Yiqun Tang · Jie Zhou · Ping Yang ·
Jingjing Yan · Nianqing Zhou

Groundwater Engineering

图书在版编目（CIP）数据

工程地下水 = Groundwater Engineering：英文 / 唐益群等著. -- 上海：同济大学出版社，2016.10

ISBN 978-7-5608-5960-6

Ⅰ.①工… Ⅱ.①唐… Ⅲ.①基础（工程）—人工降低地下水—英文 Ⅳ.① TU46

中国版本图书馆 CIP 数据核字（2015）第 203595 号

Jointly published with Tongji University Press, Shanghai, China
ISBN: 978-7-5608-5960-6 Tongji University Press, Shanghai, China

Library of Congress Control Number: 2015952991

Groundwater Engineering
工程地下水

Yiqun Tang, Jie Zhou, Ping Yang, Jingjing Yan, Nianqing Zhou
唐益群 周洁 杨坪 严婧婧 周念清 著

责任编辑	江岱	责任校对 徐春莲	封面设计 Springer
出版发行	同济大学出版社 www.tongjipress.com.cn		
	（地址：上海市四平路 1239 号 邮编：200092 电话：021-65985622）		
经 销	全国各地新华书店		
印 刷	虎彩印艺股份有限公司		
开 本	787mm×1092mm 1/16		
印 张	26.5		
字 数	530 000		
版 次	2016 年 10 月第 1 版 2016 年 10 月第 1 次印刷		
书 号	ISBN 978-7-5608-5960-6		
定 价	180.00 元		

本书若有印装质量问题，请向本社发行部调换 版权所有 侵权必究

Preface

The largest largest source of freshwater lies in underground water. Rapid economic and construction development makes engineering geological and environmental problems of groundwater more serious. In many cases, such as soil deformation or pit bottom bursting in foundation pit excavation, land subsidence by engineering dewatering, quicksand, piping, or sand liquefaction in underground construction, stability problem in bedrock area, corrosion of concrete and steel bar, etc., groundwater always plays a crucial role. These engineering geological, hydrogeological problems or construction disasters have been paid a substantial amount of attention by researchers and engineers. A lot of new knowledge about groundwater engineering has been accumulated over the past decades. Combined with the engineering practice experience and the summary of construction lessons, the prevention or alleviation of engineering geological and environmental problems relevant to groundwater must be of greater significance and emergence.

The authors have been involved in the teaching and research work on groundwater engineering for many years. For a textbook, it is an achievement on the summary of previous basic knowledge and our practical engineering experience. It also plays an important role as a most applicable education material for both senior undergraduate and graduate students. The integration of theory and practice makes it a professional textbook for related students. Moreover, it can provide valuable references for technical staff and managers of engineering construction.

Involving several disciplines of engineering geology, hydrogeology, and geotechnical engineering, this book mainly covers the general field of groundwater from an engineering perspective, based on new research results in China and abroad. The first two chapters provide theoretical aspects, such as basic theory in groundwater and parameter calculation in hydrogeology. The large main part introduces the problem caused by groundwater and dewatering construction design, including geological problem and prevention caused by groundwater, construction dewatering, engineering wellpoint dewatering method, dewatering well and drilling, groundwater dewatering in foundation pit engineering, and groundwater engineering in bedrock area. Chapter 9 presents approaches in computer modeling for groundwater engineering. Finally an introduction to groundwater corrosion on concrete and steel is discussed in Chap. 10 as supplementary material.

I am grateful to have an excellent group of authors such as Prof. Yiqun Tang, Ph. D. Jie Zhou, graduate student Jingjing Yan, Associated Prof. Ping Yang, Prof. Nianqing Zhou, Associated Prof. Jianxiu Wang, and Assistant Prof. Guo Li.

Specifically, Chaps. 1 and 3–6 are written by Yiqun Tang, Jie Zhou, and Jingjing Yan. Chapter 2 is organized by Ping Yang. Chapter 7 is mainly revised by Guo Li. In Chap. 9, Jianxiu Wang has given the most contribution, while in Chap. 10, Jie Zhou and Tang have made great efforts.

The re-edition, organization, and revision of the whole book have been done by Yiqun Tang, Jie Zhou, Ping Yang, and Jingjing Yan. I am also very grateful to the graduate students Chen Tang and Ph.D. Jie Zhou. They made a special effort on the graphic drawing and processing work. The case study and exercises are organized by Prof. Yiqun Tang, Associated Prof. Ping Yang, Ph.D. Jie Zhou, and graduate students Jie Xu and Kai Sun.

The completion of the book was supported by the National Key Technologies R&D Program of China through Grant No. 2012BAJ11B04, 12th five-year teaching material planning program, and pilot program of comprehensive reform on major higher education teaching quality and teaching reform project by the Ministry of Education.

I also express our appreciation here since some basic material and knowledge is referred from Handbook of hydrogeology of water-supply. Some notation has been specifically marked in relevant texts. Some reference could not be correctly found due to the long-time missing record. I apologise in case of minor inaccuracies, which authors have not noticed. It should be noted that the copyright holder of the materials on land subsidence data of Tokyo (Figs. 5.35–5.37) could not be traced with proper credit, we would appreciate any information that could enable us to do so.

The experiments in this book were mostly conducted in the key laboratory of geotechnical and underground engineering at Tongji University, Ministry of Education. Ph.D. Qi Liu has done us a favor during the experimental design. All the authors are appreciated for this.

This book will be an essential handy reference for industrial and academic researchers working in the groundwater field and can also serve as a lecture-based course material to provide fundamental and practical information for both senior undergraduate and graduate students, who will need to work in the fields of geology engineering, hydrogeology, geotechnical engineering, or to conduct related research.

Shanghai, China
June 2015

Prof. Yiqun Tang

Contents

1 Groundwater 1
   1.1 Basic Concepts of Groundwater 1
      1.1.1 Geological Occurrence of Groundwater 1
      1.1.2 Hydraulic Properties of Earth Materials
and Groundwater 4
      1.1.3 Aquifers 8
   1.2 Types of Groundwater 9
      1.2.1 Buried Conditions 9
      1.2.2 Aquifer Characteristic 12
   1.3 Groundwater Movement 13
      1.3.1 Basic Concepts 13
      1.3.2 Linear Seepage Principles 20
      1.3.3 Nonlinear Seepage Principles 29
      1.3.4 Flow Nets 30
   1.4 Exercises 34

2 Hydrogeological Parameters Calculation 35
   2.1 Hydrogeological Tests 35
      2.1.1 Pumping Test 36
      2.1.2 Water Pressure Test 49
      2.1.3 Water Injection Test 56
      2.1.4 Infiltration Test 57
   2.2 Measurement of Groundwater Table, Flow Direction
and Seepage Velocity 61
      2.2.1 Measurement of Groundwater Table 61
      2.2.2 Measurement of Groundwater Flow Direction 61
      2.2.3 Measurement of Seepage Velocity 62
   2.3 Capillary Rise Height Determination 64
      2.3.1 Direct Observation Method 64
      2.3.2 Water Content Distribution Curve Method 64

	2.4	Pore Water Pressure Determination	67
		2.4.1 Pore Water Pressure Gauge and Measurement Methods	67
		2.4.2 Calculation Formulas	67
	2.5	Hydrogeological Parameters Calculation in Steady Flow Pumping Test	69
		2.5.1 Calculation of Hydraulic Conductivity	69
		2.5.2 Calculation of Radius of Influence	73
		2.5.3 Case Study	80
	2.6	Hydrogeological Parameters Calculation in Unsteady Flow Pumping Test	83
		2.6.1 Transmissibility, Storage Coefficient, and Pressure Transitivity Coefficient Calculation for Confined Aquifer	83
		2.6.2 Transmissibility, Storage Coefficient, Leakage Coefficient, and Leakage Factor Calculation for Leaky Aquifers	89
		2.6.3 Specific Yield, Storage Coefficient, Hydraulic Conductivity and Transmissibility Calculation of Unconfined Aquifer	90
	2.7	Other Methods for Hydrogeological Parameters Calculation	95
		2.7.1 Transmissibility and Well Loss Calculation	95
		2.7.2 Calculation of Transmissibility Coefficient and Water Storage Coefficient by Sensitivity Analysis Method Based on Pumping Test Data	98
		2.7.3 Hydrogeological Parameter Optimization Based on Numerical Method and Optimization Method Coupling Model	102
	2.8	Case Study	109
	2.9	Exercises	111
3	**Groundwater Engineering Problem and Prevention**		**113**
	3.1	Adverse Actions of Groundwater	113
		3.1.1 Suffosion	113
		3.1.2 Pore-Water Pressure	113
		3.1.3 Seepage Flow	114
		3.1.4 Uplift Effect of Groundwater	114
	3.2	Suffosion	114
		3.2.1 Types of Suffosion	114
		3.2.2 Conditions of Suffosion	115
		3.2.3 Prevention of Suffosion	115
	3.3	Piping and Prevention	116
		3.3.1 Piping	116
		3.3.2 Conditions of Piping	117

		3.3.3	Prevention of Piping..............................	117
		3.3.4	Case Study..	118
	3.4	Quicksand and Prevention................................		119
		3.4.1	Quicksand..	119
		3.4.2	Causes of Quicksand................................	120
		3.4.3	Conditions of Quicksand............................	121
		3.4.4	Determination of Quicksand.........................	121
		3.4.5	Quicksand in Foundation Pit........................	123
		3.4.6	Quicksand in the Caisson...........................	125
		3.4.7	The Prevention and Treatment of Quicksand..........	126
	3.5	Liquefaction of Sands and Relevant Preventions...........		128
		3.5.1	Liquefaction.......................................	128
		3.5.2	The Factors Affecting Liquefaction.................	130
		3.5.3	Evaluation of Liquefaction Potential................	135
		3.5.4	Anti-Liquefaction Measurement......................	143
	3.6	Pore-Water Pressure Problems............................		144
		3.6.1	The Influence of Pore-Water Pressure on Shear Strength...	144
		3.6.2	Instantaneous and Long-Term Stability in Foundation Pit in Saturated Clay...............................	147
	3.7	Seepage...		150
		3.7.1	The Stability of Foundation Pit with Retaining Wall Under Seepage Condition...........................	150
		3.7.2	The Stability of Slope Under Seepage Condition.....	154
	3.8	Piping and Soil Displacement in Foundation Pit Bottom....		155
		3.8.1	Piping in the Foundation Pit.......................	155
		3.8.2	Soil Displacement..................................	157
		3.8.3	The Foundation Pit Bottom Stability Encountering Confined Water Pressure............................	158
		3.8.4	The Measurements of Foundation Pit Piping..........	160
	3.9	Exercises...		164
4	**Construction Drainage**..			165
	4.1	Summary..		165
	4.2	Open Pumping Methods...................................		168
		4.2.1	Open Ditches and Sump Pumps........................	168
		4.2.2	Multilayer Open Pumping from Ditches and Sumps..	170
		4.2.3	Deep Ditches Pumping...............................	171
		4.2.4	Combined Pumping...................................	172
		4.2.5	Dewatering by Infrastructure.......................	173
		4.2.6	Open Pumping in Sheet Pile Supporting System.......	173
	4.3	Calculation on Open Pumping Amount......................		174
		4.3.1	Formulas...	174
		4.3.2	Empirical Method...................................	179

	4.4	The Common Section of the Ditches in Foundation Pit.	180
	4.5	The Calculation of the Power of Pumps in Requirement	180
	4.6	The Performance of Common Pumps	180
	4.7	Case Study	181
	4.8	Exercises	182

5 Wellpoint Dewatering in Engineering Groundwater 183

- 5.1 Light Wellpoint Dewatering 186
  - 5.1.1 Range of Application 186
  - 5.1.2 Major Equipment 186
  - 5.1.3 Wellpoint Arrangement 189
  - 5.1.4 Wellpoint Construction Processes 193
  - 5.1.5 Parameter Calculation 197
  - 5.1.6 Choice of Filter Screen and Sand Pack 209
- 5.2 Ejector Wellpoint 211
  - 5.2.1 Scope of Application 211
  - 5.2.2 Major Equipment and Working Principles 213
  - 5.2.3 Design of the Pumping Device Structure 215
  - 5.2.4 Layout of Ejector Wellpoint and Attention for Construction 218
- 5.3 Tube Wellpoint 218
  - 5.3.1 Scope of Application 218
  - 5.3.2 Major Equipment and Working Principles 219
  - 5.3.3 Construction Method 220
- 5.4 Electroosmosis Wellpoint 221
  - 5.4.1 Scope of Application 221
  - 5.4.2 Major Equipment and Working Principles 221
  - 5.4.3 Key Points and Attention of Construction 222
- 5.5 Recharge Wellpoint 223
  - 5.5.1 Working Principles 223
  - 5.5.2 Key Points and Attentions of Construction 223
- 5.6 Monitoring of Wellpoint Dewatering 224
  - 5.6.1 Flow Observation 224
  - 5.6.2 Water Table Observation 225
  - 5.6.3 Pore Water Pressure Measurement 225
  - 5.6.4 Total Settlement and Layered Settlement Observation 225
  - 5.6.5 Earth Pressure Measurement 225
- 5.7 Design Cases of Dewatering Projects 226
  - 5.7.1 Ejector Wellpoint Case 226
  - 5.7.2 Tube Wellpoint Case 229
- 5.8 Common Issues of Wellpoint Dewatering Methods and Their Solutions 234
  - 5.8.1 Light Wellpoint 234
  - 5.8.2 Ejector Wellpoint 235
  - 5.8.3 Tube Wellpoint 238

	5.9	Impact of Wellpoint Dewatering on the Environment and the Prevention 240
		5.9.1 Ground Deformation Near a Dewatering Wellpoint ... 240
		5.9.2 Mechanism of Settlement Caused by Dewatering 243
		5.9.3 Impact of Changes in Groundwater Level on Soil Deformation 247
		5.9.4 Differences Between Load Consolidation and Osmotic Consolidation 248
		5.9.5 Relationship Between Settlement Rate and Groundwater Pressure 249
		5.9.6 Calculation of Wellpoint Dewatering Influence Range and Ground Settlement 251
		5.9.7 Precautions of Adversely Affects on Environment Caused by Wellpoint Dewatering 255
	5.10	Case Study 259
	5.11	Exercises 260

6 Dewatering Well and Requirements of Drilling Completion 263
    6.1   Structural Design of Dewatering Well 263
           6.1.1 Determination of Well Pipe, Depth and Diameter of Drilling 263
           6.1.2 Design of Filter in Well Pipe 265
    6.2   Technical Requirements of Dewatering Well Completion 271
           6.2.1 Water Sealing Requirement for Drilling 271
           6.2.2 Demands of Drilling Flushing Fluid 271
           6.2.3 Requirements of Drilling Inclination 271
    6.3   Well Washing 271
           6.3.1 Mechanical Methods for Well Washing 272
           6.3.2 Chemical Methods for Well Washing 276
    6.4   Case Study 279
    6.5   Exercises 281

7 Dewatering Types in Foundation Pit 283
    7.1   Types and Effect of Dewatering in Foundation Pit 283
           7.1.1 Effects of Dewatering in Foundation Pit Construction 283
           7.1.2 Different Types of Dewatering in Foundation Pit Construction 284
    7.2   The Seepage Properties of Dewatering in Foundation Pit 284
           7.2.1 Water-Proof Curtain 284
           7.2.2 Length of Filter 288
           7.2.3 Vertical Hydraulic Conductivity of Aquifer 288

	7.3	The Classification and Characteristics of Dewatering in Foundation Pit.................................	289
		7.3.1 The First Class	289
		7.3.2 The Second Class	290
		7.3.3 The Third Class...........................	291
		7.3.4 The Fourth Class	292
	7.4	Dewatering Design of Foundation Pit Engineering	292
		7.4.1 Design for the First Class Dewatering..............	292
		7.4.2 Design for the Second Class Dewatering............	293
		7.4.3 Design for the Third Class Dewatering.............	294
		7.4.4 Design for the Fourth Class Dewatering	295
	7.5	Case Study	296
		7.5.1 Case 1—The Second Class Foundation Dewatering Engineering of Small Area and Large Drawdown.....	296
		7.5.2 Case 2—The Third Class Foundation Dewatering Engineering of Large Drawdown and Double Aquifers.................................	301
	7.6	Exercises	310
8	**Engineering Groundwater of Bedrock Area**		311
	8.1	Concepts and Classifications of Groundwater in Bedrock Area	311
		8.1.1 Concept of the Bedrock Groundwater..............	311
		8.1.2 Classification of the Bedrock Groundwater	311
	8.2	Forming Conditions, Characteristics, and Storage Regularities of the Bedrock Fissure Water	316
		8.2.1 Forming Conditions of the Bedrock Fissure Water	316
		8.2.2 Characteristics of the Bedrock Fissure Water.........	317
		8.2.3 Occurrence Regularity of the Bedrock Fissure Water	317
		8.2.4 Flow Regularity of the Bedrock Fissure Water........	318
	8.3	Groundwater Seepage Model of Fractured Rock Mass	320
		8.3.1 Dual Model of Fracture-Pore....................	320
		8.3.2 Non-Dual-Medium Model......................	323
	8.4	Three-Dimensional Numerical Model for Bedrock Fissure Water...	327
		8.4.1 Equivalent Three-Dimensional Model..............	328
		8.4.2 One-Dimensional Model.......................	328
		8.4.3 Water Catchment Corridor Model	329
	8.5	Project Types and Instances of Bedrock Fissure Water	329
		8.5.1 Groundwater of the Rock Slope Engineering.........	329
		8.5.2 Groundwater of the Tunnel Project................	349
	8.6	Exercises	356

9 Numerical Simulation of Engineering Groundwater... 357
- 9.1 Basic Principle ... 357
  - 9.1.1 Finite Difference Method ... 357
  - 9.1.2 Finite Element Method ... 360
  - 9.1.3 Boundary Element Method ... 363
- 9.2 Numerical Simulation of Foundation Pit Dewatering ... 363
  - 9.2.1 Analysis of Prototype ... 364
  - 9.2.2 Three-Dimensional Numerical Modeling of FDM ... 368
  - 9.2.3 Three-Dimensional Numerical Simulation of FDM ... 372
  - 9.2.4 Settlement Calculation ... 377
  - 9.2.5 Effects and Analysis ... 379
- 9.3 Case Study ... 380
- 9.4 Exercises ... 380

10 Groundwater Pollution and Corrosivity Assessment ... 381
- 10.1 Groundwater Quantity Analysis ... 381
  - 10.1.1 Groundwater Quantity Analysis Representation Methods ... 381
  - 10.1.2 Groundwater Quantity Analysis Contents ... 384
  - 10.1.3 Water Sample Requirements ... 385
- 10.2 Groundwater Pollution ... 387
  - 10.2.1 Concepts of Groundwater Pollution ... 387
  - 10.2.2 Pollutants, Pollution Sources, and Pollution Paths or Ways ... 388
  - 10.2.3 Investigation and Monitoring of Groundwater Pollution ... 391
- 10.3 Groundwater Corrosion Evaluation ... 392
  - 10.3.1 Groundwater Corrosive Effects to Concrete ... 392
  - 10.3.2 Groundwater Corrosive Effects to Steel ... 400
- 10.4 Case Study ... 401
- 10.5 Exercises ... 403

Bibliography ... 405

Chapter 1
Groundwater

Groundwater is the subsurface water in soil pore spaces or in the fractures of rock formations. As the component of soil and rock mass, it plays important role on the engineering behaviors of earth materials (soils or rocks), and it is also an essential part of engineering environment. Its geological occurrence and seepage flow influence the strength, deformation, stability, and durability of structures. Therefore, groundwater is a significant element in geotechnical engineering or foundation engineering area. In this regard, any soil or rock engineering evaluation without considering groundwater is not available and comprehensive. In China, during the discipline development and project practice in geotechnical engineering, the knowledge of groundwater is far behind the international developed level.

1.1 Basic Concepts of Groundwater

1.1.1 Geological Occurrence of Groundwater

Water in soils or rocks occurs in many kinds of forms. According to the physical and chemical properties, they are aqueous vapor, film adsorbed water, free water (gravitational water), capillary water, and water in bulk (water in the solid state of aggregation, such as ice).

1.1.1.1 Aqueous-Vapor Water

In soil engineering, this part of subsurface water occupies the voids in the soil or rock above the groundwater table which is called phreatic water or soil moisture in the unsaturated zone. It can be moved from the atmosphere into the voids, or can be formed by evaporation of liquid water. The aqueous-vapor water can flow as air and also can migrate from high humidity to low humidity places. It controls the moisture distribution in soil or rock mass to some extent.

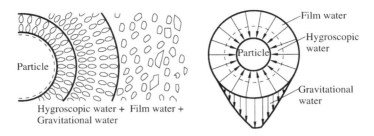

Fig. 1.1 Absorbed water and gravitational water

1.1.1.2 Adsorbed Water

The surfaces of loose soil particles carry a net negative charge. It attracts cations in the water. This electrostatic adsorption forces become larger as soil is finer. Water molecules act as a bar magnet with positive and negative charges in two ends. Under the electrostatic attraction, they are clustered and can be attracted rigidly around the surfaces of the individual soil particles, to form a very thin hull of film of water. This water film is the adsorbed water, or called bound water as well (Fig. 1.1).

Depending on the force of electrostatic adhesion on the particle surface, the adsorbed water can be also divided into strongly adsorbed water and loosely adsorbed water. The strongly adsorbed water is called hygroscopic moisture. The hygroscopic moisture film is known to be bound or attached rigidly to the soil particles with an immense physical force up to about 10,000 atmospheres. Thus this hygroscopic soil moisture film is densified akin to solid state with high density of 2 g/cm^3. It has large viscosity and elasticity. Hygroscopic moisture is not in union with the groundwater. Therefore, it does not take part in the fluctuation of the groundwater table, or does it transmit hydrostatic pressure. It can only be removed by drying the soil particles at +105 °C, resulting as aqueous vapor. The loosely bound water is known as film moisture. It is slightly away from particles as a hull or film upon the layer of the hygroscopic moisture film. It is composed of the main part of the water film. Its density is the same as the free water but with large viscosity. Film moisture does not transmit externally applied hydrostatic pressure, nor can be affected by gravity, but in case of upward migration, it is stressed, however, in the sense of soil moisture tension. This kind of moisture translocates very slowly. It moves in the form of a liquid film from points of higher potentials (heat, electric) to lower ones, from greater concentrations to smaller ones, and from points of thicker films to thinner films (Fig. 1.2).

Providing the film moisture is greatly stressed, and it can be removed from the particle surfaces and transformed into gravitational water. Therefore, during the exploitation of confined aquifer in loose sediment, the film moisture of soils within the embedded clay layer or aquitard may be transformed into gravitational water. It must be paid attention since the water quality and quantity may both have some influence.

Fig. 1.2 Film movement of loosely bound water

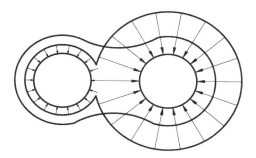

1.1.1.3 Capillary Water

Capillary water is that soil moisture which is located within the interstices and voids of capillary size of the soil above the groundwater table. Capillary movement in soils is the movement of the soil moisture through minute pores between the soil particles. The minute pores serve as capillary channels through which the soil moisture rises above the groundwater table. The rise takes place and the liquid is held by means of a force called the surface tension force of the meniscus at the top of the water column in a capillary tube, or by surface tension forces plus the effect of gravity. Capillary water is hydraulically and continuously connected to the groundwater table or to a perched groundwater table, and can be raised against the force of gravity. For capillary of rise in soil to exist, this height is called the capillary rise height, thus capillary-saturated zone between groundwater table and the plane of meniscus is known as closed capillary fringe. It contains no air and the thickness depends mainly on the fineness of the soil particles. The larger the pore size, the lesser the height of rise or lesser the capillary fringe.

Between the aeration zone and the closed capillary fringe, there is the so-called open capillary fringe, i.e., the air-containing capillary zone in a perched groundwater. In addition, pore corner or neck moisture is the annular moisture wedge held by the concave meniscus or rather surface tension forces, in the angularities formed by the points of contact of the soil particles. Capillary water can transmit static water pressure and is able to be absorbed by plant roots.

Capillary water cannot be drained away by means of drainage systems installed within the capillary fringe, but it can be controlled by lowering the groundwater table. The drainage system must be installed in the groundwater to pull it down together with the capillary fringe, thus controlling the capillary height to which the capillary water can rise. Capillary water can be removed from soils by drainage only when the quantity of water present in the soil is in excess of that retained by surface tension forces.

1.1.1.4 Gravitational Water

Gravitational water is the water which is in excess of the amount of moisture the soil can retain. It translocates as a liquid and it can be drained away by the force of gravity. It transmits hydraulic pressure.

1.1.1.5 Solid Water (Water in Bulk)

When the soil temperature is below the freezing point of water, the pore water within soils is frozen as solid. The solid water is mainly distributed in the mountains above the snow line and some cold regions, where the shallow groundwater exists as solid water throughout the year.

Aqueous water, adsorbed water, capillary water, and gravitational water are vertically distributed in the shallow subsurface soil. When drilling a well in the loose deposits, initially there exists aqueous water and adsorbed water in the dry soils. Subsequently, wet soil can be observed. It reveals the existence of capillary water. Much deeper, water can be found to flow into the well and forms the groundwater surface. This is the gravitational water (Fig. 1.3).

As shown in Fig. 1.3, along the soil profile, area from stable groundwater table to soil surface is known as zone of aeration, including aqueous water and adsorbed water; and some perched gravitational water and open capillary water fringe followed; and then some capillary water close to the groundwater surface. Below the groundwater surface, it is the main zone of saturation, where gravitational water is mainly located (Fig. 1.3).

1.1.2 Hydraulic Properties of Earth Materials and Groundwater

1.1.2.1 Specific Storage (Water Storativity)

In a saturated porous medium that is confined between two transmissive layers of rock or clay, water will be stored in the pores of the medium by a combination of

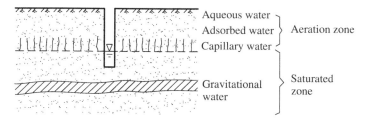

Fig. 1.3 The vertical profile of different water occurrences

1.1 Basic Concepts of Groundwater

two phenomena; there are water compression and aquifer expansion. As water is forced into the system at a rate greater than it is being extracted, the water will compress and the matrix will expand to accommodate the excess. In a unit volume of saturated porous matrix, the volume of water that will be taken into storage under a unit increase in head, or the volume that will be released under a unit decrease in head, is called specific storage μ_s. It is shown in Eq. (1.1):

$$\mu_s = \rho g(\alpha + n\beta) \tag{1.1}$$

where

α is aquifer compressibility; ρ is fluid density; n is porosity; g is a gravitational acceleration; β is water compressibility.

This unit has the dimension of 1/L and is quite small, usually 0.0001 or less. The storage efficient of an aquifer, or simply, the storativity μ^* is given as $\mu^* = b\mu_s$, where b is the saturated thickness of the aquifer. Storativity is defined as the volume of water per unit aquifer surface area taken into or released from storage per unit increase or decrease in head, respectively. It is a dimensionless quantity. In confined aquifers, the value of storativity ranges from 0.005 to 0.00005.

1.1.2.2 Specific Retention (Water Retentivity)

In unconfined porous media, gravity drainage will proceed until the forces of surface tension and molecular attraction to the matrix grains become equal to the force of gravity. Water retention refers to water capacity after gravity drainage. Under the influence of gravity, the water retained in the pores includes adsorbed water and partial perched capillary water or pore capillary water. Specific retention S_t is used to evaluate the ability of water retention. It is the ratio of the volume of water retained in the pores to the total matrix volume (transmissive layer), and it can be expressed as decimal or fraction as follows:

$$W_m = \frac{V_m}{V} \quad \text{or} \quad W_m = \frac{V_m}{V} \times 100\% \tag{1.2}$$

where W_m is the specific retention, decimal or fraction; and V_m is the water volume retained in pores under gravity drainage, m^3.

According to the modes of water retention, it can be divided into capillary water, specific retention, and adsorbed water-specific retention. Generally, the adsorbed water-specific retention is used, which is the ratio of maximum water capacity retained in pores to the total matrix volume. It depends on the particle size. In common, the smaller the pore size, the larger the specific surface area and the higher the amount of adsorbed water; so that the larger the specific retention. Table 1.1 presents the specific retention values of loose soils.

Table 1.1 The specific retention values of loose soils

Soil type	Coarse sands	Medium sands	Fine sands	Very fine sands	Loam	Clay
Particle size (mm)	2–0.5	0.5–0.25	0.25–0.1	0.1–0.05	0.05–0.002	<0.002
Adsorbed water-specific retention (%)	1.57	1.6	2.73	4.75	10.8	44.85

1.1.2.3 Specific Yield (Water Yield)

Water yield refers to the water capacity draining from a saturated porous matrix under gravity, in unconfined porous media. Water releases from the saturated porous media under the influence of gravity when decreased in water head. In unconfined porous media, that is, where there is no overlying confining cover, storage of water in its upper part is defined as specific yield μ. This is the ratio of the volume of water that drains from a saturated porous matrix under the influence of gravity to the total volume of the matrix, per unit drop in the water table. Specific yield is normally much greater than specific storage, as water released from elastic storage leaves the pores still saturated. Specific yield is often in the range of 0.2–0.3, or three to four orders of magnitude greater than elastic storage.

$$\mu = \frac{V_g}{V}, \quad \text{or} \quad \mu = \frac{V_g}{V} \times 100\,\% \qquad (1.3)$$

where V_g is the water amount released from a saturated porous matrix, per unit drop in water table.

Specific yield first depends on the void sizes of rock or soil, and then is the amount of voids. The values of different loose soils are presented in Table 1.2.

In the upper parts of an unconfined porous medium, where elastic storage is not significant, the sum of specific yield and specific retention equals porosity $\mu + W_m = n$.

When an unconfined porous medium is very thick, the lower parts of the medium may also contain water under elastic storage, owing to the increase of pressure and consequent water compressibility and matrix expansion with increasing depth. In this case, the total storativity of the medium is expressed as $\mu^* = \mu + b\mu_s$.

Field capacity is used to describe essentially the same phenomena as specific retention, but it is normally used in agricultural soil moisture studies. It is a function not only of specific retention, but also the evaporation depth and the unsaturated permeability of the soil.

1.1 Basic Concepts of Groundwater

Table 1.2 The specific yield values of loose soils

Soil type	Clay	Sand loam	Silt sand	Fine sand	Medium sand	Coarse sand	Gravel	Fine gravel	Medium gravel	Coarse gravel
Specific yield (%)	2	7	8	21	26	27	25	25	23	22

1.1.2.4 Coefficient of Permeability (Water Permeability)

Permeability is defined as the property of a porous material which permits the passage or seepage of fluids such as water. Various soils have different permeabilities. Theoretically, all soils are more or less porous, in practice the term "permeable" is applied to soils which are porous enough to permit the flow of water through such a soil. Conversely, soils which permeate with great difficulties are termed "impermeable." Generally, the permeability depends on the resistance to flow offered by the soil, through which the flow takes place. The resistance to the flow depends upon the type of soil, size, and shape of the soil particles (rounded, angular or flaky), the degree of packing (density of soil), and size and geometry of the voids. Also, it is relevant to temperature of water (viscosity and surface tension effects). In addition, coarse-textured soils (such as gravel, sand, etc.) are more pervious than fine-textured soils (silt, clay). The smaller the grain particles and pore voids, the poorer the permeability. Although clay has relatively high porosity, up to 50 %, the pores are occupied by absorbed water. It is really hard for the movement of free water. Thus clay is called as impermeable layer.

1.1.3 Aquifers

Groundwater occurs in many types of geologic formations; those known as aquifers are of most importance. An aquifer may be defined as a formation that contains sufficient saturated permeable material to yield significant quantities of water to wells and springs. This implies an ability to store and to transmit water; unconsolidated sands and gravels are a typical example. Generally, aquifers are really extensive with well-developed inter-connecting voids and good permeability, such as various sands, gravel, and hard rock with fissure and karst caves. Aquifers may be overlain or underlain by a confined bed, which may be defined as a relatively impermeable material stratigraphically adjacent to one or more aquifers. Clearly, there are several types of confining beds:

Aquiclude—a saturated but relatively impermeable material that does not yield appreciable quantities of water to wells, such as clay.

Aquifuge—a relatively impermeable formation neither containing nor transmitting water, such as solid granite.

Aquitard—a saturated but poorly permeable stratum that impedes groundwater movement and does not yield water freely to wells, and may transmit appreciable water to or from storage zone, such as sandy clay.

An aquifer first should be a permeable stratum. It is the saturated part of a permeable stratum below the groundwater table, and the unsaturated part can be a permeable stratum without containing water. Thus, a permeable stratum can be an aquifer, such as alluvial gravel aquifer, or unyielding permeable stratum, such as talus sandy loam. It can be one part as the aquifer below the groundwater table and the other part is unyielding permeable stratum above the groundwater table (Fig. 1.4).

Fig. 1.4 Aquifer and permeable stratum

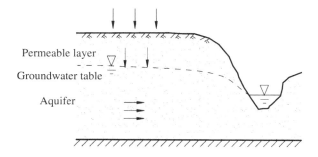

An aquifer or aquiclude is relatively defined. The boundary between these two terms is not so determined and clear. Actually, they relatively exist. Such as the embedded silt sand layer within the river bed alluvial coarse sand layers, it can be regarded as aquiclude. While if it is embedded in the clay, it must be considered as aquiclude. Therefore, there are different meanings under different geological circumstances.

The relativity of aquifers is also reflected on real value of the water yielding amount, i.e., whether it can meet the actual needs of exploitation or whether it harms the mining engineering projects. Red sand mudstone is an example. It has small yielding capacity. Compared with the gravel pore water or limestone karst water, it is so rare that could not make any sense for the water supply and mining filling water. So, here it can be regarded as aquiclude. However, as for the rural area, it is lack of water mostly, where drilling well for water can solve the domestic water supply and also as part of the irrigation water, this red sand mudstone can be meaningful aquifer.

In addition, transformation between aquifer and aquiclude happens in the layers of aquitard, such as sandy clay. Usually, it is a good aquiclude, and when this kind of soil is located in deep underground with large hydraulic gradient, leakage recharge may also occur and provide appreciate quantities water to become an aquifer.

1.2 Types of Groundwater

1.2.1 Buried Conditions

The buried condition refers to occurrence of all the aquifers in the subsurface soil geological profile. Based on this, the groundwater can be divided into three types: perched water, phreatic water, and confined water.

1.2.1.1 Perched Water

The aeration zone, also termed the vadose zone, unsaturated zone, is the part of earth between the land surface and the top of the phreatic zone, i.e., the position at which the groundwater is at atmospheric pressure. Hence, the aeration zone extends from the top of the ground surface to the water table. When local aquitard exists in the aeration zone, some gravitational water accumulated above a perch water table; this kind of water is called the perched groundwater. Runoff water, seeping into the soil, may also be trapped in depressions in pocket of moraine clay located below ground surface in permeable sand, thus forming a perched groundwater. The amount of groundwater accumulated depends upon the season and rate of evaporation from the depression in the direction of ground surface. Generally, the perched groundwater is accumulated in rain seasons and gradually dried in dry seasons. When the distribution area is very small and could not be often supplied, the water amount could not be retained all over the year. The perched groundwater table fluctuates obviously due to the small water amount. It can only be the small water supply in the water-deficient area or temporary water supply. At the same time, the contamination circumstance should be paid attention since the really short path from the surface water supply.

1.2.1.2 Phreatic Water

The water in an unconfined aquifer, which has a free upper groundwater table, is called phreatic water (Fig. 1.5). There is no upper confining bed in phreatic water or only local upper confining bed (if it has). The upper surface in such a zone is called phreatic groundwater table. The distance from this phreatic groundwater table to lower confining bed is called the phreatic aquifer thickness. The distance from the ground surface to the phreatic groundwater table is the phreatic buried depth.

Since the phreatic water is connected directly with the aeration zone, the water amount within the whole range of phreatic zone can be supplied by the atmospheric precipitation, surface water, or condensated water. The phreatic water has free

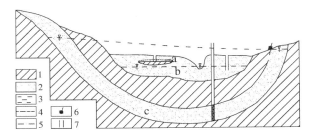

Fig. 1.5 Phreatic water, confined water, and perched water. *1* aquiclude; *2* aquifer; *3* saturated zone; *4* Phreatic water table; *5* confined pressure piezometric level; *6* spring; *7* well, solid line means no water entry along the wall; *a* perched water; *b* phreatic water; *c* confined water

groundwater table without confining pressure. The movement of the groundwater in this zone along a slope is downward, because it is subjected to the gravitational force only. The drainage of phreatic water has two ways: one is the runoff to the appropriate terrains, such as spring, drainage exit, and converging into surface water. This is called runoff discharge. Second is evaporation into the atmosphere through aeration zone or plant roots.

1.2.1.3 Confined Water

A confined aquifer is a water-bearing stratum that is confined or overlain by an impermeable layer that does not transmit water in any appreciable amount. They probably are few truly confined aquifers, because tests have shown that the confining strata, or layers, although they do not readily transmit water, over a period of time contribute large quantities of water by slow leakage.

Confined water is confined groundwater under hydrostatic or pressure head (a permeable water-bearing soil layer or aquifer sandwiched between impermeable zones above and below it). The top aquitard of confined aquifer is the upper confined bed or confined bed. Bottom aquitard is called lower confined bed. The distance between these two confined beds is the thickness of confined aquifer.

Water pressure resistance is an important feature of confined aquifer. Figure 1.6 shows synclinal basin bedrock. The central part of the aquifer is buried beneath the impermeable layer. Two ends expose at the surface. The water supply is provided from higher exposure and discharge at the other side. Water from the recharge area flowing into the confined area is subjected to the confining pressure due to the top aquitard. Conversely, the water pressure is conducted on the upper confined bed as

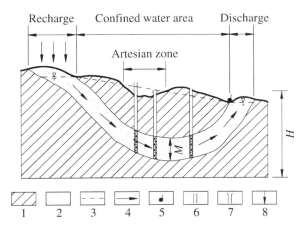

Fig. 1.6 Confined water. *1* confining bed; *2* aquifer; *3* groundwater table; *4* groundwater flow direction; *5* spring (confined spring); *6* borehole (*dash line* is the water-entry part); *7* artesian exposure; *8* precipitation recharge; *H*—confined pressure head; *M*—thickness of the aquifer

well. To confirm the confining pressure of water is not difficult. When borehole is drilled into the confined aquifer, the water table will raise to a certain location higher the top aquitard. The height which exceeds the hydrostatic level is the confined water head. The static water level in the borehole is the piezometric level at this location. It is higher above the surface, thus the water is artesian when drilled to exposure. So the confined water is also known as artesian water.

1.2.2 Aquifer Characteristic

Based on the aquifer medium types, the groundwater can be divided into pore water, fissure water, and karst water.

The void space in the aquifer is the storage site and transport channel for groundwater. Thus the void characteristics determine the storage, transport, and accumulating properties.

1.2.2.1 Pore Water

Pore water is distributed in a variety of loose Quaternary sediments. The main feature is good uniformity and continuity of water quantity in distribution. There exists good hydraulic connection within the same aquifer characterized by consistent groundwater table.

1.2.2.2 Fissure Water

Fissure water is the groundwater located in the fissured bedrock. The distribution and accumulation is relevant to the fissure development and mechanical properties of rock. The water amount can be very huge in the well-developed fissured rock. Conversely, it is very rare. Thus there may be large fluctuation in the water amount even in the same rock structure. Nonuniformity is the feature of fissure water in distribution. This property may make great difference on the aquifer yielding amount even two very close boreholes.

1.2.2.3 Karst Water

The groundwater retained and transported in the soluble karst rock voids is called the karst water. It can be phreatic water or confined water, depending on the buried condition.

The spatial distribution of karst water varies greatly, even more uneven than the fissure water. It can be accumulated in a karst cave to form a water-rich region. Or in some other place, it can flow away along the karst pore channels, to make serious water shortage.

1.2 Types of Groundwater

Table 1.3 Groundwater categories

Aquifer medium type	Pore water	Fissure water	Karst water
Aeration zone water	Vadose water, seasonal perched gravitational water above local aquitard (perched groundwater); perched capillary water and gravitational water	Seasonal gravitational water and capillary water in the shallow fissured rock	Seasonal gravitational water in upper karst channels of exposed karst formation
Phreatic water	Water in variety of shallow loose deposits	Water exposed in all kinds of shallow fissured rock	Water exposed on the surface of karst formations
Confined water	Shallow water retained in the loose deposits in the Mountain basins and plains	Water in all kinds of fissured rock covered by the synclinal structure, structural basin, fault rock block	Water in all kinds of Karst formations covered by the synclinal structure, structural basin, fault rock block

In the combination of these two classifications, groundwater can have nine categories as shown in Table 1.3.

1.3 Groundwater Movement

Groundwater exists in the voids of rock or soil mass in a variety of meanings (hygroscopic water, film water, capillary water, gravitational water, etc.). Except hygroscopic water, other types of water are all involved in the activities of aeration zone and saturated zone. Even though the film water could not move under gravity, it can transmit hydrostatic pressure and move under a certain high water head. Loam and clay layers can be aquitard under high water head difference. Previous research is mainly concentrated in the movement of gravitational water in saturated zone. In practice, some problems about the groundwater movement in aeration zone (even hygroscopic water) should be paid attention.

1.3.1 Basic Concepts

1.3.1.1 Hydraulic Head

Considering a representative element volume (REV) A (Fig. 1.7) under the groundwater table in the soil, where all the pores are hydraulically connected and

Fig. 1.7 Groundwater head in soil

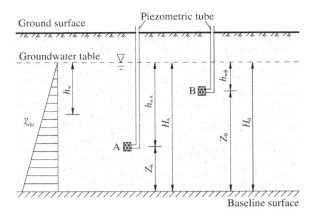

saturated, the water in A has hydrostatic pressure u_w. If one piezometric tube is connected into A, the water surface will rise to a certain height until the weight of the water in tube equals to u_w, i.e.,

$$h_w = \frac{u_w}{\gamma_w} \tag{1.4}$$

or

$$u_w = \gamma_w \cdot h_w \tag{1.5}$$

where γ_w is the unit weight of water, kN/m^3; h_w is the height from A to the piezometric tube water surface, which is usually known as piezometric water head, m; and u_w is the hydrostatic pressure, also called pore water pressure, kN/m^2.

Here, three water heads should be distinguished: pressure head h_w, elevation head Z, and the total hydraulic head H. Elevation head refers to the distance of the reference point above a datum plane (normally mean sea level, Fig. 1.7). Total head H is defined as the sum of the elevation head and the pressure head, i.e.,

$$H = h_w + Z \tag{1.6}$$

Generally, the water flows from high water head to low water head, where the water head refers to the total head, neither the pressure head nor elevation head. In Fig. 1.7, though $h_{wA} > h_{wB}$, $Z_B > Z_A$, there is no groundwater movement due to $H_A = H_B$. When considering the pore water pressure u_w, the pressure head should be paid attention, since its value can be negative or positive, depending on the position. When at the groundwater level, $h_w = 0$ thus $u_w = 0$. u_w linearly changes along the depth. The total head is also termed as piezometric head since the velocity head can generally be ignored due to the really slow water movement.

Fig. 1.8 One-dimensional seepage experimental device

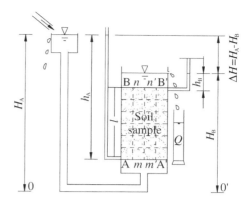

1.3.1.2 Hydrodynamic Pressure

Resistance exists in the water flow from the soil particles. Conversely, the soil particles are definitely exerted pressure when water flows through. The total pressure conducted on particle skeleton per unit is termed as hydrodynamic pressure G_D (kN/m³). The experimental device in Fig. 1.8 is taken as an example.

When there is no seepage between points A and B ($\Delta H = 0$), saturated soil unit AA'B'B is selected as object to consider the conducted force. As shown in Fig. 1.9a, F represents the force applied on the bottom surface of copper mesh AA'B'B. According to the equilibrium of forces, F equals to the effective weight of AA'B'B $\gamma' Al$. It reflects that the force conveyed to underlying soil is the effective weight.

When $\Delta H > 0$, there exists downward seepage (Fig. 1.9b). Compared to Fig. 1.9a, there is an additional water pressure $\gamma_w \Delta HA$. This part water pressure force is generated by the water head difference. When the water seepage flows from

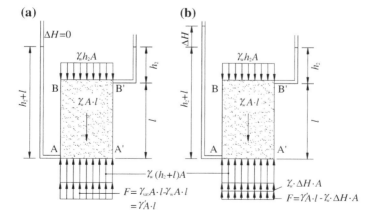

Fig. 1.9 Forces on the saturated REV

AA′ to BB′, the seepage force is consumed totally by the particle resistance. The resistance of particle exerted on water equals to the force applied on the particle skeleton by water flow, i.e., hydrodynamic pressure:

$$G_D = \frac{\gamma_w \cdot \Delta H \cdot A}{l \cdot A} = \gamma_w \cdot I \qquad (1.7)$$

Therefore, the hydrodynamic pressure is proportional to the hydraulic gradient. The direction is the same with water flow. The unit is kN/m³.

In addition, F refers to the effective contact force between upper and lower interface. From the equilibrium of forces,

$$F = \gamma' A \cdot l - \gamma_w \cdot \Delta H \cdot A \qquad (1.8)$$

where the first part on equation's right-hand side is the effective weight of AA′B′B and the second is the additional uplift force.

If $F > 0$, it means the soil REV is still on the copper mesh AA′B′B, contacted. If $F < 0$, it means the soil REV is uplifted without contacting on AA′B′B. This is the seepage failure. When the soil upper and lower parts are apart away from each other, the soil mass is unstable and piping or quicksand may occur. $F = 0$ is the critical situation. From Eq. (1.8), it can be derived

$$I_c = \frac{\Delta H}{l} = \frac{\gamma'}{\lambda_w} \approx 1 \qquad (1.9)$$

where I_c is the critical hydraulic gradient. In practice, the requirement $I < I_c$ should be meet to ensure the safety, and some safety factor should be ensured as well.

The above is the case of upward seepage. If the seepage is downward, the hydrodynamic pressure has the same direction with weight; and the seepage can only increase the force between water flow and soil particle skeleton: $F = F = \gamma' Al + \gamma_w \Delta HA$. This circumstance is favorable to the stability.

Such as the dewatering in pit (Fig. 1.10), the water flow is upward in the foundation pit. The seepage stability should be checked. Form the flow net, it can be easily seen that the most dangerous place is close to the deep end of sheet pile

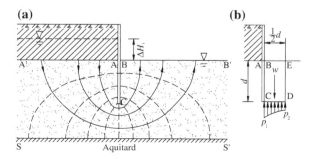

Fig. 1.10 Seepage stability checking

wall, where it is the largest hydraulic gradient. Usually, there are two types of verifications which should be made:

1. The hydraulic gradient at the water flow exposure $I = \frac{\Delta H_i}{l_{min}}$, where ΔH_i is the water head difference of the shadow area in Fig. 1.10a, l_{min} is the shortest seepage path in the area, $K_s = \frac{I_c}{I}$, and the safety factor is designed to no smaller than 2.0.
2. The stability of the whole area BCDE (Fig. 1.10b). The width can be selected as half of depth $d/2$, where d is the buried depth of sheet pile wall. The safety factor can be calculated as

$$K_s = \frac{\text{downward effective weight } W'}{\text{upward seepage force } P} = \frac{\gamma' \frac{1}{2} d^2}{\left(\frac{P_1 + P_2}{2}\right) \frac{1}{2} d} \quad (1.10)$$

where P_1 and P_2 are, respectively, the seepage forces at point C and D (Fig. 1.10b). They can be evaluated through equipotential lines. Generally, the safety factor is required as $K_s \geq 1.5$–2.0.

Seepage failure may result in catastrophe engineering accidents. Moreover, suffosion or piping phenomena is also one kind of seepage failure. Even though the entire soil mass is stable, the fine particles are taken away from the coarse particles and if this circumstance continuously happens, the pore void will be enlarged a lot. The flow velocity increases greatly. Some serious damage will happen. Particularly, in the noncohesive soil with uniformity coefficient $\mu_u > 10$, piping can occur under small hydraulic gradient (0.3–0.5). Thus, to prevent seepage failure (piping, quicksand, or boiling sand), the designation should try to minimize the hydraulic gradient; and some additional filter layer should be added at the flow exposure place when necessary.

1.3.1.3 Permeation and Seepage

Groundwater permeation is defined as water movement in voids of soils or rocks. In variety of the size, shape, and connectivity of voids in earth materials, complicated and tortuous stream channels are formed accordingly (Fig. 1.11). Even though they are the same void but different locations, the flow directions and velocities of groundwater must be different, in which groundwater in the void center flows faster, while in the places contacting particles, it moves slowly. Permeation is the real water flow existing in the earth materials. It is characterized by the discontinuity along the whole cross section of aquifer. From the aspects of theory and practice, there are great difficulties of the study on the specific circumstance. Therefore, in need of practical engineering, a hypothetical flow model is proposed to replace the real flow action. First, the tortuosity of water flow is neglected by just considering the main groundwater flow direction. Second, the groundwater is regarded as flowing through the entire cross section without particle skeletons (Fig. 1.12). It is called seepage.

Fig. 1.11 Schematic of permeation

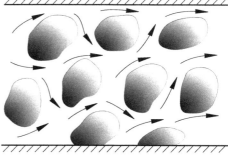

Fig. 1.12 Schematic of seepage

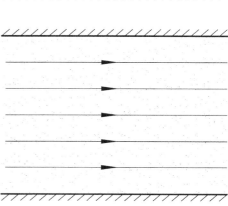

1.3.1.4 Laminar Flow and Turbulent Flow

Groundwater is not static in the saturated earth materials (such as soils or rocks). It flows from places of high water table to low water table. According to the observation and experimental verification, groundwater flow has two basic states, i.e., laminar flow and turbulent flow.

When groundwater moves continuously with paralleling streamlines, this type of flow condition is called of laminar flow (Fig. 1.13). Conversely, in the circumstance

Fig. 1.13 Groundwater movement of laminar flow. *1* soil particle; *2* absorbed water film

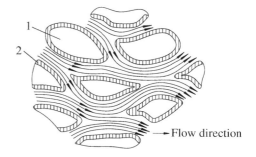

1.3 Groundwater Movement

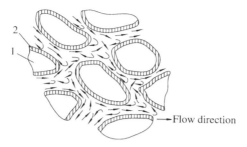

Fig. 1.14 Groundwater movement of turbulent flow. *1* soil particle; *2* absorbed water film

of turbulent flow, groundwater moves discontinuously with endless accelerations, decelerations, and changes in direction, as shown in Fig. 1.14.

Research results indicate that laminar flow always occurs when the flow velocity is much smaller. There is a critical value to distinguish between laminar and turbulent flow. When the flow velocity exceeds this value, the flow state changes from laminar to turbulent. Groundwater generally moves very slowly within earth materials. Mostly, the movement of groundwater can be regarded as laminar flow. Turbulent flow can be found in rocks such as basalt and limestone that contain large underground openings, or large karst caves.

1.3.1.5 Steady Flow and Unsteady Flow

The flow characteristics can be described by the variation of motion elements, including dynamic pressure, velocity, acceleration, etc. Providing the flow movement is just the function of space domain with no change occurring in time domain. This type of flow is called steady flow. As shown in Fig. 1.15a, when water level in the tank is kept identical, the motion elements of water flow from the hole of the tank wall are relevant to the location, but rarely change with time. This condition is defined as steady flow.

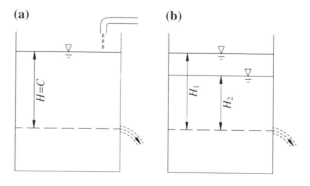

Fig. 1.15 Steady and unsteady flows

In the case of Fig. 1.15b, the water flow is not only controlled by the location, but also varies with time. It is the unsteady flow condition. There is no water supply in the flow condition shown in Fig. 1.15b. The water head decreases with time, which is resulted in the variation of each other motion element.

1.3.2 Linear Seepage Principles

1.3.2.1 Darcy's Law

The flow movement has three states, i.e., laminar, turbulent, or combined flows. Laminar flow usually occurs in the permeation of groundwater in soil pores or rock fissures. Turbulent flow always takes place in the underground cave or large rock fissures. It has characteristics of eddies and swirls, with interlacing streamlines. In some circumstances, these two conditions simultaneously arise.

Most natural underground flow is regarded as laminar flow. French hydraulic engineer, Henry Darcy, investigated the water flow more than a century ago. In the statement, the flow rate through porous media is proportional to the length of the flow path, and is known as universally as Darcy's law [shown in Eq. (1.11)]:

$$Q = KA \frac{h}{L} \quad (1.11)$$

where Q is the flowing rate, m³/d; K is the hydraulic conductivity, a constant that serves as a measure of the permeability of the porous medium, m/d; A is the cross-sectional area of water flow, m²; L is the distance of seepage path between these two flow cross sections, m; and h is the water head loss between these two cross sections, m, $h = H_1 - H_2$; $\frac{h}{L}$ is the hydraulic gradient, noted by I, which represents the head loss per unit length along seepage path.

Expressed in general terms

$$Q = -KA \frac{dh}{dl} \quad (1.12)$$

or simply Darcy velocity

$$v = \frac{Q}{A} = -K \frac{dh}{dl} \quad (1.13)$$

where $\frac{dh}{dl}$ is the hydraulic gradient (dimensionless). The negative sign indicates that the flow of water is in the direction of decreasing head. Equation (1.13) states, Darcy's law in its simplest form, that the seepage velocity v is proportional to the hydraulic gradient. When $I = 1$, i.e., $K = v$, namely hydraulic conductivity equals the hydraulic gradient, cm/s or m/d.

Fig. 1.16 Cross section of groundwater (AB phreatic water surface; A'B' aquitard layer)

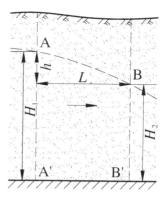

According to Eq. (1.13), the value of seepage velocity of soils is equivalent of the hydraulic conductivity under hydraulic gradient of 1. It depends on two factors. One is the permeability of soils (the amount of K) and the other is the hydraulic conditions (the amount of I) (Fig. 1.16).

Here, two aspects should be noted. First, the seepage velocity is referred to as the Darcy velocity, which assumes that water flow occurs through the entire cross section of soils without regarding to solids and pores. Actually, the flow is limited only to the pore space. The seepage path is complicated and tortuous. The real flow cross-sectional area is smaller than A, so that real average interstitial velocity is greater than Darcy velocity v. In practice, the average flow amount through the entire soil is taken much more care. Thus the apparent velocity v with cross-sectional area A and seepage path L is convenient and useful. Second, Darcy's law is applicable in sands or other soils with much smaller grain particles. Because in large pore voids (such as gravels, pebble, or karst caves), too fast flow velocity, the irregular flow paths of eddies and swirls associated with turbulence occur first in the larger pore space. The head loss varies approximately with the velocity rather than linearly. Velocity in laminar flow is proportional to the first power of the hydraulic gradient, and it seems reasonable to believe that Darcy's law applies to laminar flow in porous media.

Seepage velocity (Darcy velocity) is not the real flow velocity of groundwater (u). The cross-sectional area A is the entire soil, and not the area of pore. To obtain the real average velocity u, the flow is limited only to the pore space as

$$u = \frac{Q}{A'} = \frac{Q}{A \cdot n} \qquad (1.14)$$

where n is the porosity of soils, %.

In conjunction Eq. (1.13) with Eq. (1.14), it can be found that the real average velocity u is

Fig. 1.17 Seepage velocity and hydraulic gradient

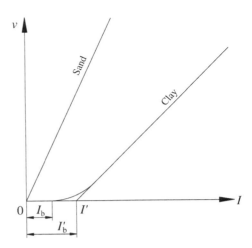

$$u = \frac{v}{n} \quad (1.15)$$

Because n is always smaller than 1, in relative order of magnitude, real average velocity is greater than seepage velocity.

When groundwater flows in sands, Darcy's law is reasonable, as shown in Fig. 1.17 investigated by the experiments. The velocity is proportional to the first power of the hydraulic gradient. But in clays, Eq. (1.13) is not so applicable. Around mineral particle surfaces, there are absorbed water films, which obstruct or block the pore channels for water flowing. Meantime, they could not be ignored. Results show that smaller hydraulic gradient could not resist the adhesion of absorbed water film, so that the water could not flow through these pores. Until the hydraulic gradient is larger than a certain critical value I_b (yield hydraulic gradient), clay can be permeable (see Fig. 1.17). If the intercept of the linear part of the curve in clay is I'_b (threshold hydraulic gradient), Eq. (1.13) can be expressed as the hydraulic conductivity

$$v = K(I - I'_b) \quad (1.16)$$

1.3.2.2 Validity of Darcy's Law

In applying Darcy's law, it is important to know the range of validity within which it is applicable. Because velocity in laminar flow is proportional to the first power of the hydraulic gradient, it seems reasonable to believe that the Darcy's law applies to laminar flow in porous media. Thus sometimes Darcy's law is called laminar seepage principle. Since 1940s, many experiments have revealed that not all the

1.3 Groundwater Movement

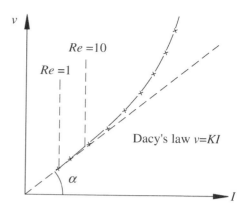

Fig. 1.18 Seepage velocity and hydraulic gradient

laminar flows have the characteristics of linear relationship between velocity and hydraulic gradient. Jacob and Bell investigated the relations between seepage velocity and hydraulic gradient (Fig. 1.18). The Reynolds number, which expresses the dimensionless ratio of inertial to viscous forces, serves as a criterion to distinguish between laminar and turbulent flow. Hence, by analogy, the Reynolds number can be employed to establish the limit of flows described by Darcy's law, corresponding to the value where the linear relationship is no longer valid.

Reynolds number is expressed as

$$Re = \frac{\rho v D}{\mu} \quad (1.17)$$

where ρ is the fluid density; v is the velocity; D is the diameter of cross section; and μ is the dynamic viscosity of the fluid. Experimental results shown in Fig. 1.18 show that Darcy's law is valid for $Re < 1$ and does not depart seriously up to $Re = 10^1$. It represents an upper limit to the validity of Darcy's law. A range of values rather than a unique limit must be stated because as inertial forces increase, turbulence occurs gradually. For fully developed turbulence, the head loss varies approximately with the second power of the velocity rather than linearly. When flow velocity of groundwater is much slower, the flow movement is mainly controlled by viscous forces. The influence of inertial forces can be ignored. Darcy's law is applicable. As the velocity increases, the water flow has continuously variable velocity and acceleration. The inertial forces are proportional to the second order of velocity. Darcy's law could not be available any more.

Since the shape, size, and orientation of pores in soils are quite complicated, they vary in large range. The state transition of laminar to turbulent occurs in some pores, while other pores may not change. Thus this transition from linear laminar flow, to nonlinear laminar flow, turbulent flow, develops gradually, without apparent limit. Fortunately, most natural underground flow occurs with $Re < 1$, so Darcy's law is applicable.

1.3.2.3 Hydraulic Gradient

Hydraulic gradient is expressed as the dimensionless ratio of water head loss along the seepage path to corresponding seepage length. During the water flow through soil particles, the head loss is resulted in energy consumption being lost by frictional resistance dissipated as heat energy. It is defined as potential loss. Therefore, hydraulic gradient can be understood as the energy consumption of frictional resistance per unit length along seepage path.

1.3.2.4 Hydraulic Conductivity

The permeability of a rock or soil defines its ability to transmit a fluid. For practical work in groundwater engineering, where water is the prevailing fluid, hydraulic conductivity K is employed. A medium has a unit hydraulic conductivity if it will transmit in unit time a unit volume of groundwater at the prevailing kinematic viscosity through a cross section of unit area, measured at right angles to the direction of flow, under a unit hydraulic gradient. The units are

$$K = -\frac{v}{dh/dl} = -\frac{\text{m/d}}{\text{m/m}} = \text{m/d} \qquad (1.18)$$

which indicates that hydraulic conductivity has the same units of velocity.

Hydraulic conductivity of soils can be determined by a variety of techniques, including calculation from formulas, laboratory methods, tracer tests, auger hole tests, and pumping tests of wells.

First, they are several factors, which influence the permeability.

1. Grain size distribution. Generally, coarse, uniform, and smooth grains have large K value. For clean sands (including no fines), the hydraulic conductivity can be estimated by

$$K = 100-150\,(d_{10})^2 \qquad (1.19)$$

where d_{10} is the effective grain diameter, smaller than which the accumulative weight percentage is summed up to 10 %. When sands contain fines, the hydraulic conductivity decreases sharply, as the finer content increases.
2. Degree of density. The denser the soil, the smaller the hydraulic conductivity. Experimental results show, as for sand, that the K value is proportional to the second power of the void ratio, while in clay, the exponential index is larger. Because of the thickness of the absorbed water film, the empirical relationship can hardly be established.
3. Saturation. The higher the saturation, the larger the hydraulic conductivity it is. It is mainly due to the existence of air, which would decrease the flow cross-sectional area, or even obstruct the small pore spaces.

4. Soil structure. Fine-grained soils always have complex soil structure. Once it is disturbed, the shape, size, and distribution of previous flow cross section will correspondingly change. The resulting hydraulic conductivity must be different. The reconstituted or compaction soil samples always have smaller hydraulic conductivity compared to undisturbed soil.
5. Soil texture. The hydraulic conductivity is also greatly influenced by soil texture, such as, if there is a thin sand interbed imbedded in the clay layer, it must be resulted that the horizontal hydraulic conductivity is much greater than the vertical value, even by tens of orders. Therefore, in the laboratory methods for determining the hydraulic conductivity, most representative soil sample is most important. Sometimes, field pumping tests are very necessary for the hydraulic conductivity of natural soil layers.
6. Water temperature. Experimental results show that the hydraulic conductivity is also related to the properties of groundwater, including unit weight γ_w and coefficient of viscosity η(Pa·s). Under different temperatures, γ_w rarely changes; but η varies a lot. Higher temperature is resulted in smaller coefficient of viscosity, correspondingly larger hydraulic conductivity. K and $1/\eta$ almost has a linear relationship. Thus, the hydraulic conductivity (K_T) under temperature T (°C) should be amended to standard hydraulic conductivity value (K_{10}) under the temperature of 10 °C:

$$K_{10} = \frac{\eta_T}{\eta_{10}} K_T \qquad (1.20)$$

where η_T and η_{10} are the coefficient of viscosity under the temperature of T °C and 10 °C (All the values can be checked in Physical handbook). As the temperature is 5 °C, $\frac{\eta_T}{\eta_{10}} = 1.161$. While it is 0.773, under temperature is 20 °C. Apparently, the influence of temperature could not be neglected.

The temperature of underground water is usually stable at 10 °C, so generally the value under 10 °C is regarded as standard criterion. Some other countries take 15 °C or 20 °C for standard.

Second, in the laboratory, hydraulic conductivity can be determined by a permeameter, in which flow is maintained through a small sample of material, while measurement of flow rate and head loss is made. The constant-head and falling-head types of permeameters are simple to operate and widely employed.

The constant-head permeameter shown in Fig. 1.19 can measure hydraulic conductivity of consolidated or unconsolidated formations under low heads. Water enters the medium cylinder from the bottom and is collected as overflow after passing upward through the materials. From Darcy's law, it follows that the hydraulic conductivity can be obtained from

$$K = \frac{VL}{Ath} \qquad (1.21)$$

Fig. 1.19 The constant-head permeameter

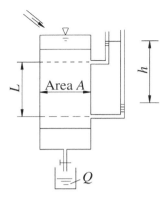

where V is the flow volume in time t, and other dimensions A, L, and h are shown in Fig. 1.19. It is important that the medium be thoroughly saturated to remove entrapped air. Several different heads in a series of tests provide a reliable measurement.

A second procedure utilizes the falling-head permeameter illustrated in Fig. 1.20. Here, water is added to the tall tube; it flows upward through the cylindrical sample and is collected as overflow.

The test consists of measuring the rate of fall of the water level in the tube. The hydraulic conductivity can be obtained by noting that the flow rate Q in the tube

$$dQ = A_1 dh/dt \qquad (1.22)$$

Fig. 1.20 The falling-head permeameter

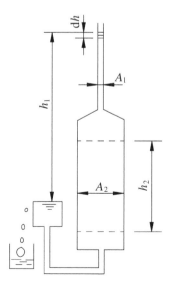

1.3 Groundwater Movement

must equal the amount through the sample, which by Darcy's law is

$$dQ = A_2 Kh/L \, dt \tag{1.23}$$

After equaling and integrating from t_1 to t_2,

$$A_1 dh/dt = A_2 Kh/L dt$$
$$A_1 \int_{t_1}^{t_2} \frac{dh}{h} = \frac{KA_2}{L} \int_{t_1}^{t_2} dt$$
$$A_1 \ln \frac{h_1}{h_2} = \frac{KA_2}{L}(t_2 - t_1) \tag{1.24}$$
$$K = \frac{A_1 L}{A_2 t} \ln \frac{h_1}{h_2}$$

where L, A_1, and A_2 are shown in Fig. 1.20, and $(t_2 - t_1)$ is the time interval for the water level in the tube to fall from h_1 to h_2.

Permeameter results may bear little relation to actual field hydraulic conductivities. Undisturbed samples of unconsolidated materials are difficult to obtain, while disturbed sample experience changes in porosity, packing, and grain orientation, which modify hydraulic conductivities. So one or even several samples from aquifer may not represent the overall hydraulic conductivity of an aquifer. Variations of several orders of magnitude frequently occur for different depths and locations in an aquifer.

Example 1:
A field sample of medium sand is tested to determine the hydraulic conductivity using a constant-head permeameter with a head difference of 83 mm (h). The permeameter has length of 200 mm (L) and a diameter of 75 mm (D). In 1 min, 71.6 cm³ of water is collected at the outlet. Determine the hydraulic conductivity of the sample.

Solution:
Equation (1.21) is used to compute the hydraulic conductivity in a constant-head permeameter test:

$$K = \frac{QL}{Ath} = \frac{4 \times 71.6 \times 20}{\pi \times 7.5^2 \times 8.3 \times 60} = 6.5 \times 10^{-2} \, \text{cm/s}$$

Further question: what should the maximum allowable piezometric head difference be for a series of tests?

Example 2:
A field sample of silty sand with a cross-sectional area of 44.18 cm³ is tested to determine the hydraulic conductivity using a falling-head permeameter with a cross-sectional area of 1.77 cm³ and the initial head of 130 cm. Over a period of 135 s, the head in the tube falls to 80 cm. Estimate the hydraulic conductivity of the sample.

Table 1.4 Estimation of hydraulic conductivity

Soil type	Permeability	K (cm/s)	Soil type	Permeability	K (cm/s)
Boulders, cobbles, gravels	Very good	$>1 \times 10^{-1}$	Silty clay	Poor	1×10^{-5}–10^{-6}
Sands	Good	1×10^{-2}–10^{-3}	Clay	Very poor	$<1 \times 10^{-7}$
Sandy clay	Medium	1×10^{-3}–10^{-4}			

Solution:
Equation (1.24) is used to compute the hydraulic conductivity in a falling-head permeameter test:

$$K = \frac{A_1 L}{A_2 t} \ln \frac{h_1}{h_2} = \frac{1.77 \times 15}{44.18 \times 135} \ln \frac{130}{80} = 2.16 \times 10^{-3} \text{ cm/s}$$

Third, some empirical values of common soil layers are presented in Table 1.4. It can be employed in case of lacking specific relevant information.

1.3.2.5 Intrinsic Permeability

To avoid confusion with hydraulic conductivity, which including the properties of groundwater, an intrinsic permeability k may be expressed as

$$k = \frac{K\mu}{\rho g} \quad (1.25)$$

where K is the hydraulic conductivity; μ is the dynamic viscosity; ρ is the fluid density; and g is the acceleration of gravity, m/s². Thus, intrinsic permeability possesses units of area. Because values of k in Eq. (1.25) are usually very small in units of m², it is always used in square micrometers $(\mu m)^2 = 10^{-12}$ m².

1.3.2.6 Transmissivity

The term transmissivity T is widely employed in groundwater engineering. It may be defined as the rate at which water of prevailing kinematic viscosity is transmitted through a unit width of aquifer under a unit hydraulic gradient. It follows that

$$T = KM = (\text{m/day})(\text{m}) = \text{m}^2/\text{day} \quad (1.26)$$

where M is the saturated thickness of the aquifer.

Example 3:

A leaky confined aquifer is overlain by an aquitard that is also overlain by an unconfined aquifer. The estimated recharge rate from the unconfined aquifer into the confined aquifer is 0.085 m/year. Piezometric head measurements in the confined aquifer show that the average piezometric head in the confined aquifer is 6.8 m below the water table of the unconfined aquifer. If the average thickness of the aquitard is 4.30 m, find the vertical hydraulic conductivity K_v of the aquitard. What type of material could this possibly be?

Solution:

Equation (1.18) is used to compute the hydraulic conductivity in a constant-head permeameter test:

$$K_v = -\frac{v}{dh/dl} = -\frac{2.329 \times 10^{-4}}{(6.8/4.30)} \text{ m/day} = 1.473 \times 10^{-4} \text{ m/day}$$
$$= 1.705 \times 10^{-7} \text{ cm/s}$$

From Table 1.4, the aquitard is composed of clay.

1.3.3 Nonlinear Seepage Principles

When the Reynolds number is larger than 1–10, it must be a turbulent flow. The seepage velocity could not be linear to the hydraulic gradient any more. Presently, there is no commonly used nonlinear motion equation. Most familiar is P. Forchheimer Equation:

$$I = av + bv^2 \tag{1.27}$$

or

$$I = av + bv^m \; (1.6 \leq m \leq 2) \tag{1.28}$$

where a and b are constants determined by experiments. When $a = 0$, Eq. (1.28) can be changed to

$$v = KI^{\frac{1}{2}} \tag{1.29}$$

This is called Chezy's law, i.e., the seepage velocity is proportional to the square root of hydraulic gradient.

At the beginning, the combined state of laminar and turbulent flow is mentioned. There is no apparent limit between these two states of flow movement. Rum Gail proposed a combined flow state, in which laminar and turbulent flows both exist, as follows:

$$v = KI^{\frac{1}{m}} \qquad (1.30)$$

where *m* is the liquidity index, 1–2. When *m* = 1, Eq. (1.30) is converted as Darcy's law. While as 2, it is equivalent to Chezy's law. As 1 < *m* < 2, inertial forces could not be ignored anymore and it plays some parts on the movement of groundwater flow.

1.3.4 Flow Nets

Groundwater flow through soils can often be described approximately in a relatively simple way by a flow net. For specific boundary conditions, flow lines and equipotential lines can be mapped in two dimensions to form a flow net. The two sets of lines form an orthogonal pattern of small squares.

The purpose of mapping a flow net is aimed to visibly investigate the seepage path. More importantly, it can calculate the seepage amount and also could determine the water head at each location in the flow net. In practical engineering, many two- or three-dimensional conditions of seepage can be encountered. In these cases, mapping a flow net is a very effective way. As shown in Fig. 1.21, in the construction of foundation pit dewatering, it is a two-dimensional seepage problem around the underground diaphragm wall. It is most convenient to calculate the seepage amount and head loss by the flow net contour.

Before mapping the flow net, there are two basic conditions should be met.

1. First, flow lines should strictly reflect the flow directions. This property is determined by the definition of flow line and equipotential line. Specifically, each tangential direction of an arbitrary point in the flow line represents the direction of seepage velocity. In Fig. 1.22, point m is the crossing point of the flow line 1-1 and equipotential line a-a, where the slope is

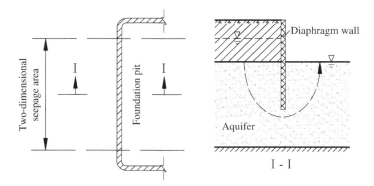

Fig. 1.21 Two-dimensional seepage problem

1.3 Groundwater Movement

Fig. 1.22 Portion of an orthogonal flow net formed by flow and equipotential lines

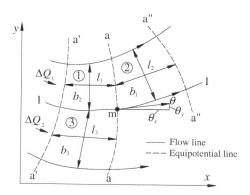

$$\left(\frac{dy}{dx}\right)_{\text{flow-line}} = \frac{v_y}{v_x} \qquad (1.31)$$

The equipotential line is formed by the points with the same water head H, which is along the equipotential line aa, $\Delta H = 0$ between each point. In two-dimensional steady seepage flow, $H = f(x, y)$ has no relationship with z, t. Thus, it has

$$\Delta H = \frac{\partial H}{\partial x} dx + \frac{\partial H}{\partial y} dy = 0 \qquad (1.32)$$

According to Darcy's law along x, y directions,

$$v_x = K \cdot I_x = K \frac{\partial H}{\partial x} \qquad (1.33)$$

$$v_y = K \cdot I_y = K \frac{\partial H}{\partial y} \qquad (1.34)$$

After substituting into Eq. (1.32),

$$\frac{v_x}{K} dx + \frac{v_y}{K} dy = 0 \qquad (1.35)$$

Thus the slope of equipotential line aa can be

$$\left(\frac{dy}{dx}\right)_{\text{equipotential-line}} = -\frac{v_x}{v_y} \qquad (1.36)$$

In conjunction Eq. (1.36) with Eq. (1.31),

$$\left(\frac{dy}{dx}\right)_{\text{flow-line}} \left(\frac{dy}{dx}\right)_{\text{equipotential-line}} = -1 \qquad (1.37)$$

Therefore, it confirms that flow and equipotential lines are always orthogonal.

2. Second, each small square formed by two sets of orthogonal lines has the same $\frac{b_i}{l_i}$ value (shown in Fig. 1.22, in small orthogonal square i, b_i is the average distance of flow line and l_i is the average distance of equipotential line). For convenient calculation, the flow volume ΔQ and head loss ΔH in each small square are identical.

Investigating the seepage a ①, ②, ③, from Darcy's law, it has

$$\Delta Q_1 = K \cdot \frac{\Delta H_1}{l_1} \cdot b_1 \times 1 = K \cdot \frac{\Delta H_2}{l_2} \times b_1 \times 1 \qquad (1.38)$$

$$\Delta Q_3 = K \cdot \frac{\Delta H_3}{l_3} \cdot b_3 \times 1 \qquad (1.39)$$

where ΔH_1 is the water head loss of equipotential lines from a'a' to aa; ΔH_2 is the water head loss of equipotential lines from aa to a"a".

From Eqs. (1.38) and (1.39), providing

$$\frac{b_1}{l_1} = \frac{b_2}{l_2} = \frac{b_3}{l_3} = \cdots = \frac{b_i}{l_i} \qquad (1.40)$$

Then,

$$\begin{aligned} \Delta H_1 &= \Delta H_2 = \cdots = \Delta H_i \\ \Delta Q_1 &= \Delta Q_2 = \cdots = \Delta Q_i \end{aligned} \qquad (1.41)$$

The ratio value of $\frac{b_i}{l_i}$ can be arbitrary in mapping a flow net. Generally, $\frac{b_i}{l_i}$ is set to be 1. Each small seepage area is close to a small square.

The steps for mapping a flow net consist of four main parts.

1. Plot the contour of structure and soils according a certain mapping scale (Fig. 1.23).
2. Determine the boundary conditions, such as acb and ss' are flow lines. If no flow crosses an impermeable boundary, flow lines must parallel it. Similarly, if now flow crosses the water table, it becomes a bounding flow surface, or called equipotential lines, such as a'a and bb'.
3. Try to plot several apparent flow lines (almost paralleling, noncrossing smooth, and gentle lines. Because water always flows along the shortest seepage path, the flow direction changes along the smallest slope of flow lines). Flow lines should be orthogonal with the inlet and outlet water surfaces. They also should be paralleling to impervious surfaces.

1.3 Groundwater Movement

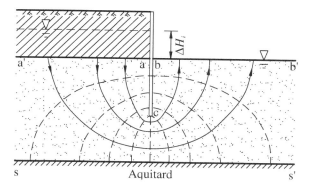

Fig. 1.23 Mapping the flow net (trial-and-error method)

4. Add some equipotential lines (they must be orthogonal with flow lines. Each small seepage area is close to square). Sometimes, this process should be tried for several times before success.

Described as above, the most used method to plot a flow net is trial-and-error method. Besides this, the model test (such as model of sands in a water tank), electric model test, or in some simplified cases, the differential equation governing flow can be solved to obtain the flow net. Most complicated conditions can only be analyzed by numerical model. About the specific descriptions about these methods, groundwater dynamics can be referred. In engineering, the trial-and-error method is most convenient. The accuracy also meets the requirement.

From the flow net, some information and parameters can be deduced. First is the flow rate Q.

Providing that there are N_f small segments in equipotential lines, the flow rate is $Q = N_f \cdot \Delta Q$; N_D in flow lines, so that the head loss in each small orthogonal square along the flow lines $\Delta H_i = \frac{\Delta H}{N_D}$. Thus

$$\Delta Q = K \frac{\Delta H_i}{l_i} b_i = K \Delta H_i Q$$
$$Q = N_f \Delta Q = N_f K \frac{\Delta H}{N_D} = K \frac{N_f}{N_D} \Delta H \quad (1.42)$$

As shown in Fig. 1.23, $N_f = 4$, $N_D = 10$, based on the values of ΔH and K, the flow rate in unit length along the foundation pit can be estimated.

Second is the water head H and hydraulic gradient I at each location. According to $\Delta H_i = \frac{\Delta H}{N_D}$, i.e., along the direction of flow, if groundwater moves n equipotential lines, the water head decreases $n \cdot \Delta H_i$. In Fig. 1.23, the total water head at each location along equipotential line aa' is H_a, so the water heads of all the crossing points along the flow lines with equipotential line aa' are, respectively, $H_a - \frac{1}{N_D} \Delta H$, $H_a - \frac{2}{N_D} \Delta H, \ldots$. In the last, equipotential line is bb', $n = N_D = 10$, and the water heads of all points in this line are the same as $H_a - \Delta H = H_b$. The water head of each point between two equipotential lines is derived by linear interpolation. Therefore, providing the water head between two arbitrary points, the relevant hydraulic gradient I can be estimated. It can be easily seen that the denser the equipotential lines, the larger the hydraulic gradients are.

1.4 Exercises

1. How many types of groundwater, according to the physical and mechanical properties?
2. What parameters reflect the water-physical properties of groundwater?
3. What is aquifer, aquiclude? Is aquiclude completely impermeable?
4. Based on the occurrence condition, what types are there in groundwater?
5. How to distinguish the pressure head, potential head, total head?
6. What is the dynamic water pressure?
7. What is the validity of Darcy's law?
8. What factors influence the hydraulic conductivity?
9. What is the flow net? And how it makes?

Chapter 2
Hydrogeological Parameters Calculation

Hydrogeological parameters of aquifer are the essential and crucial basic data in the designing and construction progress of geotechnical engineering and groundwater dewatering, which are directly related to the reliability of these parameters.

There are three types of hydrogeological parameters that reflect the hydraulic properties of aquifer, as follows:

The first type is the parameters that represent the properties of aquifer. Hydraulic conductivity (K) and transmissibility (T) represent the aquifer's permeability. The water reserving capacity is represented by the specific yield (μ) in unconfined aquifer and storage coefficient (μ^*) in confined aquifer. The rate of water head conduction is represented by groundwater table conductivity in unconfined aquifer and pressure transitivity in confined aquifer, which are both a.

The second type parameters show the interaction of aquifers after dewatering, including leakage coefficient (σ) and leakage factor (B).

The third type parameters refer to the capacity of water exchange between aquifers and the external environment. It includes parameters that refer to the receiving capacity of external recharge and the degree of water loss. The former includes infiltration coefficients (α) of precipitation, river and irrigation, and the latter mainly for coefficient of phreatic evaporation.

There are many methods in hydrogeological parameter calculation. Laboratory tests and pumping and injection tests are the most common methods in geotechnical engineering design and construction. With the data of long-term groundwater observation, hydrogeological parameters can also be back calculated by analytical and numerical solutions and optimization method.

2.1 Hydrogeological Tests

In geotechnical engineering, hydrogeological in situ tests include pumping test, recharge test, infiltration test, injection test, water pressure test, connection test, groundwater flow direction and velocity test et al. These tests are used to calculate hydrogeological parameters and find out the hydraulic connection between different aquifers and between groundwater and surface water. Hydrogeological and geotechnical engineering design and construction conditions should be considered when selecting test method.

2.1.1 Pumping Test

Pumping test is one of the most common geotechnical engineering investigation methods in finding out the permeability and calculating the parameters of aquifers. Different types of pumping tests are applied in different engineering programs according to their objectives and hydrogeological conditions.

Pumping tests can be divided into three types according to the operation and the number of wells, shown in Table 2.1.

2.1.1.1 Objective, Task, and Types of Pumping Test

1. Objective and task of pumping test

Pumping test is on the basis of well flow theory. During this test, groundwater is pumped out through the main well and the change of flow rate in observation wells is measured. Meanwhile, the variation of state and distribution of seepage field in the time and space is also measured. Pumping test is aimed at finding out the hydrogeological condition of engineering construction field, quantifying the water amount of pumping wells and aquifers, calculating the hydrogeological parameters and finally providing a basis for groundwater solution program.

The main tasks of pumping test are as follows:

(1) Measure the variation of drawdown with the change of discharge of wells or drilling holes, then calculate the unit inflow and estimate the maximum yielding water of the aquifer.
(2) Determine the hydrogeological parameters of aquifer, including hydraulic conductivity, transmissibility, specific yield, storage coefficient, pressure transitivity, leakage factor, and influence radius et al.
(3) Measure the shape of cone of depression, and its expanding progress.
(4) Find out the hydraulic connection between different aquifers and between groundwater and surface water.

Table 2.1 Pumping test classification and applied range

Type	Applied range
Simple pumping test in drillings or exploration wells	Rough estimate of the hydraulic conductivity of aquitard
Pumping test without observation well	Preliminary determination of hydraulic conductivity
Pumping test with observation wells	Accurate determination of hydraulic conductivity

(5) Determine the aquifer boundary condition, including its location and properties.
(6) Conduct pumping simulation to provide necessary data for well-group design, which includes determining reasonable distance and diameter of wells, drawdown and the flux of water.

2. Types of pumping test

According to different classification principles, pumping tests can be classified as follows:

(1) Steady flow pumping test and unsteady flow pumping test, according to groundwater flow state on the basis of well flow theory.

- (a) Steady flow pumping test is an early common method, which requires the test must last for a long time after meeting the stable flow and drawdown. Steady flow theory is used in calculation of aquifer's parameters, such as hydraulic conductivity, influence radius, etc. However, groundwater flows are mostly unsteady in nature; only the areas which have abundant and stable water supply can form a relatively steady seepage field. Therefore, its application is limited.
- (b) Unsteady flow pumping test has been used universally since 1970s in our country. It requires the water discharge or water table to remain constant. Generally, it is the water discharge flux that remains constant or staged constant and the water table changes with time. The duration of the unsteady flow pumping test is determined by s-lg$t$ curve. If the aquifer has an infinite recharge boundary, then pumping can be terminated after an inflection point appears on the curve. While if the aquifer has a constant head boundary, impermeable boundary, or leakage recharge, there are generally two inflection points.

  The results of unsteady flow theories and formulas can be more accurate than steady flow theories, and so the former has a wider application. It can calculate more parameters, such as transmissibility, specific yield, storage coefficient, pressure transmission coefficient, leakage factor and so on. Also it can determine the simple boundary conditions and take full advantages of all the information provided throughout the whole pumping process. However, the calculation is much more complex that needs higher technical standards for observation. Generally, for the early unsteady stage and later steady stage, relevant formulas are applied, respectively, to calculate the parameters in different stages.

(2) Single well pumping test and multiwells pumping test, depending on whether there is observation well(s).

- (a) Single well pumping test is the pumping test that only has one pumping well, which also known as main well, and has no observation well. It is simple and less expensive, but not very accurate, which makes it suit for preliminary investigation stage. The main well is usually set at the place where is rich in groundwater. Aquifer's water abundance, permeability, and the relationship between pumping discharge and drawdown can be found through single well pumping test.

(b) Multiwells pumping test is the pumping test that has a pumping well and one or more observation well(s). It has a wider application. It can determine not only the hydraulic conductivity and pumping discharge, but also the anisotropy of hydraulic conductivity, the radius and shape of the depression cone, the width of supply area, the reasonable well spacing, interference coefficient, and the hydraulic connection between groundwater and surface water. Besides, seepage velocity test also can be taken during the pumping test. This kind of pumping test costs a lot, but the results of which are more accurate. Therefore, it is more used in detailed investigation stage than preliminary investigation stage. In the area which has the value of water supply, at least one group of multi-wells pumping test should be taken.

(3) Fully penetrating well pumping test and partially penetrating well pumping test according to the type of pumping well.

Generally, fully penetrating well pumping test is the primary choice, for its comprehensive well flow theory. Only in the condition that the aquifer is thick and homogeneous, or in the specialized study of filter's effective length, the partially penetrating well pumping test is adopted.

(4) Layering pumping test and combination pumping test according to aquifer's condition involved in test.

(a) Layering pumping test is the pumping test that conducted the test for separate aquifers to determine each aquifer's hydrogeological characters and parameters.

(b) Combination pumping test is the pumping test that tests several layers of aquifers in one pumping well. The results reflect the average value of those aquifers' hydrogeological parameters. In the condition that the layers are not numerous, the approximate value of each aquifer's parameters can be determined by recharging the well layer by layer and conducting combination pumping test accordingly.

3. Arrangement of main well and observation wells

Main well should be considered arranging in the following locations: the main water source aquifer, aquifer with large thickness and abundant water, the possible connection part between surface water and groundwater, fault or karst-concentrated zone, the representative control region, such as boundaries of different sections and aquifers.

The design of observation wells in the plane and profile layout depends on the test tasks, accuracy, feature size, aquifer's character, as well as data processing and calculation methods and other factors. If only to eliminate "well loss" or "water jump" effects, just one observation well near the pumping well need to be arranged. If to obtain reliable hydrogeological parameters, one to four rows of observation wells can be arranged according to aquifer's character and groundwater flow condition, shown in Tables 2.2, 2.3 and Fig. 2.1.

2.1 Hydrogeological Tests

Table 2.2 Distance between main well and observation wells

Aquifer's characters	Hydraulic conductivity (m/day)	Groundwater type	Distance (m)			Influence radius (m)
			First well	Second well	Third well	
Hard with developing fissures	>60	Confined	15–20	30–40	60–80	>500
		unconfined	10–15	20–30	40–60	
Hard with slight developing fissures	60–20	Confined	6–8	10–15	20–30	150–250
		unconfined	5–7	8–12	15–20	
Pure cobble, gravel and coarse-medium sand	>60	Confined	8–10	15–20	30–40	200–300
		unconfined	4–6	10–15	20–25	
Cobble and gravel with fine particles	60–20	Confined	5–7	8–12	15–20	100–200
		unconfined	3–5	6–8	10–15	
Anisotropic sand	20–5	Confined	3–5	6–8	10–15	80–150
		unconfined	2–3	4–6	8–12	

Table 2.3 Arrangement of observation lines

Aquifer's characters		Arrangement of observation lines	Graph
Homogeneous and isotropic	Small water gradient	One line that is perpendicular to groundwater flow direction	Figure 2.1 (1)
	Large water gradient	Two lines that are perpendicular and parallel to groundwater flow direction	Figure 2.1 (2)
Heterogeneous and anisotropic	Small water gradient	Two lines that are perpendicular to groundwater flow direction and one line that is parallel to groundwater flow direction	Figure 2.1 (3)
	Large water gradient	Two lines that are perpendicular to groundwater flow direction and two lines that are parallel to groundwater flow direction	Figure 2.1 (4)

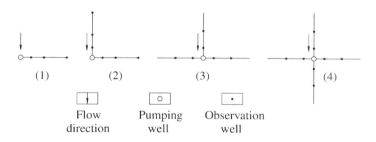

Fig. 2.1 Arrangement of observation wells

The number, distance, and depth of observation wells depend on the test task, accuracy, and pumping type. There should be no less than three observation wells arranged in one line to figure out the shape of the depression cone. For parameter calculations, only two observation wells in one line are needed for a steady pumping test, and usually three wells for an unsteady pumping test to take full use of all observation data. If the test task is to find out the hydraulic connection or boundary characters, the observation wells should not be less than two.

The distance between observation wells should be small near the main well and became larger far from the main well. The distance between the main well and the closest observation well depends on the permeability of aquifer and the drawdown, which can be several meters to 20 m on the principle of in favor of controlling the shape of depression cone and avoiding the turbulence and 3D flow around the observation well. For unsteady flow pumping tests, observation wells should be evenly distributed on a logarithmic axis and ensure the observation of the initial water table changes. The empirical distance data of observation wells can be found in the relevant handbooks.

The depth of observation wells generally is required to be 5–10 m deep in tested aquifers, except for thin aquifers. If the aquifer is heterogeneous, the depth and the filter's position of the observation wells should be in accord with the main well.

2.1.1.2 Technical Requirements for Pumping Tests

1. Steady flow pumping test
  (1) Drawdown
  Generally, at least three drawdowns should be made to determine the relation between water discharge and drawdown (Q-s curve), which can judge the correctness of tests and indicate the water discharge. While, only one drawdown is enough if the maximum drawdown is <1 m in the following conditions: the requirement to test accuracy is not very high, the test is taken in a secondary aquifer, the water discharge is too small (<0.1 L/s m), and the pumping equipment has limitations. If the Q-s relation has been determined and the correctness of pumping tests can be ensured, only two time tests need to be taken out. This is because that there are no more than two unknown coefficients in Q-s relation and the type of Q-s curve can be determined by two times pumping tests using the coefficient n in $\frac{Q_2}{Q_1} = \sqrt[n]{\frac{s_2}{s_1}}$, where $n < 1$ is unmoral, $n = 1$ is linear type, $1 < n < 2$ is exponential type, $n = 2$ is parabolic type, $n > 2$ is logarithmic type. Although this method can save one test workload process, it has poor reliability.

The maximum drawdown is mainly determined by the test purpose. When calculating the parameters, the drawdown should be smaller to avoid turbulence and 3D flow. When calculating for groundwater resource evaluation and dewatering, the drawdown should be able to extrapolate to the design requirements. When determining the boundary properties and hydraulic connection, the drawdown should be large enough to fully expose the problems, for the impermeability of some layers is related to the waterhead difference on both sides of boundary. The maximum

2.1 Hydrogeological Tests

drawdown ($s_{\max}$) can be 1/3–1/2 of aquifer's thickness in unconfined aquifer, and can be the distance between static water table and the aquifer's roof. The rest drawdowns can be evenly distributed ($s_1 = s_{\max}/3$, $s_2 = s_{\max}/2$), which is convenient for drawing Q-s curve. The minimum drawdown and the difference of each two drawdowns usually is no <1 m. In geotechnical engineering, construction design, or groundwater dewatering design, formal pumping test conducted three times, and the difference between each drawdown is more appropriate than 1 m.

(2) Stable duration time

Stable duration time refers to the time that the pumping test lasts after the seepage field reaches approximate stabilization. The time from the beginning of pumping to steady seepage field depends on the groundwater type, aquifer's parameters, boundary and recharge conditions, and drawdown value. This time is longer when it is in unconfined or leakage aquifers, or in the condition of poor water recharge or large drawdown. The duration time is different in different investigation stage, test purpose, and aquifer condition. Generally, it should meet the requirement of test reliability. It is easier to find out the slight and trending change and the false stability that is caused by temporary recharge.

The stable duration time does not need to be long when calculating parameters, usually <24 h. In other conditions usually it will be 48–72 h. No matter what the test purpose is, it should not <2–4 h of the farthest observation well.

Generally, the stable stage is reaching when the variation of water table in pumping well is <1 % of drawdown. If the drawdown is small, the limitation is 3–5 cm. When pumped with air compressor, the variation of water table in main well allows up to 20–30, and 2–3 cm in observation well, but no trending change is allowed. The variation of water discharge should not exceed 5 %.

(3) Water table and water discharge

Natural stable water table should be observed before pumping. Water table should be observed hourly. The water table that does not change in 2 h or only changes 2 cm in 4 h is the stable water table. If the natural water table fluctuates, the average value is desirable as the natural stable water table, or eliminating the interference effects.

During pumping, the water table and water discharge should be measured at the same time. The interval time for observation should be close first and loose afterward, for example 5–10 min first and 15–30 min afterward, which should e according to the specific requirements.

When pumping is stopped or broken off, recovery water table should be measured with the same interval time. The standards for stable water table judgment is the same with above. If there is difference between natural and recovery stable water table, the drawdowns should be amended by the weighted arithmetic average of the difference regarding the time.

2. Unsteady flow pumping test

Unsteady flow pumping test can be divided into constant-flow test and constant-drawdown test. The former is much used in practice. The latter is used when in artesian well or modeling dewatering or groundwater mining.

(1) Water discharge and water table

The requirements for water discharge and water table measuring is the same with steady flow pumping test. It should be especially noticed that the flow or water table should be constant from the beginning to the end of pumping.

During pumping, the water table and water discharge should be measured at the same time. When pumping is stopped or broken off, recovery water table should be measured. The interval time in unsteady flow pumping test should be smaller than it in steady flow pumping test, especially in first 10–30 min. For example, it could be observed at 1, 2, 3, 4, 6, 8, 10, 15, 20, 25, 30 min, then at every 30 min.

(2) Stable duration time

The stable duration time for unsteady flow pumping test also depends on test tasks and purposes, hydrogeological conditions, test type, water discharge, and calculation method. It has big differences in different pumping test, which also has no uniform regulations. For the parameter calculation alone, the duration time usually does not exceed 48–96 h in our country. However, the variation is 6–600 h according to global data, and 48–96 h is the most choice.

If the aquifer is a borderless confined aquifer, curve-matching and linear graphic methods are in common use. The former only requires the early pumping data, while the latter needs pumping data for two pairs of log-periods. These mean that the total pumping time needs to be three pairs of log-period, which is 1000 min, about 17 h. So the pumping usually lasts for 1–2 days. If there are more than one observation wells, all of them should meet the above requirements. If the water discharge is ladder-like distributed, the last ladder should also continue to meet the above requirements.

In leakage flow, if inflected point method is used in parameter calculation, the duration time should be long enough to judge the maximum drawdown. If linear graphic method is used, the duration time can be shorter. If the data of steady stage is used, it also should meet the requirements for steady flow.

If the test purpose is to determine the boundary location and character, the duration time should be long enough to finish the job. For example, if there is constant head boundary, steady stage should be reached; for linear impermeable boundary, the second line segment in s-$\lg t$ curve should occur and the pumping generally lasts more than 100 min. Some impermeable boundary can be permeable when waterhead difference is high enough, so the duration time should ensure that the water table drawdown near boundary value reaches a predetermined value.

The test duration could be long in following circumstances: using large group wells pumping test to determine the boundary property, using the hydrogeological numerical method to calculate the parameters of heterogeneous area, and modeling water supply and unwatering.

3. Measurement of water temperature and weather temperature

Water temperature and weather temperature should be measured every 2–4 h. Other groundwater physical properties should be recorded if necessary.

4. Water sampling

At the end of pumping test, water samples should be taken for full chemical analysis, bacteria analysis, or other special analysis. The sample for chemical analysis should be no <2000 mL and analyzed in one week after sampling. As for bacteria analysis, 500 mL sample is needed, which should be sealed with wax and analysed within 6 h after sampling. Special analysis should be taken according to requirements.

2.1.1.3 Test Equipment and Appliances

Test equipment mainly refers to pumping equipment, such as water pump. Test appliances include flowmeter, water table indicator, water thermometer, and timer. Besides, drainage should be constructed and communication tools should be set.

1. Pumping equipment

There are many types of pumping equipment, in which the horizontal centrifugal pump, deep-well pump, and air compressor.

(1) Horizontal centrifugal pump

Centrifugal pump has a simple structure and small size, which is easy to handle and adjust the flux. It can pump large quantities of water that are even mixed with a mass of sand, but the pumping head is small, only 5–9 m. It is commonly used in shallow well pipes and volume water or group wells.

(2) Deep-well pump

The main advantage of deep-well pump is that it can pump deep water evenly. However it is hard to adjust the flux, and not suit for water with high sand content. It can be used in wells that the water table is more than 10 m and with less sand.

(3) Air compressor

Air compressor has simple structure and can be easily handling and can pump water with a mass of sand. It is not affected by the slight curve of pipe well. However, the efficiency of the air compressor is only 15–25 %, which leads too much wasted power. It is not able to pump evenly and stably, and cannot run a long time. Sometime it cannot meet the engineering needs, so it is not suitable for large-scale pumping test work.

However, it is usually used in drilling washing. In order to save cost and time, it is also used in pumping test after washing work.

(4) Other pump types

There are many other pump types, which can be chosen according to specific conditions. For example, axial flow pump is suitable for volume and shallow water while jet pump and rob pump are suitable for the opposite condition; water hammer pump is suitable for the condition that small flux and less energy. Submersible pump is suitable for deep water and low sand content.

In short, the choice of pump depends on static groundwater table, designed outlet water, dynamic water table, well diameter, sand content, and other requirements. Generally, the pumping water should be more than the designed outlet water.

2. Utensils for flow measurement

(1) Weir box

Weir box is the most common flowmeter. A triangle weir box is suitable for small flow as shown in Fig. 2.2, and a trapezoid weir box is suitable for mass flow. Usually, weir box is made of steel, but in the group well pumping test it can be made of brick or wood for the amount of temporary weir box is too much.

Flow calculation formulas for triangle weir box are as follows:
when

$$H = 0.021 - 0.200 \text{ m}: \quad Q = 1.4 H^{\frac{5}{2}} \tag{2.1}$$

when

$$H = 0.301 - 0.350 \text{ m}: \quad Q = 1.343 H^{\frac{5}{2}} \tag{2.2}$$

when

$$H = 0.201 - 0.300 \text{ m}: \quad Q = \frac{1}{2}(1.4 + 1.343) H^{\frac{5}{2}}$$

where H is the water head, in m, which is measured by steel ruler; the ruler is 0.8–1.0 m far from overflow plate, and its zero point and the crest of weir are in the same horizontal line; Q is the water flow, m^3/s.

Fig. 2.2 Triangle weir box

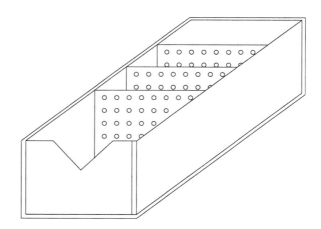

2.1 Hydrogeological Tests

(2) Orifice flowmeter

The principle of orifice flowmeter is to set a thin-walled hole with a certain diameter near the end of the outlet pipe and measure the waterhead of the two sides of orifice or of the position at a certain distance from the orifice if the flowmeter is at the end of water pipe. The waterhead is only dependent on flow velocity, if the diameters of water pipe and orifice are determined. So the quantity of flow can be calculated from that waterhead. There are two types of orifice flowmeters as shown in Figs. 2.3 and 2.4. The orifice flowmeter is portable and accurate, but not suitable for air compressors.

The following formula can be used to calculate flow in unit time:

$$Q = 0.0125 E d^2 \sqrt{\frac{H}{1000}} \text{ (water temperature : } 1-20\,°C) \qquad (2.3)$$

where Q is the quantity of flow, m^3/h; d is the diameter of orifice, mm; H is the waterhead difference, mm; E is the coefficient determined by the diameters of

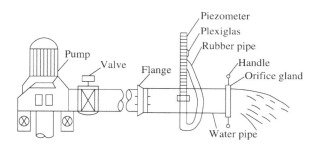

Fig. 2.3 Installation of orifice flowmeter

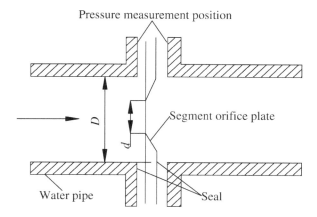

Fig. 2.4 Segment orifice flowmeter

orifice and water pipe, as well as the connection method of orifice. If the flange plate is used, then:

$$E = \frac{k}{\sqrt{1-B^4}} = 0.606 + 1.25(B-0.41)^2 \quad (2.4)$$

$$B = d/D \quad (2.5)$$

where k is the drainage coefficient.

(3) YKS-1 impeller orifice instantaneous flowmeter

The flow velocity, which is used to calculate the flow, can be measured by the impeller speed. The impeller speed is measured by electronic device. This type of flowmeter is small, light, and easy to use, which however is also not suitable for air compressors.

(4) Water meter

It is used together with centrifugal pump or deep-well pump. The water should be clear and there should be no sand or mud in it to keep the water meter work normally. The measurement error is ±2–3 %.

3. Water table indicator

The common water table indicators include electronic ones and float-type ones. The former ones indicate the water table by an ammeter, a bulb, or a loudspeaker. Recently, the pressure indicators and capacitor-based indicators are getting recognized. All the above-mentioned types belong to contact measurement. The no-contact ultrasonic water gauge is a new type with bright prospects.

2.1.1.4 Comprehensive Analysis of Pumping Test Data

1. Site data analysis

During the pumping test, the water table and flow should be observed and recorded carefully. Besides, following diagrams should be drawn to know the test progress, find out anomaly, and lay a foundation for indoor data statistic.

(1) Steady flow pumping test

(a) Draw water discharge versus time and drawdown versus time curves for main well.

The normal curve is drawn in Fig. 2.5. At the beginning of pumping, the values of drawdown and water discharge are all big and unstable. Over time, they become stable. According to the changing trend of these curves, the start and end of the stable phase can be determined reasonably.

(b) Draw drawdown versus time curves for observation wells if there is, such as s_1 curves for OW1, OW2 et al. in Fig. 2.5.

(c) Draw flow versus drawdown curves ($Q = f(s)$ curves).

Draw the point that represents a certain flow under certain second stable drawdown. Connect all the points to get the flow versus drawdown curve, as shown in Fig. 2.6. The meanings of these curves are:

2.1 Hydrogeological Tests

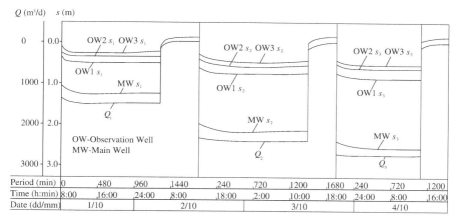

Fig. 2.5 Water discharge versus time and drawdown versus time curves

Fig. 2.6 $Q = f(s)$ curves

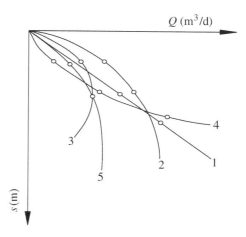

Curve 1—the curve for confined groundwater.
Curve 2—the curve for unconfined groundwater, confined–unconfined groundwater, or confined groundwater that is influenced by a 3D flow or turbulent flow, or by the resistance of well wall and filter.
Curve 3—the curve for groundwater with deficient water supply or the flow cross-section is blocked during pumping.
Curve 4—if the pump faucet is at the same position with filter, this curve indicates that pumping is affected by a 3D flow or turbulent flow, which makes it correct; if the pump faucet is above filter, this curve indicates that the results of pump test are wrong and the test should be redone.
Curve 5—this curve refers that under a certain drawdown s, the pump flow Q will be constant; this curve occurs when the drawdown is too large.

The $Q = f(s)$ curves can be used to understand the hydraulic characteristics of aquifer and yield capacity of dilled hole, to predict the maximum yield quantity, and to verify whether the results of pumping test are correct or not.

(d) Draw unit pumping-flow versus drawdown curves ($q = f(s)$ curves)

Connect the points that refer to the drawdown with a certain unit pumping-flow of the same drill hole will get the unit pumping-flow versus drawdown curve, as shown in Fig. 2.7. The meanings of these curves are the same with Fig. 2.6.

(e) Draw water table recovery curves

The method is same with the draw drawdown—time curves.

If the pumping test is normal, the curve should be rising linearly at first, and then the rise becomes slow, and finally turns horizontal. The wavy curves indicate that the observation results are wrong.

The water table recovery curves can be used to estimate the groundwater type and permeability performance of the stratum. If the water table recovers quickly, it may be the confined aquifer or strong permeable stratum. Conversely, if the water table recovers slowly, it is usually an unconfined aquifer or aquitard.

(2) Unsteady flow pumping test

> (a) Draw drawdown versus time curves (s-t curves) with the same method referred above. If the unsteady flow pumping test time is short, then magnify the time scale of abscissa. If the data include main well and observation wells, the s-t curves of them can be drawn in the same figure.
> (b) Draw drawdown versus logarithmic time curves (s-lg$t$ curves).
> (c) Draw double logarithmic curves of drawdown versus time (lg$s$-lg$t$ curves).
> (d) Draw double logarithmic curves of observation well drawdown versus distance to main well (lg$s$-lg$r$ curves).
> (e) Draw water table recovery curves with logarithmic time (s'-lg$(1 + t_p/t')$ curves), where s' is the remaining drawdown, m; t_p is the time from the start of pumping to the end of pumping, min; t' is the time of water table recovery, starting from the end of pumping, min.

Fig. 2.7 $q = f(s)$ curves

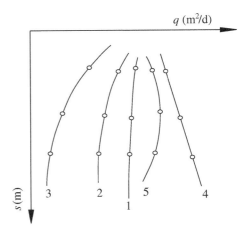

2. Indoor data analysis

(1) Draw a comprehensive result figure of pumping test, which includes: geological drilling histogram, technical structure graph of drill hole construction, Q-t, s-t curves, Q-s curves, q-s curves, table of pumping test results, table of water quality analysis, and drill hole layout plan.
(2) Calculate the hydrogeology parameters of aquifer: based on the data of steady flow pumping test and/or unsteady flow pumping test, calculate the hydrogeology parameters with multiple method and fill the summary sheet.
(3) Estimate the maximum flow of the drill hole.
(4) Write the work summary of pumping test which includes: the purposes and principles of pumping test, test method, test process, major achievements, abnormal phenomena during test and their solutions, quality analysis, and conclusions and so on.

2.1.2 Water Pressure Test

2.1.2.1 Test Purposes

The purposes for water pressure test are: exploring the fissured properties and permeability of rock and soil layers; calculating the parameters, such as unit water sucking amount (ω); providing bases for relevant design.

2.1.2.2 Test Types

Water pressure tests can be divided into following types:

1. Multistage water pressure test, synthesized water pressure test, and one-stage water pressure test according to test stages.
2. One-point water pressure test, three-point water pressure test, and multipoint water pressure test according to the number of flux-pressure relationship point.
3. Low-pressure test and high-pressure test according to pressure degree.
4. Water column pressure test, gravity flow water pressure test, and mechanical water pressure test according to pressure source, which are shown in Figs. 2.8, 2.9 and 2.10.

2.1.2.3 Main Parameters

1. Steady flux
It refers to the steady flux that is pressed into the field under certain hydrogeological conditions and pressure.

Fig. 2.8 Water column pressure test

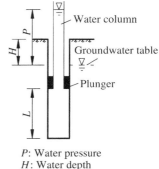

P: Water pressure
H: Water depth
L: Test segment length

Fig. 2.9 Gravity flow water pressure test

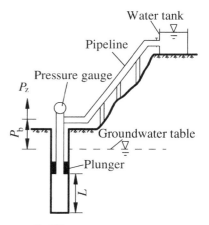

P_z: Water pressure
P_b: Pressure gauge reading
L: Test segment length

Keep the pressure constant and measure the flux every 10 min. When it meets one of following criterions, the water flow can be considered as stable, according to the *Code of Water Pressure Test in Borehole for Water Resources and Hydropower Engineering (SL31-2003)*:

(1) The difference between maximum and minimum value of four consecutive readings is <10 % of final reading Q_L, which is $Q_{max} - Q_{min} < Q_L/10$.
(2) The flow is reducing gradually till four consecutive readings are all <0.5 L/min, that is 0.5 L/min > Q_1 > Q_2 > Q_3 > Q_4.
(3) The flow is increasing gradually till four consecutive readings are no longer increase.
   In simple water pressure test, it can be lower than above standards.

Fig. 2.10 Mechanical water pressure test

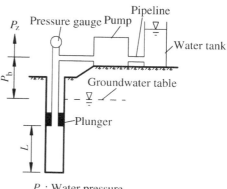

P_z: Water pressure
P_b: Pressure gauge reading
L: Test segment length

2. Pressure stage and pressure value
 (1) Total test pressure

Total pressure for water pressure test refers to the average pressure that acts on the test section. It is measured with the height of water, that is "1 m water height" = 0.98 N/cm² = 9.8 kPa ≈ 1 N/cm². The total pressure can be calculated by the following formula:

$$P = P_b + P_z + P_s \tag{2.6}$$

where P is the total test pressure, N/cm²; P_b is the reading of pressure gauge, N/cm²; P_z is the pressure of water column, N/cm²; P_s is the pressure loss that in single-pipe column plunger from pressure gauge to the bottom of plunger, N/cm².

 (2) Zero line (0–0 line) and pressure loss

Water column pressure refers to the pressure of water from zero line to the middle of pressure gauge. Therefore, the zero line (0–0) for pressure calculation should be determined first. There are three conditions, as follows:

 (a) When the water table is below the test section, the 0–0 line is the horizontal line through 1/2 of the test section, shown in Fig. 2.11.
 (b) When the water table is in the test section, the 0–0 line is the horizontal line through 1/2 of the test section that is above the water table, shown in Fig. 2.12.
 (c) When the water table is above the test section, the 0–0 line is the water table, as shown in Fig. 2.13.

The pressure is measured from water table, which should be determined before test.

Standards for stable groundwater table are as follows:

If the natural groundwater table is not affected by outer factors, or changes little, it can be determined by the average value of 2–3 times observation.

Fig. 2.11 P_z water pressure when the water table is below the test section

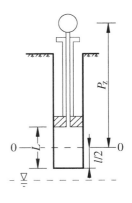

Fig. 2.12 P_z Water pressure when the water table is in the test section

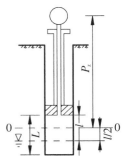

Fig. 2.13 P_z Water pressure when the water table is above the test section

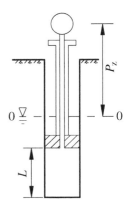

If the groundwater table changes, the stable water table is observed by following steps. At the initial observation stage, the interval time should be short, and then observed every 10 min. When the water table is no longer changed, or the rate of change of the three consecutive readings of water table is <1 cm/min (that is 10 cm/10 min), the last measured water table could be considered as stable water table.

2.1 Hydrogeological Tests

If the initial water table is higher than the stable water table in drilling, it will gradually decrease to stable, shown in Fig. 2.14. The stable standard is: $H_2-H_1 \leq$ 10 cm, $H_3-H_2 \leq$ 10 cm and the decreasing rate is <1 cm/min.

If the initial water table is lower than the stable water table in drilling, it will gradually increase to stable, shown in Fig. 2.15. The stable standard is: $H_1-H_2 \leq$ 10 cm, $H_2-H_3 \leq$ 10 cm and the decreasing rate is <1 cm/min.

The pressure loss P_s can occur in following the conditions: uniform diameter, sudden change of diameter (to bigger or smaller).

(a) Pressure loss in uniform diameter

The water pressure loss when flowing in uniform diameter can be calculated as:

$$\Delta P_{s1} = 0.49 \lambda \cdot \frac{l}{d} \cdot \frac{v^2}{g} \tag{2.7}$$

where ΔP_{s1} is the pressure loss in the uniform diameter pipe, N/cm^2; l is the length of pipe, m; d is the inner diameter of pipe, m; v is the velocity of water, m/s; g is the

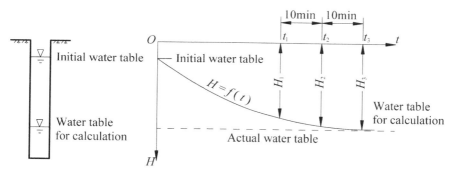

Fig. 2.14 Duration curve of water table decreasing

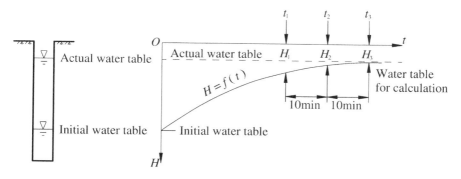

Fig. 2.15 Duration curve of water table increasing

Table 2.4 Resistance coefficients

d_2/d_1	0.1	0.2	0.4	0.6	0.8
α	0.5	0.42	0.33	0.25	0.15

Notes d_1 is the larger diameter and d_2 is the smaller diameter

acceleration of gravity, 9.81 m/s^2; λ is the friction coefficient, 0.02–0.03 for steel pipe.

(b) Pressure loss in sudden expansion diameter pipe

$$\Delta P_{s2} = 0.49 \cdot \frac{(v_1 - v_2)^2}{g} \tag{2.8}$$

where ΔP_{s2} is the pressure loss in the sudden expansion diameter pipe, N/cm^2; v_1 is the velocity of water in small diameter segment, m/s; v_2 is the velocity of water in large diameter segment, m/s.

(c) Pressure loss in sudden reduction diameter pipe

$$\Delta P_{s3} = 0.49\alpha \cdot \frac{v^2}{g} \tag{2.9}$$

where ΔP_{s3} is the pressure loss in the sudden reduction diameter pipe, N/cm^2; v is the velocity of water in small diameter segment, m/s; g is the acceleration of gravity, 9.81 m/s^2; α is the resistance coefficient, see Table 2.4.

In engineering investigation, usually choose one-point water pressure test, and the water pressure is 30 N/cm^2.

(3) Length of test segment

The length of test segment usually is 5 m.

If the rock core is intact ($\omega = 0.01$ L min^{-2} m^{-2}), it can be lengthen, but not longer than 10 m. For tectonic fracture zones, karst segments, sand and gravel layers with strong permeability, the length should be determined by specific condition. If the length of rock core is <20 cm, it can be included into the test segment. For tilt test drilling, the test length is the actual length of the drilling.

2.1.2.4 Test Data Compilation

1. Test data reliability judgment

The reliability of one-point water pressure test data depends on the quality of drilling and pressure process. Ensure the data reliability by following test programs: drill with clean water → wash the test drilling → set plunger → observe stable water table → press water, keep the pressure constant and read Q → error check → loosen and pull out the plunger.

2.1 Hydrogeological Tests

2. Test outcome and application

(1) Unit water sucking amount ω

The major outcome of water pressure test is unit water sucking mount (ω), which can be calculated as:

$$\omega = \frac{Q}{LP} \qquad (2.10)$$

where ω is the unit water sucking amount, L/min m^2; Q is the steady packing flow in drill hole, L/min; L is the length of test section, m; P is the total applied pressure, N/cm^2.

The decimal of ω is limited to 0.01.

The ω got from water pressure test usually less than the real value, so it unsafe to use it in engineering design.

(2) Estimate the hydraulic conductivity K according to ω

If the distance from the lower end of test section to the aquifer's bottom is larger than the length of test section, the aquifer's hydraulic conductivity K can be estimated as:

$$K = 0.527\omega \lg \frac{0.66 L}{r} \qquad (2.11)$$

where K is the hydraulic conductivity, m/day; L is the length of test section, m; r is the radius of drill hole or filter, m; ω is the unit water sucking amount, L/min m^2.

If the distance from the lower end of test section to the aquifer's bottom is less than the length of test section, the aquifer's hydraulic conductivity K can be estimated as:

$$K = 0.527\omega \lg \frac{1.32 L}{r} \qquad (2.12)$$

The meanings of symbols are the same with above.

(3) Relations between unit water sucking amount and rock fracture

The relations between unit water sucking amount and rock fracture coefficient are shown in Table 2.5.

Table 2.5 Relations between unit water sucking amount and rock fracture coefficient

Unit water sucking amount (L/min m^2)	Fracture coefficient	Rock evaluation
<0.001	<0.2	Complete
0.001–0.01	0.2–0.4	Relatively complete
0.01–0.1	0.4–0.6	Some fracture
0.1–0.5	0.6–0.8	More fracture
>0.5	>0.8	Cracked

2.1.2.5 Test Equipment and Demands

1. Pipeline: inner pipe use steel and outer pipe use rubber.
2. Water supply equipment: if the water pressure test is for geological investigation, it's better to adopt gravity flow type water pressure test.
   The flow of the pump should be no <100 L/min under 150 N/cm^2 pressure, and the flow pressure should be stable. The pump should be with an agile and reliable valve.
3. Pressure gauges: the pressure gauge should be qualified with an accuracy no <2.5°; the working pressure is usually in the 1/3–1/4 measuring range; when lightly knock the pressure gauge during working, the pointer change should be no more than 2 % of the measuring range; the pointer can return to zero when stop loading.
4. Flow measurement: measuring cylinder, water meter.
5. Water table measurement: measurement bell and plumb; electronic water table indicator.

2.1.3 Water Injection Test

Drilling water injection test is a simple measurement method for aquifer permeability in field, the principles of which is similar with pumping test.

Drilling water injection test is usually used in following conditions: (1) the water table too deep to be pumped; (2) the rock and soil layer is dry.

The testing apparatus are shown in Fig. 2.16.

Inject water in drilling continuously and constantly to form constant water table. The duration of stable time depends on test purposes and requirements, which is

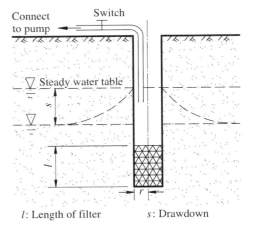

Fig. 2.16 Schematic of drilling water injection test

l: Length of filter s: Drawdown

usually 2–8 h. This kind of test can be used in calculating the hydraulic conductivity (K) and unit water sucking amount (ω).

According to engineering experience, in horizontal aquifer with huge thickness, K can be calculated by following formulas:
When $l/r \leq 4$:

$$K = \frac{0.08Q}{rs\sqrt{\frac{l}{2r} + \frac{1}{4}}} \tag{2.13}$$

when $l/r > 4$:

$$K = \frac{0.366Q}{ls} \lg \frac{2l}{r} \tag{2.14}$$

where l is the length of filter, m; Q is the constant injection water amount, m³/day; s is the waterhead in drilling, m; r is the radius of drilling or filter, m.

The hydraulic conductivity that calculated above is 15–20 % less than it calculated by pumping test formulas.

The K_1 for single layer and K for double layers can be calculated in two tests. Since $KL = K_1 l_1 + K_2 l_2$, so that $K_2 = (Kl - K_1 l_1)/l_2$.

If the water table is deep and the medium is uniform, and in the condition that $50 < h/r < 200$ and the water height in drilling is higher than 1 m, K can be calculated by Eq. (2.15):

$$K = 0.423 \frac{Q}{h^2} \lg \frac{2h}{r} \tag{2.15}$$

where h is the water height in drilling, m.

The error of K that calculated by Eq. (2.15) is <10 %.

2.1.4 Infiltration Test

Infiltration test is a simple method that is taken in trial pit to measure the hydraulic conductivity of vadose zone in field. The most common methods are trial pit method, single-loop method, and double-loop method as shown in Table 2.6.

2.1.4.1 Test Method

1. Trial pit method

Trial pit method is the test that conducted in trial pit, which is 30–50 m deep. The shape can be square, whose length of side is 30 cm, or can be round with the

Table 2.6 Infiltration test method (from Handbook of Engineering Geology 1992)

Test method	Test sketch map	Advantages and disadvantages	Notes
Trial pit method		1. Simple device; 2. Effected by lateral penetration, leading to low accuracy	When there is anti-seepage measurement on the wall of round pit, $F = \pi r^2$. When there is no anti-seepage measurement, $F = \pi r(r + 2Z)$, where r is the radius of pit bottom, Z is the thickness of water in pit
Single-loop method		1. Simple device; 2. Lateral penetration is not considered, leading to low accuracy	
Double-loop method		1. Simple device; 2. Effect of lateral penetration is excluded, so that the results are accurate	

37.75 cm diameter. The water table is 3–5 m beneath the pit bottom. Lay a layer of gravel sand with a thickness of 2 cm. Control the flow continuous balancing and the water a constant thickness (10 cm) since test starting. When the injection water amount is stable and then lasts for 2–4 h, the test can be completed.

If it is the coarse sand, gravel or cobble layer that tested, the water thickness (Z) should be kept in 2–5 cm. When $(H_k + Z + l)/l \approx 1$, hydraulic conductivity can be calculated by following formula: $K = \frac{Q}{F} = v$, where H_k is the capillary pressure head (in m), can be found in Table 2.7; l is the water penetration depth when test completes, which can be determined after excavation or by water content analysis.

The sketch map of the trial pit method is shown in Fig. 2.17. This method is usually used in sand, which is not much influenced by capillary pressure. As for clay, the result is usually on high side.

2. Single-loop method

Embed an iron loop that has a height of 20 cm, a diameter of 37.70 cm and an area of 1000 cm^2 on the pit bottom. At the start of the test, control the water column more than 10 cm high in the loop by Mariotte bottle. When the infiltration amount

2.1 Hydrogeological Tests

Table 2.7 Capillary pressure head (H_k) of different soil type (from Handbook of Engineering Geology 1992)

Soil type	H_k (m)	Soil type	H_k (m)
Silty clay (SC)	1.0	Fine clayed sand (SM)	0.3
Clay (CLS)	0.8	Silty sand	0.2
Clayed silt (CL)	0.6	Fine sand	0.1
Sandy silt (MLS)	0.4	Medium sand	0.05

Notes The H_k values in above table are always lower

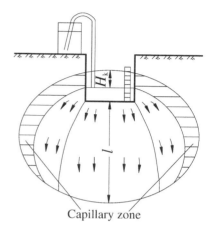

Fig. 2.17 Sketch map for trial pit method applied in clay

Q is constant, the test is complete and the infiltration rate can be calculated by Eq. (2.16), which is equal to the hydraulic conductivity of soil layer.

$$v = \frac{Q}{F} = K \qquad (2.16)$$

In addition, infiltration rate can be calculated by following steps: measured the infiltration amount in a certain period of time (e.g. 30 min); compute the average infiltration rate value; draw the infiltration rate duration curve, shown as Fig. 2.18.

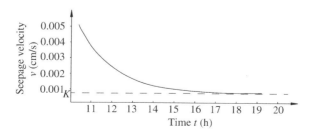

Fig. 2.18 Seepage velocity duration curve in infiltration test

It can be found that the infiltration rate decreases with time and tend to be a constant, which can be considered as hydraulic conductivity.

3. Double-loop method

Embed two iron loops on the bottom of the trial pit. The outer one has a diameter of 0.5 m and inner one is 0.25 m. Keep the water table the same in both loops by Mariotte bottle. Calculate hydraulic conductivity according to the data that is obtained from the inner loop. The water in inner loop only infiltrates vertically, which can exclude the effect of lateral infiltration and makes the results more accurate.

2.1.4.2 Parameters Calculation

When the infiltration water amount is tend to be constant, the hydraulic conductivity can be calculated by following equation, in which the capillary pressure has been considered.

$$K = \frac{Ql}{F(H_k + Z + l)} \quad (2.17)$$

where Q is the constant infiltration amount, cm^3/min; F is the infiltration area of inner loop, cm^2; Z is the thickness of water in inner loop, cm; H_k is the capillary pressure head, cm; l is the infiltration depth when test complete, cm.

If the infiltration can be steady for a very long time, K can be calculated by Eq. (2.18):

$$K = \frac{V_1}{Ft_1 a_1}[a_1 + \ln(1 + a_1)] \quad (2.18)$$

$$a_1 = \frac{\ln(1 + a_1) - \frac{t_1}{t_2}\ln\left(1 - \frac{a_1 V_2}{V_1}\right)}{1 - \frac{t_1 V_2}{t_2 V_1}} \quad (2.19)$$

where V_1, V_2 are the total infiltration amount during t_1 and t_2, m^3; t_1 and t_2 are the cumulative time; F is the infiltration area of inner loop, cm^2; a_1 is the alternative factor, calculated by trial method.

2.1.4.3 Test Data Compilation

1. Draw the layout of pit plane position.
2. Draw the hydrogeological cross-sectional view and the test device.
3. Draw the penetration rate duration curve.
4. Calculate the hydraulic conductivity.
5. Organize the original recording sheets.

2.2 Measurement of Groundwater Table, Flow Direction and Seepage Velocity

2.2.1 *Measurement of Groundwater Table*

Groundwater table is the naturally relative stable water table, which means it has no obvious up or down trend during a period.

Water table can be measured by water table indicator, which should be chosen according to engineering properties, construction conditions and measurement accuracy.

2.2.1.1 Measurement Bell

Measurement bell is a common tool used in borehole and observation hole. It is a metal cylinder with a diameter of 25–40 mm and length of 50–80 mm. The top is closed, connecting with a measuring line, and its accuracy is 1–2 cm. It can make a sound after contacting with water, which may hardly to identify when water table is too low.

2.2.1.2 Battery Water Table Indicator

Battery water table indicator consists of electrodes, wires, μA ampere meter and dry battery. The accuracy is about 1 cm. It is convenient to use, making it available for all boreholes with any diameter or depth.

2.2.1.3 Auto Water Table Recorder

It adopts the clock spring principle so that it can automatically record water table. It can be used in wells with a diameter larger than 89 mm. The accuracy is about ± 1.5 cm.

2.2.2 *Measurement of Groundwater Flow Direction*

Groundwater flow direction can be measured by three-point method, shown in Fig. 2.19. Set three boreholes to create a near equilateral triangle. Measure the water table in these boreholes, and then draw the water table contour map. The direction that is perpendicular to contours and point to the descent side is the groundwater flow direction.

Fig. 2.19 Layout of drillings for groundwater direction measurement

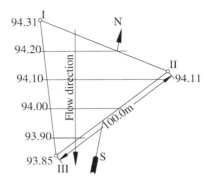

Besides, groundwater flow direction also can be measured by artificial radioisotopes method. Trickle the radioactive tracer into a single well, and then measure the concentration of the tracer around the well. The direction with the highest concentration of tracer is the flow direction.

2.2.3 Measurement of Seepage Velocity

2.2.3.1 Hydraulic Gradient Method

Measure the hydraulic gradient between adjacent water table contours on water table contour map. Groundwater flow velocity can be calculated by Darcy's law:

$$v = KI \qquad (2.20)$$

where v is the seepage velocity, m/day; K is the hydraulic conductivity of aquifer, m/day; and I is the hydraulic gradient.

2.2.3.2 Indicator or Tracer Method

Indicators and radioactive tracer can be used in in situ seepage velocity measurement. Here are some requirements: borehole should be set in a typical position of the tested aquifer; groundwater around borehole is steady laminar flow.

Set tracer injection hole and observation hole along flow direction line. Two assistant observation holes can be set to prevent tracer or indicator bypassing the main observation hole, shown as Fig. 2.20. The distance between Delivery hole (DH) and Observation hole (OH) depends on the permeability of soil or rock, shown in Table 2.8.

Draw indicator concentration versus time curve and use the time corresponding to the peak or average concentration to calculate the actual flow velocity:

Fig. 2.20 Layout of drillings for groundwater velocity measurement

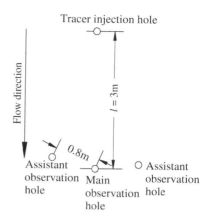

Table 2.8 Distances between injection hole and observation hole

Rock or soil type	Distance (m)
Silt	1–2
Fine sand	2–5
Coarse sand with gravel	5–15
Fractured rock	10–15
Limestone high karst degree	>50

$$u = \frac{l}{t} \tag{2.21}$$

where u is the average actual groundwater flow velocity, m/h; l is the distance between injection hole and observation hole, m; t is the time mentioned above, s.

Seepage velocity can be calculated by following formula:

$$v = nu$$

where n is the porosity.

The seepage velocity also can be tested in situ by single well artificial radioisotopes method. The common radioactive tracers include ^{3}H, ^{51}Cr, ^{60}Co, ^{82}Br, ^{131}I, ^{137}Cs et al. According to the tracer's concentration versus time curve, the average actual flow velocity u can be calculated by following formula, shown in Fig. 2.21.

$$U = \frac{V}{st} \ln\left(\frac{C_0}{C}\right) \tag{2.22}$$

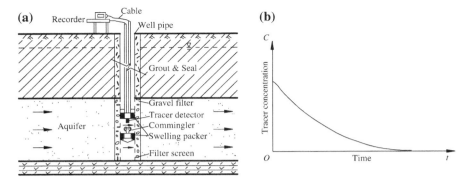

Fig. 2.21 Single well test. **a** Single well dilution method. **b** Dilution duration curve of indicator

where C_0 and C are tracer's concentration of time $T = 0$ and $T = t$ respectively, μg/L; t is the time duration of observation, h; s is the vertical cross section area of isolated part that water pass through, m²; V is water volume in isolated part, m³.

Single well method is shown in Fig. 2.21.

Other measurement methods are shown in Table 2.9.

2.3 Capillary Rise Height Determination

Methods for capillary rise height determination are as follows:

2.3.1 Direct Observation Method

This method could apply to silty soil and clayed soil, which has a large capillary rise height. A boundary of wet and dry soils can be easily observed in trial pit, the distance from which to water table is the capillary rise height.

2.3.2 Water Content Distribution Curve Method

2.3.2.1 Plastic Limit Test Method

This method could apply to silty soil and clayed soil. Soil samples are taken every 15–20 cm above the water table, and their water content and plastic limit are tested. Then two curves that water content and plastic limit against depth are obtained, whose intersection depth to water table is the capillary rise height.

2.3 Capillary Rise Height Determination 65

Table 2.9 The measurement methods for actual groundwater flow velocity (from Handbook of Engineering Geology 1992)

Method	Principle	Indicator			Operation	Identification	Notes
		Name	Distance between IH and OH	Amount			
Chemical method	Determine the time that salt appear and its concentration change through chemical analysis	NaCl	>5 m	10–15 kg	There are two ways for tracer delivery: (1) Put the tracer cylinder, which has a conic valve at bottom, to predetermined depth and open the valve to make the solution spill into the delivery hole; (2) Bury the cylinder, which has a taphole that connects a rubber hose to ground, into predetermined depth, then pull the solution in through rubber hose. Sample the water from observation holes by sample collector that no more than 50 cm² cross-sectional area	Determine the concentration of chlorine by titration. Titrate the solution until the color of water sample turns into brownish red and not fade	(1) Of all these methods, the best one is taking the nitrate as indicator. Its advantages include: high sensitivity, less disturbance, simple and easy operation, good repeatability, low cost, and easy to get. However, the $NaNO_2$ is unstable and has certain toxicity. By comparison, $NaNO_3$ has low toxicity and low sensitivity, and needs for rectify of NO_3^- concentration. (2) The preparation method of testing powder for NO_2^- and NO_3^- concentration test: porphyrize 100 g $BaSO_4$, 75 g citric acid, 4 g sulfanilic acid and 2 g α-naphthylamine; then mix them uniformly, store the mixture in brown bottle and keep dry. For NO_3^- concentration test, mix another 10 g $MnSO_4$ and 4 g zinc powder. The NO_2^- and NO_3^- concentration of standard colorimetric solution is 0.001 mg/L. (3) When taking the nitrate as indicator, it is must to pre-sample one bottle of testing water, for comparison in case there is an exception during test
		$CaCl_2$	3–5 m	5–10 kg			
		NH_4Cl	<3 m	3–5 kg			
		$NaNO_2$	>5 m	Concentration of NO_2^- <1 mg/L		Method for NO_2^- concentration test: first, compound testing powder; then dissolve 0.5 g testing powder into 50 mL water sample; compare its color with standard solution	
		$NaNO_3$	>5 m	Concentration of NO_3^- <5 mg/L		The method for NO_3^- concentration test is the same with above. But the result is the total concentration of NO_2^- and NO_3^-. This result subtract the concentration of NO_2^- will get the concentration of NO_3^-	
Colorimetric method	Use the concentration change of reagent color to determine the time that is taken by the reagent pass through the two holes	Alkaline water	Fluorescent yellow, fluorescent red, eosin		Every 5 cm	Incompact stratum 1–5 g	Use fluorescent colorimeter to confirm the existence of dyestuff and determine its concentration. Or pull self-confected solution with different concentration into colorimeter, and compare its color with water sample at regular time intervals
		Weak acid water	Congo red, methylene blue, aniline blue			Karst fracture stratum 1–10 g	
						10–30 g	
						10–40 g	

(continued)

Table 2.9 (continued)

Method	Principle	Indicator Name	Distance between IH and OH	Amount	Operation	Identification	Notes
Electrolytic method	Determine the electrolyte movement and distribution condition in observation hole by special electrical test equipment	Ammonium chloride			The delivery method is the same with above. A special electrode, which is isolated from the casing pipe, is put into the observation hole. The electric circuit is from the electrode and trough battery, ammeter, regulating rheostat, finally to the casing pipe of delivery hole	Measure the current intensity at regular time intervals to confirm the existence of tracer and determine its concentration	
Electricize method	Dissolve the salt into groundwater. The flow of saline water makes the electric field change near the deliver hole	Salt			Put the salt into the delivery hole at the position where correspond to the aquifer. Put the electrode A into drill hole and insert the electrode B in the ground where is $20H$ away from drill hole (H is the bury depth of aquifer), and make $MN = (2-4) H$ (electrode N is set on upstream)	The equipotential line that observed on ground changes from circle to ellipse. The flow direction of groundwater is parallel to the ellipse's long axis	
Atomic method of radioactive tracer	Determine the time that the tracer takes to pass through the observation hole by special equipment	Tritium (H^3), iodine (I^{13}), bromine (Br^{82}), sodium (Na^{23}), sulfur (S^{35}) et al.	If the velocity of groundwater is 10^{-2}–10^{-5} cm/s, then the delivery distance is 50–1 m	The intensity of radioactive source usually is 10–15 mci	Deliver the tracer into center hole, then record the radioactivity degree at regular intervals in observation holes by detect device, which takes Cr-M counter tube as probe and scaler	Take the projection value of maximum radioactive intensity on time axis as the time that the tracer passing through the observation hole	

2.3 Capillary Rise Height Determination

2.3.2.2 Maximum Molecular Water Absorption Method

This method could apply to sand, that is high column method for medium-coarse sand and water absorption medium method for fine sand. Soil samples are taken every 15–20 cm above the water table, and their maximum water absorption and natural water content are tested. Then two curves that natural water content and maximum water absorption against depth are obtained, whose intersection depth to water table is the capillary rise height.

2.4 Pore Water Pressure Determination

Pore water pressure of saturated soil foundation changes during foundation treatment and base construction. Its crucial to measure pore water pressure due to its big effect on soil deformation and stability.

Engineering projects that pore water pressure measurement should be taken and its aims are shown in Table 2.10.

2.4.1 Pore Water Pressure Gauge and Measurement Methods

Pore water pressure gauges should be chosen in accordance with measurement aims, period, and soil permeability. Their accuracy, sensitivity, and range should meet the needs. Table 2.11 shows the instrument types and their application conditions.

2.4.2 Calculation Formulas

Calculation formulas for different types of pore water pressure gauges are shown in Table 2.12.

Table 2.10 Engineering projects and measurement aims

Engineering projects	Measurement aims
Preloading foundation	Consolidation degree estimation and loading rate controlling
Dynamic consolidation	Time intervals controlling and effective influence depth determination
Prefabricated pile construction	Pilling rate controlling
Engineering dewatering	Relief well pressure monitoring and land subsidence controlling
Landslide	Landslide monitoring and treatment

Table 2.11 Pore water pressure gauges and their application conditions

Pore water pressure gauge		Application conditions
Riser pipe pressure gauge (open type)		Hydraulic conductivity $>10^{-4}$ cm
Water pressed pressure gauge (hydraulic type)		Low hydraulic conductivity
		Accuracy >2 kPa
		Period of measurement <1 month
Electric pressure gauge	Vibration string type	All kinds of soil
		Accuracy <2 kPa
		Period of measurement >1 month
	Differential transformer type	All kinds of soil
		Accuracy <2 kPa
		Period of measurement >1 month
	Resistance type	All kinds of soil
		Accuracy <2 kPa
		Period of measurement <1 month
Pneumatic pressure gauge (air pressure type)		All kinds of soil
		Accuracy >10 kPa
		Period of measurement <1 month
Piezo-cone static penetration apparatus		All kinds of soil
		Short period of measurement

Table 2.12 Calculation formulas for different types of pore water pressure gauges

Type	Calculation equation	Symbols
Hydraulic type	$u = P_a + \rho_w h$	u—Pore water pressure, kPa
		P_a—Gauge reading, kPa
		h—Distance between pore water pressure gauge and the base table of piezometer, cm
		ρ_w—Water density, g/cm^3
Air pressure type	$u = c + aP_a$	c, a—Calibration constants of pressure gauge
Vibration string type	$u = K(f_0^2 - f^2)$	K—Sensitivity coefficient of pore water pressure gauge, measured in kPa/Hz^2 for vibration string type and kPa/µε for resistor type
		f_0—Frequency of pore water pressure gauge at zero, Hz
		f—Frequency of pore water pressure gauge after pressed, Hz
Resistor type	$u = K(\varepsilon_1 - \varepsilon_0)$	ε_1—Reading of pore water pressure gauge after pressed, µε
		ε_0—Reading of pore water pressure gauge before pressed, µε
Differential resistor type	$u = (A - A_0)K$	A—Initial reading, V
		A_0—Measured value, V
		K—Calibration coefficient, kPa/V

2.5 Hydrogeological Parameters Calculation in Steady Flow Pumping Test

Hydraulic conductivity K and conductivity coefficient T can be calculated by steady flow formulas with test data. Hydraulic conductivity represents the aquifer's permeability, which equal to seepage velocity when hydraulic gradient $I = 1$. Hydraulic conductivity relates to properties of both aquifer and liquid. Conductivity coefficient $T = KM$ (M represents aquifer's thickness).

2.5.1 Calculation of Hydraulic Conductivity

When single well pumping test reaching to a steady state, the drawdown s and water discharge Q can be measured. Generally, three group data, which are s_1 and Q_1, s_2 and Q_2, s_3 and Q_3 should be got.

2.5.1.1 Dupuit Formula

Dupuit formula can be applied to hydraulic conductivity calculation for homogeneous, isopachous, and infinite horizontal extent aquifer with a fully penetrating pumping well.
Confined aquifer:

$$K = \frac{Q}{2\pi M s_w} \ln \frac{R}{r_w} \tag{2.23}$$

Unconfined aquifer:

$$K = \frac{Q}{\pi \left(H_0^2 - h_w^2\right)} \ln \frac{R}{r_w} \tag{2.24}$$

where K is the aquifer's hydraulic conductivity, LT^{-1}; Q (in L^3T^{-1}) and s_w (in L) are water discharge and drawdown, respectively, when pumping test reaching to a steady state; R is the radius of influence of the pumping well, L; r_w is the radius of the pumping well, L; M is the aquifer's thickness, L; H_0 is the natural water table of unconfined aquifer, L; h_w is the water table in pumping well when pumping test in unconfined aquifer reaching to a steady state, L.

Formulas should be chosen correctly according to hydrogeological conditions, boundary conditions and well structure.

2.5.1.2 Three-Dimensional Single Well Formula

Dupuit formula does not consider 3D flow near pumping well, so the drawdown it used is lager, which lead to the calculated K is usually less than actual value. Three-dimensional formula is an amendment of Dupuit formula, as Eq. (2.25).

$$s = \frac{Q}{2\pi KM}\ln\frac{R}{r_w} \pm \frac{Q^2}{8\pi^2 r_w^4}\left(\frac{f}{6M^2 D}Z^3 + \frac{Z^2}{M^2} - \frac{fM}{24D} - \frac{1}{3}\right) \quad (2.25)$$

where D is the diameter of the pumping well, L; f is the friction coefficient of the filter tube, which is equal to 64/Re in laminar flow; g is the acceleration of gravity, m/s²; Z is the distance between the bottom of filter tube to the bottom of the aquifer, L. Other symbols are the same with that of Dupuit formula.

For ease of use, two parts of the former formula can be represented by A and C, respectively, that are:

$$A = \frac{Q}{2\pi KM}\ln\frac{R}{r_w}$$

$$C = \frac{1}{8\pi^2 r_w^4}\left(\frac{f}{6M^2 D}Z^3 + \frac{Z^2}{M^2} - \frac{fM}{24D} - \frac{1}{3}\right)$$

So the 3D formula can be expressed as:

$$s = AQ \pm CQ^2 \quad (2.26)$$

or

$$s = s_w \pm CQ^2 \quad (2.27)$$

where s_w is the drawdown in pumping well, L; s is the amendment drawdown considered 3D flow, L.

A and C are constants when size of pumping well is determined. That means s and Q has a parabolic relation and A, C can be obtained by graphic method.

Making $\varepsilon = \frac{s}{Q}$ as unit flow drawdown, then s-Q parabola will simplify to ε-Q line as follow, shown in Fig. 2.22.

$$\varepsilon = A + CQ \quad (2.28)$$

where A is the intercept; C is the slope, which can be rewritten as:

$$C = \frac{s_{i+1}/Q_{i+1} - s_i/Q_i}{Q_{i+1} - Q_i} \quad (2.29)$$

Fig. 2.22 ε-Q curve

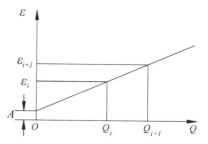

where s_i and s_{i+1} are the drawdowns for the ith pumping and $(i + 1)$th pumping, respectively; Q_i and Q_{i+1} are the water discharge for the ith pumping and $(i + 1)$th pumping, respectively.

Steps for hydraulic conductivity calculation using single well pumping data are as follows:

1. Draw s_w (or Δh_w^2)-Q curve

Draw s_w-Q curve for confined aquifer or Δh_w^2-Q curve for unconfined aquifer according to steady flow pumping test data. Here, $\Delta h_w^2 = H_0^2 - h_w^2$.

There are three types of s_w (or Δh_w^2)-Q curves in engineering practice, shown in Fig. 2.23.

2. Calculate K

(1) If s_w (or Δh_w^2)-Q curve is a straight line, as line (1) in Fig. 2.23, that means the groundwater flow is two dimensional and $C = 0$ in Eq. (2.26). So hydraulic conductivity K can be calculated by following formulas:

Fully penetrating well in confined aquifer:

$$K = \frac{Q}{2\pi s_w M} \ln \frac{R}{r_w}$$

Fully penetrating well in unconfined aquifer:

$$K = \frac{Q}{\pi(H_0^2 - h_w^2)} \ln \frac{R}{r_w} = \frac{Q}{\pi \Delta h_w^2} \ln \frac{R}{r_w} \quad (2.30)$$

Fig. 2.23 s_w (or Δh_w^2)-Q curve

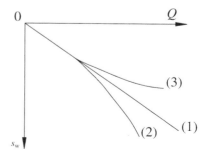

Dupuit formula that be used in unconfined aquifer should meet the need that the water gradient of cone of depression <1/4. If the drawdown in unconfined aquifer is less than one tenth of the aquifer's thickness, formulas for confined aquifer can be used and M will be replaced with H_0, which represents the aquifer's thickness.

For partially penetrating pumping well, corresponding formulas should be chosen.

(2) Curves (2), (3) in Fig. 2.23 represent three-dimensional groundwater flow, so that ε-Q curve should be drawn as shown in Fig. 2.22, where $\varepsilon = \frac{s}{Q}$ in confined aquifer and $\varepsilon = \frac{\Delta h_w^2}{Q}$ in unconfined aquifer. Intercept A can be obtained from it and then hydraulic conductivity K can be calculated as follows:

Fully penetrating well in confined aquifer:

$$K = \frac{Q}{2\pi AM} \ln \frac{R}{r_w} \tag{2.31}$$

Fully penetrating well in unconfined aquifer:

$$K = \frac{1}{\pi A} \ln \frac{R}{r_w} \tag{2.32}$$

2.5.1.3 Steady Pumping Test with Observation Wells

Hydraulic conductivity can be calculated in steady pumping test with observation wells, which is usually more than two. Specific steps are as follows:
1. Draw s_w (or Δh^2)-lg$r$ curve curves according to pumping test data, shown in Fig. 2.24, where $\Delta h^2 = H_0^2 - h^2$.
2. Calculate K by following formula:
Fully penetrating well in confined aquifer:

$$K = \frac{Q}{2\pi M(s_1 - s_2)} \ln \frac{r_2}{r_1} = \frac{2.3Q}{2\pi M} \frac{1}{m_r} \tag{2.33}$$

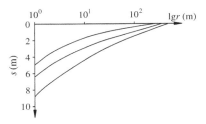

Fig. 2.24 s_w (or Δh_w^2)-lg$r$ curve

where (r_1, s_1) and (r_2, s_2) are the coordinate values of two random points of the straight-line portion in s-$\lg r$ curve; m_r is the intercept of the straight-line portion, $m_r = \frac{s_1 - s_2}{\lg r_1 - \lg r_2}$.

Fully penetrating well in unconfined aquifer:

$$K = \frac{Q}{\pi(\Delta h_1^2 - \Delta h_2^2)} \ln \frac{r_2}{r_1} = \frac{2.3Q}{\pi} \frac{1}{m_r} \tag{2.34}$$

where $(r_1, \Delta h_1^2)$ and $(r_1, \Delta h_2^2)$ are the coordinate values of two random points of the straight-line portion in $\Delta h^2 - \lg r$ curve;

$$\Delta h^2 = H_0^2 - h^2$$

where H is the thickness of unconfined aquifer before pumping, L; h is the height of water column that from aquifer's bottom to water surface in observation well, L; m_r is the intercept of the straight-line portion, $m_r = \frac{\Delta h_1^2 - \Delta h_2^2}{\lg r_2 - \lg r_1}$.

2.5.1.4 Choice of Formulas and Empirical Values

Formulas for hydraulic conductivity calculation can be chosen from Tables 2.13, 2.14, 2.15, 2.16 and 2.17, according to different engineering condition. Empirical values can be chosen from Tables 2.18, 2.19, and 2.20.

2.5.2 Calculation of Radius of Influence

Radius of influence of pumping well is one of the aquifer's original data in hydraulic conductivity calculation. It can be determined by steady pumping test with observation wells. There are two kinds of methods, as follows:

2.5.2.1 Graphing Method

Draw s (or Δh^2)-$\lg r$ curves according to pumping test data, shown in Fig. 2.24. Lengthen the straight-liner segment to $\lg r$ axial, and the intersection R is the radius of influence of pumping well in hypothetical cylindrical aquifer. According to the conception of "reference influence radius", R is a constant that does not change with water discharge. So all s (or Δh^2)-$\lg r$ curves should intersect in one point, which is R. If they did not, reasons should be found, for example the change of recharge condition (Fig. 2.25).

Table 2.13 Partially penetrating well in unconfined aquifer (filter submerged) (from Handbook of hydrogeological investigation of water supply 1977)

Graphs	Formulas	Application condition
	$K = \dfrac{0.366Q}{ls_w} \lg \dfrac{0.66l}{r_w}$	1. Filter installed in the middle of aquifer 2. $l < 0.3H$ 3. $c = (0.3 - 0.4)H$ 4. No observation well
	$K = \dfrac{0.16Q}{l(s_w - s_1)}\left(2.3 \lg \dfrac{0.66l}{r_w} - \text{arsh}\dfrac{l}{2r_1}\right)$	Condition 1, 2, and 3 is the same with above; 4. Have one observation well
	$K = \dfrac{0.366Q(\lg R - \lg r_w)}{(s_w + l)s}$	1. Filter installed in the middle of aquifer 2. No observation well
	$K = \dfrac{0.366Q(\lg r_1 - \lg r_w)}{(s_w - s_1)(s - s_1 + l)}$	Condition 1, 2, and 3 is the same with above 4. Have one observation well
	$K = \dfrac{0.73Q(\lg R - \lg r_w)}{s_w(H + l)}$	1. Filter installed near the bottom of aquifer 2. No observation well

2.5.2.2 Formula Method

Steps of influence radius calculation with data of steady pumping test are as follows:
1. Draw s (or Δh^2)-$\lg r$ curves according to pumping test data.
2. Choose two points A ($\lg r_1$, s_1) and B ($\lg r_2$, s_2) or A ($\lg r_1$, Δh_1^2) and B ($\lg r_2$, Δh_2^2) in straight-liner segment, then calculate R by following formulas:

2.5 Hydrogeological Parameters Calculation in Steady Flow Pumping Test

Table 2.14 Partially penetrating well in unconfined aquifer (filter unsubmerged) (from Handbook of hydrogeological investigation of water supply 1977)

Graphs	Formulas	Application condition
	$K = \dfrac{0.73Q}{s_w \left[\dfrac{l+s_w}{\lg \frac{R}{r_w}} + \dfrac{l}{\lg \frac{0.66l}{r_w}} \right]}$	1. Filter installed near the top of aquifer 2. $l < 0.3H$ 3. Aquifer has a large thickness
	$K = \dfrac{0.16Q}{l'(s-s_1)} \left(2.3 \lg \dfrac{1.6l'}{r_w} - \text{arsh} \dfrac{l'}{r_1} \right)$ where $l' = l_0 - 0.5(s+s_1)$	1. Filter installed near the top of aquifer 2. $l < 0.3H$ 3. $s < 0.3l$ 4. Have one observation well; radius $r_1 < 0.3$
	$K = \dfrac{0.73Q}{s_w \left[\dfrac{l+s_w}{\lg \frac{R}{r_w}} + \dfrac{2m}{\frac{1}{2a}\left(2\lg \frac{4m}{r_w} A\right) - \lg \frac{4m}{R}} \right]}$ where m is the distance between the middle of filter and the aquifer bottom; A depends on l/m	1. Filter installed near the top of aquifer 2. $l > 0.3H$ 3. No observation well
	$K = \dfrac{0.366Q(\lg R - \lg r_w)}{H_1 s_w}$ where H_1 is the distance between bottoms of filter and aquifer	No observation well
	$K = \dfrac{0.366Q}{l s_w} \lg \dfrac{0.66l}{r_w}$	1. Pumping under river 2. Filter installed in the middle or near the top of aquifer 3. $c > \dfrac{l}{\ln \frac{l}{r_w}}$ (Usually $c < (2\text{–}3)$ cm)

Fully penetrating well in confined aquifer:

$$\lg R = \frac{s_1 \lg r_2 - s_2 \lg r_1}{s_1 - s_2} \tag{2.35}$$

Table 2.15 Fully penetrating well in confined aquifer (from Handbook of hydrogeological investigation of water supply 1977)

Graphs	Formulas	Application condition
(Pumping well, Observation well 1, Observation well 2)	$K = \frac{0.732Q}{(2H-s)s} \lg \frac{R}{r}$	No observation well
	$K = \frac{0.732Q}{(2H-s-s_1)(s-s_1)} \lg \frac{r_1}{r}$	Have one observation well
	$K = \frac{0.732Q}{(2H-s_1-s_2)(s_1-s_2)} \lg \frac{r_2}{r_1}$	Have two observation well

Table 2.16 Fully/partially penetrating well in confined aquifer (from Handbook of hydrogeological investigation of water supply 1977)

Graphs	Formulas	Application condition
	$K = \frac{0.366Q}{l} \lg \frac{al}{r}$ $a = 1.6$ Гиринский $a = 1.32$ В.Д.Бабушкин	1. In confined or unconfined aquifer 2. Filter is next to the top or bottom of aquifer 3. $l/r \geq 5$ and $l < 0.3M$
	$K = \frac{0.366Q}{Ms} \left[\frac{1}{2a} \left(2\lg \frac{4M}{r} - A \right) - \lg \frac{4M}{R} \right]$ $a = l/M$	1. In confined aquifer 2. Filter is next to the top of aquifer 3. $l/r > 5$ and $l > 0.3M$ (by Muskat formula)
(Pumping well, Observation well 1, Observation well 2)	$K = \frac{0.366Q}{Ms} \lg \frac{R}{r}$	Dupuit formula No observation well
	$K = \frac{0.366Q}{M(s-s_1)} \lg \frac{r_1}{r}$	Dupuit formula Have one observation well
	$K = \frac{0.366Q}{M(s_1-s_2)} \lg \frac{r_2}{r_1}$	Dupuit formula Have one observation well

2.5 Hydrogeological Parameters Calculation in Steady Flow Pumping Test

Table 2.17 Water table recovery method (from Handbook of hydrogeological investigation of water supply 1977)

Graphs	Formulas	Application condition	Notes
	$K = \dfrac{1.57 r_w (h_2 - h_1)}{t(s_1 + s_2)}$	Flat bottom well and trial pit with large diameter in confined aquifer	Obtain a series of k that related to water table recovery time, and draw k-t curve, in which the constant value of hydraulic conductivity can be determined, as shown in following figure. K_0: The final stable value of K
	$K = \dfrac{r_w (h_2 - h_1)}{t(s_1 + s_2)}$	Spherical well and trial pit with large diameter in confined aquifer	
	$K = \dfrac{3.5 r_w^2}{(H + 2r)t} \ln \dfrac{s_1}{s_2}$	Fully penetrating well in unconfined aquifer	
	$K = \dfrac{\pi r_w}{4t} \ln \dfrac{H - h_1}{H - h_2}$	1. Fully penetrating well in unconfined aquifer 2. Seepage just occurs on the bottom of well with large diameter	

Table 2.18 Empirical values of hydraulic conductivity K (from Handbook of Engineering Geology 1992)

Soil type	K (m/day)	Soil type	K (m/day)
Sandy silt	<0.05	Fine sand	1–5
Clayed silt	0.05–0.1	Medium sand	5–20
Silty clay	0.1–0.5	Coarse sand	20–50
Loess	0.25–0.05	Gravel	100–200
Silty sand	0.5–1.0	Gravel-boulder	20–150
		Boulder	500–1000

Table 2.19 Hydraulic conductivity of gravels (from handbook of engineering geology 1992)

Average particle size d_{50} (mm)	25.0	21.0	14.0	10.0	5.8	3.0	2.5
Nonuniform coefficient η (d_{50}/d_{10})	2.7	2.0	2.0	6.3	5.0	3.5	2.7
Hydraulic conductivity K (cm/s)	20.0	20.0	10.0	5.0	3.3	3.3	0.8

Table 2.20 Hydraulic conductivity of some kinds of soil (from Handbook of Engineering Geology 1992)

Soil type	K (m/day)	Soil type	K (m/day)
Clay	$<1.2 \times 10^{-6}$	Fine sand	1.2×10^{-3}–6.0×10^{-3}
Silty clay	1.2×10^{-6}–6.0×10^{-5}	Medium sand	6.0×10^{-3}–2.4×10^{-2}
Clayed silt	8.0×10^{-5}–6.0×10^{-4}	Coarse sand	2.4×10^{-2}–6.0×10^{-2}
Loess	8.0×10^{-4}–6.0×10^{-4}	Gravelly sand	6.0×10^{-2}–1.8×10^{-1}
Silty sand	6.0×10^{-4}–1.2×10^{-3}		

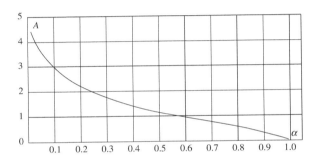

Fig. 2.25 Coefficient A-α curve

Fully penetrating well in unconfined aquifer:

$$\lg R = \frac{\Delta h_1^2 \lg r_2 - \Delta h_2^2 \lg r_1}{\Delta h_1^2 - \Delta h_2^2} \tag{2.36}$$

2.5.2.3 Choice of Formulas and Empirical Values

Table 2.21 shows some formulas for influence radius calculation. Most results are approximate values. Empirical values can be chosen from Table 2.22.

2.5 Hydrogeological Parameters Calculation in Steady Flow Pumping Test

Table 2.21 Formulas for influence radius calculation

Calculation formulas	Application conditions	Notes
$\lg R = \frac{s_1 \lg r_2 - s_2 \lg r_1}{s_1 - s_2}$	1. Confined aquifer 2. With two observation wells	Accurate (By Dupuit formula)
$\lg R = \frac{s_1(2H-s_1)\lg r_2 - s_2(2H-s_2)\lg r_1}{(s_1-s_2)(2H-s_1-s_2)}$	1. Unconfined aquifer 2. With two observation wells	Accurate (By Dupuit formula)
$\lg R = \frac{s \lg r_1 - s_1 \lg r}{s - s_1}$	1. Confined aquifer 2. With one observation well	Calculated value is larger than actual value
$\lg R = \frac{s(2H-s)\lg r_1 - s_1(2H-s_1)\lg r}{(s-s_1)(2H-s-s_1)}$	1. Unconfined aquifer 2. With one observation well	Calculated value is larger than actual value
$\lg R = \frac{2.73 KMs}{Q} + \lg r$	1. Confined aquifer 2. No observation well	Calculated value is larger than actual value
$\lg R = \frac{1.366 K(2H-s)s}{Q} + \lg r$	1. Unconfined aquifer 2. No observation well	Calculated value is larger than actual value
$R = 10s\sqrt{K}$	1. Confined aquifer 2. No observation well	Rough calculation (By W. Sihardt formula)
$R = 2s\sqrt{HK}$	1. Unconfined aquifer 2. No observation well	Rough calculation (By И.П. Кусакин formula)
$R = \sqrt{\frac{12r}{\mu}}\sqrt{\frac{QK}{\pi}}$	1. Unconfined aquifer 2. Fully penetrating well	(By Kozeny formula)
$R = 3\sqrt{\frac{KHt}{\mu}}$	Unconfined aquifer	(By Weber formula)
$R = \frac{Q}{2KHi}$	Confined aquifer	Rough calculation (By Е·Е·Керкис formula)

Symbols in this table
s_1, s_2 Drawdowns for observation wells, m
r_1, r_2 Distance between pumping well and observation well, m
r Radius of pumping well, m
H/M Thickness of unconfined/confined aquifer, m
K Hydraulic conductivity, m/day
t Time, d
μ Specific yield
i Hydraulic gradient

Table 2.22 Empirical values of influence radius R

Soil type	Primary particle diameter (mm)	Weight ratio (%)	Influence radius R (m)
Silty sand	0.05–0.1	<70	25–50
Fine sand	0.1–0.25	>70	50–100
Medium sand	0.25–0.5	>50	100–200
Coarse sand	0.5–1.0	>50	300–400
Very coarse sand	1.0–2.0	>50	400–500
Small gravel	2.0–3.0		500–100
Medium gravel	3.0–5.0		600–1500
Large gravel	5.0–10.0		1500–3000

2.5.3 Case Study

Layout of pumping test wells in Tongji University is shown in Fig. 2.26. The filters of both main well and observation wells are all drilled in the same aquifer, which contains mainly fine to medium sand with gravel and coarse sand. The buried depth

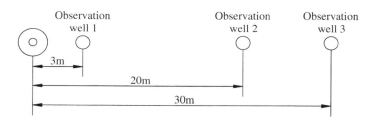

Fig. 2.26 Layout of pumping test wells

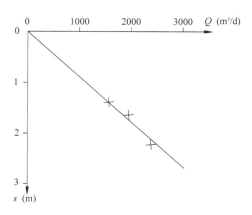

Fig. 2.27 Q-s curve

2.5 Hydrogeological Parameters Calculation in Steady Flow Pumping Test

of the aquifer is 67.7 m, and the thickness is 23.1 m. Diameters of main well and observation wells are 305 mm and 152 mm, respectively.

Three steady pumping tests were taken and the matching line that water discharge Q versus drawdown s is shown in Fig. 2.27. Other tests data are shown in Table 2.23.

2.5.3.1 Calculation of Influence Radius

1. Influence radius of pumping well is 930 m, which determined by graphing method, shown in Fig. 2.28.
2. R can also be calculated by formulas as follows:

$$\lg R_1 = \frac{s_1 \lg r_2 - s_2 \lg r_1}{s_1 - s_2} = \frac{0.52 \lg 30 - 0.31 \lg 3}{0.52 - 0.31} = 2.9532, \quad \text{and} \quad R_1 = 898 \text{ m}$$

$$\lg R_2 = \frac{0.70 \lg 30 - 0.43 \lg 3}{0.70 - 0.43} = 3.0697, \quad \text{and} \quad R_2 = 1174 \text{ m}$$

$$\lg R_3 = \frac{0.81 \lg 30 - 0.47 \lg 3}{0.81 - 0.47} = 2.8595, \quad \text{and} \quad R_3 = 724 \text{ m}$$

$$R = \frac{R_1 + R_2 + R_3}{3} = 932 \text{ m}$$

Table 2.23 Data of steady pumping tests

Test number	Water discharge Q (m³ day⁻¹)	Drawdown (m)			
		s_0 for MW	s_1 for OW1	s_2 for OW2	s_3 for OW3
1	1570	1.385	0.52	0.36	0.31
2	1954	1.635	0.70	0.47	0.43
3	2384	2.22	0.81	0.61	0.47

Fig. 2.28 s-$\lg r$ curve

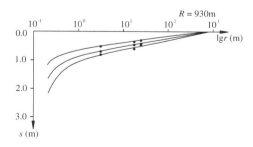

It can be found that the results of graphing method and formula method are close. So make $R = 932$ m.

2.5.3.2 Calculation of Hydraulic Conductivity K

1. Hydraulic conductivity K can be calculated by single well's test data, as follows:

$$K = \frac{Q}{2\pi Ms} \ln \frac{R}{r_0}$$

$$K_1 = \frac{1570}{2\pi \times 23.1 \times 1.385} \ln \frac{932}{0.152} = 68.11 \text{ m/day}$$

$$K_2 = \frac{1954}{2\pi \times 23.1 \times 1.635} \ln \frac{932}{0.152} = 71.81 \text{ m/day}$$

$$K_2 = \frac{2384}{2\pi \times 23.1 \times 2.22} \ln \frac{932}{0.152} = 64.53 \text{ m/day}$$

$$K = \frac{K_1 + K_2 + K_3}{3} = 68.15 \text{ m/day}$$

2. Hydraulic conductivity K can be calculated by two more wells' test data, as follows:

$$K = \frac{2.30 Q}{2\pi M} \times \frac{1}{m_r} = \frac{2.30 Q}{2\pi M} \times \frac{\lg r_3 - \lg r_1}{s_1 - s_3}$$

$$K_1 = \frac{2.30 \times 1570}{2\pi \times 23.1} \times \frac{\lg 30 - \lg 3}{0.52 - 0.31} = 118.5 \text{ m/day}$$

$$K_2 = \frac{2.30 \times 1954}{2\pi \times 23.1} \times \frac{\lg 30 - \lg 3}{0.70 - 0.43} = 114.7 \text{ m/day}$$

$$K_3 = \frac{2.30 \times 2384}{2\pi \times 23.1} \times \frac{\lg 30 - \lg 3}{0.81 - 0.47} = 111.1 \text{ m/day}$$

$$K \approx 115 \text{ m/day}.$$

It can be found that the result obtained from data of single well is less than the result got from two more wells. So using the data of more than one well is the best choice in hydraulic conductivity calculation.

2.6 Hydrogeological Parameters Calculation in Unsteady Flow Pumping Test

2.6.1 Transmissibility, Storage Coefficient, and Pressure Transitivity Coefficient Calculation for Confined Aquifer

Transmissibility T represents the ability of aquifer's water conductivity. It is the numeric equivalent of the product of hydraulic conductivity times aquifer's thickness ($T = KM$), which means it is the seepage flow under the condition of unit hydraulic gradient, unit time, and unit width.

Storage coefficient μ^* (or elastic specific yield) represents the water storage capacity of aquifer in pressurization or water yield capacity in pressure reduction. It refers to water volume that is stored or released from unit area aquifer when waterhead having a unit change. It is determined by the elastic property of water and aquifer skeleton.

Pressure transitivity coefficient (a) represents the waterhead conduction velocity of confined aquifer. It is defined as $a = T/\mu^*$.

T, μ^*, and a can be calculated according to unsteady pumping test data. If the thickness of aquifer has already been measured, the hydraulic conductivity K can also been calculated.

In horizontal, homogeneous, isotropic, and infinite extending confined aquifer without vertical recharge and thickness change, Theis formula can be written as:

$$s(r,t) = \frac{Q}{4\pi T} \int_u^\infty \frac{e^{-u}}{u} du = \frac{Q}{4\pi T} W(u) \qquad (2.37)$$

$$u = \frac{r^2 \mu^*}{4Tt}, \quad \text{or} \quad \frac{1}{u} = \frac{4Tt}{r^2 \mu^*} \qquad (2.38)$$

where $s(r, t)$ is the drawdown of certain point within influence area, m; t is the time form the beginning of pumping, h; r is the distance between calculating point and pumping well, m; $W(u)$ is the well formula which can be found from $W(u)$ table. The meaning of other symbols are the same with above.

2.6.1.1 Calculation Method

T, μ^*, and a can be calculated by curve-matching method, linear graphic method, and water table recovery method, as shown in Table 2.24. Data for these methods are got from unsteady pumping test with constant water discharge.

Table 2.24 Methods for parameters calculation of confined aquifer (From Handbook of hydrogeological investigation of water supply 1977)

Method		Calculation steps	Formulas	Graph
Curve-matching method	Drawdown (s)–time (t) curve	1. Plot the t-s data on log-log graph paper 2. Superimpose this plot on the $W(u)$-$(1/u)$ type curve sheet of the same size and scale as the t-s plot, so that the plotted points match the type curve. The axes of both graphs must be kept parallel 3. Select a match point, which can be any point in the overlap area of the curve sheets. Write down the $W(u)$, $1/u$, s and t values of match point 4. Determine T, μ^* and a	$T = \dfrac{Q}{4\pi s} W(u)$ (2.39) $\mu^* = \dfrac{4Tt}{r^2(1/u)}$ (2.40) $a = \dfrac{T}{\mu^*}$ (2.41)	Fig.2.29 s-t curve in log-log coordinate
	Drawdown (s)–distance (r) curve	1. Plot the r^2-s data on log-log graph paper 2. Superimpose this plot on the $W(u)$-u type curve sheet of the same size and scale as the t-s plot, so that the plotted points match the type curve. The axes of both graphs must be kept parallel 3. Select a match point, which can be any point in the overlap area of the curve sheets. Write down the $W(u)$, u, s and r^2 values of match point 4. Determine T, μ^* and a	$T = \dfrac{Q}{4\pi s} W(u)$ (2.39) $\mu^* = \dfrac{4Ttu}{r^2}$ (2.42) $a = \dfrac{T}{\mu^*}$ (2.41)	Fig.2.30 s-r^2 curve in log-log coordinate

(continued)

2.6 Hydrogeological Parameters Calculation in Unsteady Flow Pumping Test

Table 2.24 (continued)

Method		Calculation steps	Formulas	Graph
Linear graphic method	Drawdown (s)-time (t) line	1. Plot the s-t data on semi-log graph paper 2. Extent the line to lg$t$ axial, then get the intercept t_0 3. Determine the straight slope i, which numerically equal to Δs that corresponding to one time period ($\Delta \lg t = 1$) 4. Determine T, μ^* and a	$T = \frac{2.30Q}{4\pi \Delta s}$ (2.43) $\mu^* = \frac{2.25Tt_0}{r^2}$ (2.44) $a = \frac{T}{\mu^*}$ (2.41)	Fig.2.31 s-t curve in semi-log coordinate
	Drawdown (s)-distance (r) line	1. Plot the s-r data on semi-log graph paper 2. Extent the line to lg$r$ axial, then get the intercept r_0 3. Determine the straight slope i, which numerically equal to Δs that corresponding to one time period ($\Delta \lg t = 1$) 4. Determine T, μ^* and a	$T = \frac{2.30Q}{4\pi \Delta s}$ (2.45) $\mu^* = \frac{2.25Tt_0}{r_0^2}$ (2.40) $a = \frac{T}{\mu^*}$ (2.41)	Fig.2.32 s-r curve in semi-log coordinate
Water table recovery method		1. Plot the s'-t/t' data on semi-log graph paper, where s' is the rest drawdown, t' is the time of water table recovery, t_p is the time of pumping, and $t = t' + t_p$ 2. Determine the straight slope i, which numerically equal to Δs that corresponding to one time period ($\Delta \lg t = 1$) 3. Determine T, μ^* and a	$T = \frac{2.30Q}{4\pi \Delta s}$ (2.43) $a = 0.44 \left(\frac{r}{t_p}\right)^2 \cdot 10^{\frac{s_p}{i}}$ (2.46) where s_p is the drawdown when pumping stop. $\mu^* = \frac{T}{a}$ (2.47)	Fig.2.33 s-t/t' curve in semi-log coordinate

The biggest advantage of curve-matching method is that all test data can be used to improve the calculation accuracy. This method also has disadvantages. That is when the r is small and T is large, the steep part of observation curve appears in the first 2 min, which is difficult to measure. While the part that is easy to measure is too gentle to fit accurately. These all lead to low accuracy of parameters. The solutions are: On the one hand, using the data from initial pumping stage as much as possible. On the other hand, setting the observation wells far from pumping well when T is large, so that the drawdown can be measured accurately.

There are many advantages of linear graphic method, for example all test data can be used and the randomness of fitting curve method can be avoided. The disadvantage of this method is that the pumping time needs to be long to meet the requirement of $u \leq 0.01$, especially when T is small and r is large. Besides, the long pumping time may leads to the deviation of intercept and gradient, which will result in larger T and smaller μ^*.

The water table recovery method can avoid the interference from pumping equipment and water discharge variation during pumping period. The drawdown versus time curve of water table recovery period usually much more regular, which can improve the accuracy of parameters calculation.

Apply as many methods as possible to calculate the parameters when using unsteady pumping test data, and compare the results to choose the best one.

2.6.1.2 Case Study

The hydrogeological condition is the same as it in Sect. 2.5.3.
1. lg$s$-lg$t$ curve-matching method
lg$s$-lg$t$ curve of Observation well 2 and Observation well 3 are shown in Fig. 2.34.

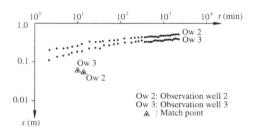

Fig. 2.34 lg$s$-lg$r$ curves

2.6 Hydrogeological Parameters Calculation in Unsteady Flow Pumping Test

Results:
Observation well 2:

$$W(u) = 1,\ 1/u = 10^3,\ s = 0.061\ \text{m},\ t = 14.1\ \text{min};$$
$$T = \frac{Q}{4\pi s}W(u) = \frac{2384}{4\pi \times 0.061} \times 1 = 3.11 \times 10^3\ \text{m}^2/\text{day};$$
$$\mu^* = \frac{4Tt}{r^2(1/u)} = \frac{4 \times 3.14 \times 10^3 \times 14.1}{20^2 \times 10^2 \times 1440} = 3.05 \times 10^{-4};$$
$$a = \frac{T}{\mu^*} = \frac{3.11 \times 10^3}{3.05 \times 10^{-4}} = 1.02 \times 10^7\ \text{m}^2/\text{day}.$$

Observation well 3:

$$W(u) = 1,\ 1/u = 10^2,\ s = 0.062\ \text{m},\ t = 10\ \text{min};$$
$$T = \frac{2384}{4\pi \times 0.062} \times 1 = 3.06 \times 10^3\ \text{m}^2/\text{day};$$
$$\mu^* = \frac{4 \times 3.06 \times 10^3 \times 10}{30^2 \times 10^2 \times 1440} = 9.44 \times 10^{-4};$$
$$a = \frac{3.06 \times 10^3}{9.44 \times 10^{-4}} = 3.24 \times 10^6\ \text{m}^2/\text{day}.$$

2. Linear graphic method

s-$\lg t$ curve of Observation well 2 and Observation well 3 are shown in Fig. 2.35.
Results:
Observation well 2:

$$T = \frac{2.30Q}{4\pi \Delta s} = \frac{2.30 \times 2384}{4\pi \times (0.50 - 0.36)} = 3.12 \times 10^3\ \text{m}^2/\text{day};$$
$$\mu^* = \frac{2.25Tt_0}{r^2} = \frac{2.25 \times 3.12 \times 10^3 \times 0.22}{20^2 \times 1440} = 2.68 \times 10^{-3};$$
$$a = \frac{T}{\mu^*} = 1.2 \times 10^6\ \text{m}^2/\text{day}.$$

Fig. 2.35 s-$\lg t$ curves

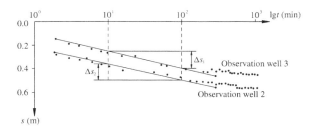

Observation well 3:

$$T = \frac{2.30 \times 2384}{4\pi \times (0.40 - 0.26)} = 3.12 \times 10^3 \text{ m}^2/\text{day};$$

$$\mu^* = \frac{2.25Tt_0}{r^2} = \frac{2.25 \times 3.12 \times 10^3 \times 0.12}{30^2 \times 1440} = 6.5 \times 10^{-4};$$

$$a = \frac{T}{\mu^*} = 4.8 \times 10^6 \text{ m}^2/\text{day}.$$

3. Water table recovery method

$s - \lg \frac{t_p + t'}{t'}$ curve of Observation well 2 and Observation well 3 are shown in Fig. 2.36.

Results:

Observation well 2:

$$T = \frac{2.30Q}{4\pi \Delta s} = \frac{2.30 \times 2384}{4\pi \times (0.276 - 0.14)} = 3.21 \times 10^3 \text{ m}^2/\text{day};$$

$$a = 0.44 \frac{r^2}{t_p} \times 10^{\frac{s_p}{i}} = 0.44 \times \frac{20^2}{0.5} \times 10^{\frac{0.61}{0.136}} = 1.08 \times 10^7 \text{ m}^2/\text{day};$$

$$\mu^* = \frac{T}{a} = \frac{3.21 \times 10^3}{1.08 \times 10^7} = 2.97 \times 10^{-4}.$$

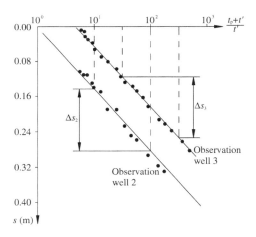

Fig. 2.36 s-$\lg \frac{t_p + t'}{t'}$ curve

2.6 Hydrogeological Parameters Calculation in Unsteady Flow Pumping Test

Table 2.25 Summary sheet for aquifer parameters

Calculation method		Data	Aquifer parameters			
			K (m/day)	T (m²/day)	a (m²/day)	μ^*
Steady flow	Single well		68.15	1.6×10^3		
	Multiple wells		115.0	2.7×10^3		
Unsteady flow	Fitting curve method	Observation well 2		3.11×10^3	1.02×10^7	3.05×10^{-4}
		Observation well 3		3.06×10^3	3.24×10^6	9.44×10^{-4}
	Linear graphic method	Observation well 2		3.12×10^3	1.2×10^6	2.68×10^{-3}
		Observation well 3		3.12×10^3	4.8×10^6	6.5×10^{-4}
	Water table recovery method	Observation well 2		3.21×10^3	1.08×10^7	2.97×10^{-4}
		Observation well 3		3.31×10^3	2.88×10^6	1.15×10^{-3}
	Average value			3.15×10^3	5.52×10^6	1.0×10^{-3}
	Range			$3.06 \times 10^3 \sim 3.31 \times 10^3$	$1.2 \times 10^6 \sim 1.08 \times 10^7$	$2.97 \times 10^{-4} \sim 2.68 \times 10^{-3}$

Observation well 3:

$$T = \frac{2.30 \times 2384}{4\pi \times (0.242 - 0.11)} = 3.31 \times 10^3 \text{ m}^2/\text{day};$$

$$a = 0.44 \times \frac{30^2}{0.5} \times 10^{\frac{0.47}{0.132}} = 2.88 \times 10^6 \text{ m}^2/\text{day};$$

$$\mu^* = \frac{T}{a} = \frac{3.31 \times 10^3}{2.88 \times 10^6} = 1.15 \times 10^{-3}.$$

Table 2.25 is the summary for hydraulic parameters.

2.6.2 Transmissibility, Storage Coefficient, Leakage Coefficient, and Leakage Factor Calculation for Leaky Aquifers

Leakage coefficient σ ($=K'/M'$) and leakage factor B represent the leakage characteristic of aquitard. Leakage recharge amount is related to hydraulic conductivity K' and thickness M' of aquitard.

Leakage coefficient σ represents the recharging amount in a unit area from aquitard to pumped aquifer under a unit waterhead difference.

Leakage factor $B = \sqrt{TM'/K'}$, where T is the transmissibility of main aquifer, K' and M' are the hydraulic conductivity and thickness of aquitard. The smaller the K' is and the bigger the M' is, the larger the B is.

Hydraulic parameters of leaky aquifers can be calculated according to unsteady pumping test data by Hantush-Jacob formula, as follows:

$$s = \frac{Q}{4\pi T} \int_u^\infty \frac{1}{y} e^{-y - \frac{r^2}{4B^2 y}} dy = \frac{Q}{4\pi T} W\left(u, \frac{r}{B}\right) \quad (2.48)$$

$$u = \frac{r^2 \mu^*}{4Tt} \quad \text{or} \quad \frac{1}{u} = \frac{4Tt}{r^2 \mu^*} \quad (2.49)$$

where T, μ^*, a, σ and B can be calculated by curve-matching method, yielding point method and tangent method, as shown in Table 2.26.

2.6.3 Specific Yield, Storage Coefficient, Hydraulic Conductivity and Transmissibility Calculation of Unconfined Aquifer

Specific yield μ refers to the volume of gravitational water that drained from unit area of unconfined aquifer when water table decreases a unit meter. Free saturation rate, also called saturation deficit, refers to the volume of water that recharge to unit area of unconfined aquifer when water table rise a unit meter. Usually, specific yield and free saturation rate are numerically same.

2.6.3.1 Bolton Formulas

Early pumping stage:

$$s = \frac{Q}{4\pi T} W(u_d, \frac{r}{D}) \quad (2.58)$$

Middle pumping stage:

$$s = \frac{Q}{4\pi T} K_0(\frac{r}{D}) \quad (2.59)$$

Late pumping stage:

$$s = \frac{Q}{4\pi T} W(u_y, \frac{r}{D}) \quad (2.60)$$

$$u_d = \frac{r^2 \mu^*}{4Tt}, \quad u_y = \frac{r^2 u}{4Tt} \quad (2.61)$$

2.6 Hydrogeological Parameters Calculation in Unsteady Flow Pumping Test

Table 2.26 Methods for parameters calculation of confined aquifer (revised from Handbook of hydrogeological investigation of water supply 1977)

Method	Calculation steps	Formulas	Graph
Drawdown (s)–time (t) curve matching method	1. Plot the t-s data on double logarithmic graphic paper 2. Superimpose this plot on the $W(u, r/B)$-$(1/u)$ type curve sheet of the same size and scale as the t-s plots, and find the best overlapping curve that the plotted points match the type curve. The axes of both graphs must be kept parallel 3. Select a match point, which can be any point in the overlap area of the curve sheets. Write down the $W(u, r/B)$, $1/u$, s and t values of match point 4. Calculate T, μ^*, a, B and σ	$T = \frac{Q}{4\pi s}[W(u, r/B)]$ (2.50) $\mu^* = \frac{4Tt[\frac{1}{u}]}{r^2}$ (2.51) $a = \frac{T}{\mu^*}$ (2.41) $B = r/[r/B]$ (2.52) $K_1/M_1 = T/B^2$ (2.53)	Fig. 2.37
Yield point method	1. Plot the t-s data on semi-log graphic paper. Determine the maximum drawdown s_{max} with extrapolation method, and calculate the drawdown of yield point $s_p = s_{max}/2$ 2. Basing on s_p, figure out the position of yield point p and the corresponding t_p 3. Draw the tangent line of the s-lg$t$ curve at the point p, and determine its slope i_p 4. According to formula $e^{(r/B)} K_0 (r/B) = 2.30 (s_p/i_p)$, determine the value of r/B and $e^{(r/B)}$ from Hantusi function table 5. Calculate T, μ^*, B and K_1/M_1 6. Verification	$B = r/[r/B]$ (2.52) $T = \frac{2.30Q}{4\pi i_p} e^{-(r/B)}$ (2.54) $\mu^* = \frac{2Tt_p}{Br^2}$ (2.55) $a = \frac{T}{\mu^*}$ (2.41) $K_1/M_1 = T/B^2$ (2.53)	Fig. 2.38
Tangent method	1. Plot the t-s data on semi-log graph paper and determine the maximum drawdown s_{max} with extrapolation method 2. Pick one point p at the s-lg$t$ curve and its coordinates are t_p and s_p 3. Draw the tangent line of the s-lg$t$ curve at the point p, and determine its slope i_p 4. According to formula $f(\delta) = 2.30 (s_{max}-s_p)/i_p$, determine the value of δ, e^{δ} and $\omega(\delta)$ from Hantusi function table 5. Calculate T, B, K_1/M_1 and μ^*; according to formula $K_0\left(\frac{r}{B} = \frac{2\pi t}{Q} s_{max}\right)$, determine the value of r/B from μ^* function table	$T = \frac{2.30Q}{4\pi i_p} e^{-\delta}$ (2.56) $B = r/[r/B]$ (2.52) $K_1/M_1 = T/B^2$ (2.53) $\mu^* = \frac{Tt_p}{B^2 \delta}$ (2.57) $a = \frac{T}{\mu^*}$ (2.41)	Fig. 2.39

Table 2.27 Bolton formulas for parameters calculation of unconfined aquifer

Method	Calculation steps	Formulas	Graph
Drawdown (s)-time (t) curve-matching method	1. Plot the t-s data on double logarithmic graph paper 2. Superimpose these plots on the standard curves of group A to find out one curve that can best match these plots. Pick one match point and record its coordinates $W(u_{\rm d}, r/D)$, $1/u_{\rm d}$, s and t, and also the r/D value of the match curve. Calculate T and μ^* 3. Superimpose the rest plots on the standard curves of group B to find out one curve that can best match these plots. Keep the r/D value constant and pick one match point, whose coordinates are $W(u_y, r/D)$, $1/u_y$, s and t. Calculate T and μ 4. When the value of s/H_0 is large, the value of T will change. Thus the drawdown should be amended by formula $s' = s - s^2/2H_0$ 5. If using Theis curves replace standard group B curves, then the time $t_{\rm wt}$ can be calculated from $\alpha t_{\rm wt} - r/D$ curve because $\alpha t_{\rm wt} = \frac{1}{4}\left(\frac{r}{D}\right)^2 \frac{1}{u_y}$	$T = \frac{Q}{4\pi[S]}\left[W(u_{\rm d}, \frac{r}{D})\right]$ (2.62) $\mu^* = \frac{4Tt}{r^2[1/u_{\rm d}]}$ (2.63) $T = \frac{Q}{4\pi[S]}\left[W(u_y, \frac{r}{D})\right]$ (2.64) $\mu = \frac{4Tt}{r^2[1/u_y]}$ (2.65) $\eta = \frac{\mu^*+\mu}{\mu}$ (2.66) $\frac{1}{\alpha} = \left[\frac{r}{D}\right]^2\left[\frac{1}{u_y}\right]$ (2.67)	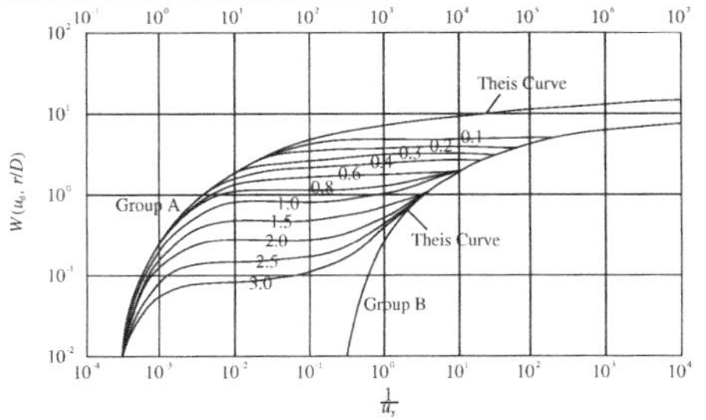 Fig. 2.40 (from Prickett, 1965)
Linear graphic method	1. Plot the s-t data on semi-log graph paper 2. Determine T, S and a by The is formulas, as shown in Table 2.19	$T = \frac{2.30Q}{4\pi i}$ (2.43) $\mu = \frac{2.25Tt_0}{r^2}$ (2.44) $a = \frac{T}{\mu}$ (2.41)	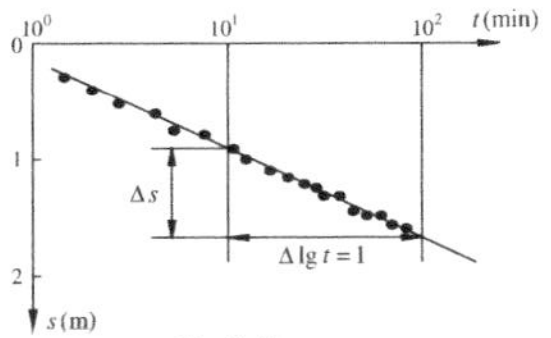 Fig. 2.41

2.6 Hydrogeological Parameters Calculation in Unsteady Flow Pumping Test

Table 2.28 Newman formulas for parameters calculation of unconfined aquifer

Method	Calculation steps	Formulas	Graph
s-t curve-matching method	1. Plot the t-s data on double logarithmic graph paper 2. Superimpose these plots on the standard group B curves to find out one curve that can best match these plots and record its β value. Pick one match point and record its coordinates s_d, t_y, s, and t, then calculate T and μ 3. Superimpose the front part of plots on the standard curves of group A to find out one curve that can best match these plots and its β value should be the same with step 2. Pick one match point, whose coordinates are s_d, t_s, s, and t. Calculate T and μ^*. The values of T calculated in step 2 and step 3 should be close 4. Calculate radial permeability K_r, ratio of anisotropy permeability K_d, vertical permeability K_z and σ	$T = \dfrac{Q}{4\pi} \dfrac{[s_d]}{[s]}$ (2.71) $\mu = \dfrac{T(t)}{r^2[t_y]}$ (2.72) $T = \dfrac{Q}{4\pi} \dfrac{[s_d]}{[s]}$ (2.73) $\mu^* = \dfrac{T(t)}{r^2[t_s]}$ (2.74) $K_r = \dfrac{T}{H_0}$ (2.75) $K_d = \beta \dfrac{H_0^2}{r^2}$ (2.76) $K_z = K_d K_r$ (2.77) $\sigma = \dfrac{\mu^*}{\mu}$ (2.78)	Fig. 2.42 (From Newman, 1972)
Linear graphic method	1. Plot the t-s data on semi-log graph paper 2. Determine the slope i_L of the linear part of s-$\lg t$ curve. Extend this linear part and get the intercept t_1 of the abscissa with $s_d = 0$. Calculate T and μ 3. Extend the horizontal part of the measured curve toward the right to intersect with the later linear part in one point, whose coordinate is t_β. Calculate $t_{y\beta}$ and β 4. If the front part of the measured curve is linear and parallel to the later linear part, then determine the slope i_E of this line and extend it to get the intercept t_E of the abscissa with $s_d = 0$. Calculate T and μ^*. The values of T that calculated in step 2 and this step should be close 5. Calculate K_r, K_d, K_z and σ	$T = \dfrac{2.30Q}{4\pi i_L}$ (2.43) $\mu = \dfrac{2.25 T t_1}{r^2}$ (2.44) $t_{y\beta} = \dfrac{T t_\beta}{\mu r^2}$ (2.79) $\beta = \dfrac{0.195}{s/\beta^{1.033}}$ (2.80) $T = \dfrac{2.30Q}{4\pi i_E}$ (2.43) $\mu^* = \dfrac{2.25 T t_E}{r^2}$ (2.40) $K_r = T/H_0$ (2.75) $K_d = \beta H_0^2 / r^2$ (2.76) $K_z = K_d K_r$ (2.77) $\sigma = t_E / t_1$ (2.81)	Fig. 2.43 (From Newman, 1973)

(continued)

Table 2.28 (continued)

Method	Calculation steps	Formulas	Graph
Water table recovery method	1. Plot the s'-t/t' data on the semi-log graph paper 2. Determine the slope i_L of the line	$T = \dfrac{2.30Q}{4\pi i_L}$ (2.43)	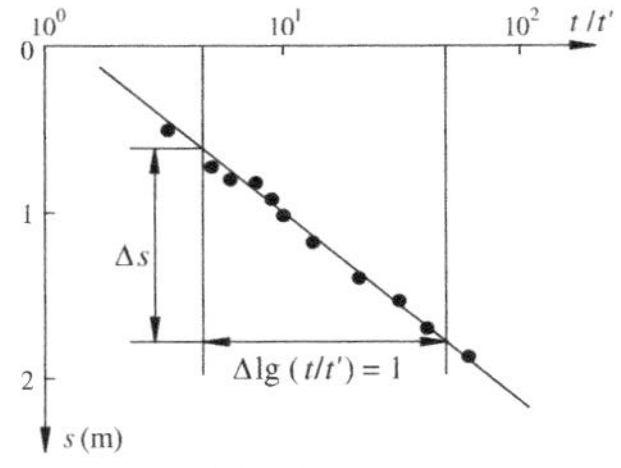 Fig.2.44

2.6 Hydrogeological Parameters Calculation in Unsteady Flow Pumping Test

Parameters of unconfined aquifer can be calculated by curve-matching method and linear graphic method according to Bolton Formula, shown in Table 2.27.

2.6.3.2 Newman Formula

Newman formula is:

$$s(r,t) = \frac{Q}{4\pi T} \int_0^\infty 4y J_0(y\beta^{1/2}) \left[\omega_0(y) + \sum_{n=1}^\infty \omega_n(y) \right] dy \quad (2.68)$$

where:

$$\omega_0 = \frac{\{1 - \exp[-t_s\beta(y^2 - \gamma_0^2)]\}\text{th}(\gamma_0)}{\{y_0^2 + (1+\sigma)\gamma_0^2 - [(y^2 - \gamma_0^2)^2/\sigma]\}\gamma_0} \quad (2.69)$$

$$\omega_n = \frac{\{1 - \exp[-t_s\beta(y^2 + \gamma_n^2)]\}\text{th}(\gamma_n)}{\{y_n^2 - (1+\sigma)\gamma_n^2 - [(y^2 + \gamma_n^2)^2/\sigma]\}\gamma_n} \quad (2.70)$$

Parameters of unconfined aquifer can be calculated by curve-matching method and linear graphic method according to Bolton Formula, shown in Table 2.28.

2.7 Other Methods for Hydrogeological Parameters Calculation

2.7.1 Transmissibility and Well Loss Calculation

Typically, well loss is ignored when using pumping test data to calculated aquifer parameters, which can lead to inaccuracy of those parameters.

This section will introduce a method that can consider the loss for transmissibility and well loss constant calculation.

2.7.1.1 Basic Theory

In confined aquifer, the Jacob approximation formula for Theis formula is:

$$s = \frac{2.30Q}{4\pi T} \lg \frac{2.25Tt}{r^2\mu^*} \quad (u<0.01) \quad (2.82)$$

It can be found from Eq. (2.82) that the value of drawdown is only related to the properties of aquifer, which is not suitable for calculation. The drawdown of pumping well mainly consists of two parts: The first part is the drawdown that related to aquifers' properties; the second part is that caused by turbulence flow around pumping well, which also know as well loss. The later is hard to be estimated and is various for different pumping wells. It can be approximately calculated by following formulas:

$$s'_w = CQ^2 \tag{2.83}$$

where s'_w is the drawdown that caused by well loss, L; C is the well loss constant, T^2K^{-5}; Q is the water discharge, L^3T^{-5}.

So the total drawdown at pumping time t is:

$$\begin{aligned} s_g &= s + s'_w \\ &= \frac{2.30Q}{4\pi T} \lg \frac{2.25Tt}{r^2\mu^*} + CQ^2 \\ &= a(b + \lg t)Q + CQ^2 \end{aligned} \tag{2.84}$$

where $a = \frac{2.30Q}{4\pi T}$; $b = \lg \frac{2.25T}{r^2\mu^*}$.

Change the water discharge for n times:

$$s_{gj} = a(b + \lg t_j)Q + CQ^2$$

$$s_{gj}/Q = A + CQ \tag{2.85}$$

where $A = a(b + \lg t_j)$; s_{gj} is the drawdown at time j, L; t_j is the time, $j = 1, 2, \ldots, m$. Formula (2.85) is a linear equation, shown in Fig. 2.45.

Fig. 2.45 $Q\text{-}s_g/Q$ lines

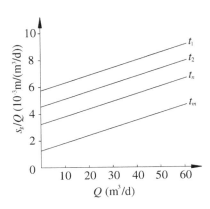

2.7 Other Methods for Hydrogeological Parameters Calculation

The intercepts that correspond to time $t_1, t_2, \ldots, t_m$ are $A_1, A_2, \ldots, A_m$, so:

$$A_m - A_j = a(b + \lg t_m) - a(b + \lg t_j)$$
$$= a \lg\left(\frac{t_m}{t_j}\right) = \frac{2.30}{4\pi T} \lg\left(\frac{t_m}{t_j}\right)$$

where $j = 1, 2, \ldots, m$.
So:

$$T = \frac{2.30 \lg(t_m/t_j)}{4\pi(A_m - A_j)} \tag{2.86}$$

Well loss constant C and transmissibility T can be calculated by Eqs. (2.85) and (2.86), respectively. Storage coefficient μ^* can be calculated by Eq. (2.84), then (2.87) can be got:

$$\lg \mu^* = \lg \frac{2.25Tt}{r_w^2} + \frac{4\pi T}{2.30}\left(CQ - \frac{s_g}{Q}\right) \tag{2.87}$$

If r is replaced by r_w, the effective radius of well r_e is no less than the actual well radius r_w, and the above formula can be rewritten as:

$$\lg \mu^* \leq \lg \frac{2.25Tt}{r_w^2} + \frac{4\pi T}{2.30}\left(CQ - \frac{s_g}{Q}\right) \tag{2.88}$$

This formula is used to estimate the range of storage coefficient.

2.7.1.2 Application Steps

Case Study

The hydrogeological condition is the same as Sect. 2.5.3. The pumping test data are shown in Tables 2.29 and 2.30.

Table 2.29 Drawdown s (m)

Flow Q (m³/day)	Observation time t (min)		
	10	40	150
1570	1.24	1.30	1.35
2335	1.92	2.0	2.06
2540	2.10	2.21	2.28
2757	2.36	2.46	2.53

Table 2.30 s_g/Q (×10^{-3} m/m³/day)

Flow Q (m³/day)	Observation time t (min)		
	10	40	150
1570	0.79	0.83	0.85
2335	0.82	0.86	0.88
2540	0.83	0.87	0.90
2757	0.86	0.89	0.92

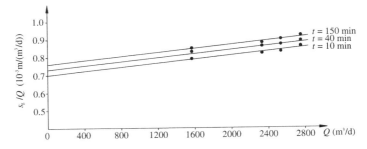

Fig. 2.46 s_g/Q-Q lines

1. Draw s_s/Q-Q lines, shown in Fig. 2.46.
2. Transmissibility calculation

$$T_1 = \frac{2.3 \lg(40/10)}{4\pi(0.735 - 0.70) \times 10^{-3}} = 3.15 \times 10^3 \text{ m}^2/\text{day}$$

$$T_2 = \frac{2.3 \lg(150/10)}{4\pi(0.77 - 0.70) \times 10^{-3}} = 3.08 \times 10^3 \text{ m}^2/\text{day}$$

3. Well loss constant calculation

$$C = \frac{(0.79 - 0.70) \times 10^{-3}}{1600} = 5.6 \times 10^{-8}$$

2.7.2 Calculation of Transmissibility Coefficient and Water Storage Coefficient by Sensitivity Analysis Method Based on Pumping Test Data

2.7.2.1 Sensitivity Analysis

When simulating an aquifer system, researchers have to determine the admissible deviation. If the admissible deviation does not affect the simulation results

2.7 Other Methods for Hydrogeological Parameters Calculation

significantly, the parameters of the actual system can be different. The admissible deviation is generally determined based on parameter variation induced into the system and the variation of characteristics in the system. These admissible deviations can be determined more effectively by using the sensitivity analysis.

In confined aquifer, according to Theis formula, there is

$$s = \frac{Q}{4\pi T} \int_u^\infty \frac{e^{-u}}{u} du$$

where

$$u = \frac{r^2 \mu^*}{4Tt}$$

So the solution of an aquifer model can be written as follows:

$$h = h(x, y, t; T, \mu^*, Q)$$

where h represents the water head.

Considering the change of one parameter, for example, T is seen as variable, when T has an increment ΔT, it has

$$h^* = h^*(x, y, t; T + T, \mu^*, Q)$$

It is assumed that the solutions of the aquifer model depend on parameter T and μ^*. T, μ^*, and Q are all independent variables. So function $h^*(x, y, t; T + \Delta T, \mu^*, Q)$ can be expanded to Taylor series. If the value of ΔT is very little, the quadratic term and high order terms can be neglected

$$h^*(x, y, t; T + \Delta T, \mu^*, Q) = h(x, y, t; T, \mu^*, Q) + U_T \Delta T \qquad (2.89)$$

$$U_T(x, y, t; T, \mu^*, Q) = \frac{\partial h(x, y, t; T, \mu^*, Q)}{\partial T} \qquad (2.90)$$

Transmissibility coefficient sensitivity $U_T(x, y, t; T, \mu^*, Q)$ will be represented by U_T.

If sensitivity U_T and the initial water head are known, the new water head caused by variation ΔT of transmissibility coefficient can be calculated by Eq. (2.89).

Similarly, the new water head caused by variation $\Delta \mu^*$ of water storage coefficient μ^* can be calculated by the following equation:

$$h^*(x, y, t; T + \Delta \mu^*, \mu^*, Q) = h(x, y, t; T, \mu^*, Q) + U_{\mu^*} \Delta \mu^* \qquad (2.91)$$

$$U_{\mu^*}(x,y,t;T,\mu^*,Q) = \frac{\partial h(x,y,t;T,\mu^*,Q)}{\partial \mu^*} \qquad (2.92)$$

Sensitivity of water storage coefficient $U_{\mu^*}(x,y,t;T,\mu^*,Q)$ $(x, y, t; T, \mu^*, Q)$ will be represented by U_{μ^*}.

Equations (2.90) and (2.92) indicate that to a given model, U_T and U_{μ^*} need to be calculated. The response of model under a variety of variation can be calculated easily by (2.89) and (2.91) without re-calculation of model equations.

Sensitivity coefficient can be obtained from Theis formula by the definition of Eqs. (2.90) and (2.92):

$$U_T = \frac{\partial s}{\partial T} = -\frac{s}{T} + \frac{Q}{4\pi T^2}\exp\left(-\frac{r^2\mu^*}{4Tt}\right) \qquad (2.93)$$

$$U_{\mu^*} = \frac{\partial s}{\partial \mu^*} = -\frac{Q}{4\pi T\mu^*}\exp\left(-\frac{r^2\mu^*}{4Tt}\right) \qquad (2.94)$$

If μ^* and T change, respectively, with the variation of $\Delta\mu^*$ and ΔT, U_T and U_{μ^*} can be obtained from Eqs. (2.93) and (2.94) can be used in Eqs. (2.89) and (2.91) to obtain the drawdown values. Data show that when $\Delta\mu^*$ and ΔT are no more than 20 % of μ^* and T, respectively, Eqs. (2.89) and (2.91) is valid.

2.7.2.2 Least-Squares Fitting

The purpose of the sensitivity analysis is to obtain the least-squares fitting of actual pumping test data to Theis formula and then to obtain the best value of μ^* and T.

T and μ^* are changed by ΔT and $\Delta\mu^*$. The new drawdown value s^* can be calculated by the following equation:

$$s^* = s + U_T\Delta T + U_{\mu^*}\Delta\mu^* \qquad (2.95)$$

The actual drawdown measured at time t is represented by $s_c(t)$. It is assumed that appropriate estimate can be made for μ^* and T, then $s(t)$ is the drawdown obtained from Theis formula by these parameters. The preliminary estimate value of μ^* and T can be changed by $\Delta\mu^*$ and ΔT. The new drawdown value s^* is calculated by Eq. (2.95). The errors dropped to minimum with the aid of the following error function, and the better fitting results compared with actual pumping test data will be obtained.

2.7 Other Methods for Hydrogeological Parameters Calculation

$$\begin{aligned}
E(\Delta T, \Delta \mu^*) &= \sum_{i=1}^{n} [s_c(t_i) - s^*(t_i)]^2 \\
&= \sum_{i=1}^{n} [s_c(t_i) - s(t_i) - U_T(t_i)\Delta T - U_{\mu^*}(t_i)\Delta \mu^*]^2 \\
&= \sum_{i=1}^{n} [s_c(t_i) - s(t_i)]^2 - 2\Delta T \sum_{i=1}^{n} U_T(t_i)[s_c(t_i) - s(t_i)] \\
&\quad - 2\Delta \mu^* \sum_{i=1}^{n} U_{\mu^*}(t_i)[s_c(t_i) - s(t_i)] + \sum_{i=1}^{n} [U_{\mu^*}^2(t_i)\Delta \mu^{*2} \\
&\quad + 2U_T(t_i)U_{\mu^*}(t_i)\Delta \mu^* \Delta T + U_T^2(t_i)\Delta T^2]
\end{aligned} \quad (2.96)$$

t_i represents any moment, when a test value of drawdown can be obtained. The error function is determined by the square sum of difference between measured values s_{μ^*} and s^*. Sensitivity coefficient U_T and U_{μ^*} are based on t_i.

The first-order derivative of ΔT and $\Delta \mu^*$ are selected, and are equal to zero. Then the errors are minimum values. The equations of ΔT and $\Delta \mu^*$ are as follows:

$$\begin{aligned}
\frac{\partial E(\Delta T, \Delta \mu^*)}{\partial \Delta T} &= -2\sum_{i=1}^{n} U_T(t_i)[s_c(t_i) - s(t_i)] + 2\Delta \mu^* \sum_{i=1}^{n} U_{\mu^*}(t_i)U_T(t_i) \\
&\quad + 2\Delta T \sum_{i=1}^{n} U_T^2(t_i) = 0
\end{aligned} \quad (2.97)$$

$$\begin{aligned}
\frac{\partial E(\Delta T, \Delta \mu^*)}{\partial \Delta \mu^*} &= \mu^* - 2\sum_{i=1}^{n} U_{\mu^*}(t_i)[s_c(t_i) - s(t_i)] + 2\Delta \mu^* \sum_{i=1}^{n} U_{\mu^*}(t_i) \\
&\quad + 2\Delta T \sum_{i=1}^{n} U_{\mu^*}(t_i)U_T(t_i) = 0
\end{aligned} \quad (2.98)$$

ΔT can be obtained by Eq. (2.97):

$$\Delta T = \left\{ \sum_{i=1}^{n} \{U_T(t_i)[s_c(t_i) - s(t_i)]\} - \left\{ \sum_{i=1}^{n} [U_{\mu^*}(t_i)U_T(t_i)] \right\} \Delta \mu^* \right\} / \sum_{i=1}^{n} U_T^2(t_i) \quad (2.99)$$

$\Delta \mu^*$ can be obtained by Eq. (2.98):

$$\Delta \mu^* = \left\{ \sum_{i=1}^{n} [U_T^2(t_i)] \sum_{i=1}^{n} \{U_{\mu^*}(t_i)[s_c(t_i) - s(t_i)]\} - \left\{ \sum_{i=1}^{n} [U_{\mu^*}(t_i) U_T(t_i)] \right\} \right.$$
$$\left. \left\{ \sum_{i=1}^{n} \{U_T(t_i)[s_c(t_i) - s(t_i)]\} \right\} \middle/ \left\{ \sum_{i=1}^{n} [U_{\mu^*}^2(t_i)] \sum_{i=1}^{n} [U_T^2(t_i)] - \left\{ \sum_{i=1}^{n} [U_{\mu^*}(t_i) U_T(t_i)] \right\}^2 \right\} \right\}$$
(2.100)

The best value (namely the best fitting results) of $\Delta \mu^*$ can be obtained from Eq. (2.100). By substitution of $\Delta \mu^*$ into Eq. (2.99), the best fitting value of ΔT can be found.

The values of ΔT and $\Delta \mu^*$ can be used to correct the first estimated values of T and μ^*. The improved values of T and μ^* are used in the program of least-squares fitting in order to obtain the new values of ΔT and $\Delta \mu^*$. This cycle continues until ΔT and $\Delta \mu^*$ are little enough to be negligible. The best values after iteration of j times can be obtained by the following equations:

$$T_j = T_{j-1} + \Delta T_{j-1} \tag{2.101}$$

$$\mu^*_j = \mu^*_{j-1} + \Delta \mu^*_{j-1} \tag{2.102}$$

The program block diagram of the calculation of hydrogeological parameters of the aquifer by sensitivity analysis is shown in Fig. 2.47.

If the initial values of T and μ^* are particularly poor, the program may not be convergent. Available data indicate that good convergence can be obtained even for the case that the initial values of μ^* and T are less or greater for two orders of magnitude.

From the above, the best transmissibility coefficient and water storage coefficient are obtained from least-squares fitting by using sensitivity analysis in order to fit Theis formula by actual pumping test data automatically. That method can be applied in more complicated hydrogeological conditions.

2.7.3 *Hydrogeological Parameter Optimization Based on Numerical Method and Optimization Method Coupling Model*

The key question of reversing hydrogeologic parameters by numerical method is to obtain a set of parameters which can objectively represent the hydrogeological characteristics of the actual aquifer. The test standard is the error between water table (head) of every node calculated by mathematical model and the actually measured water table (head).

2.7 Other Methods for Hydrogeological Parameters Calculation

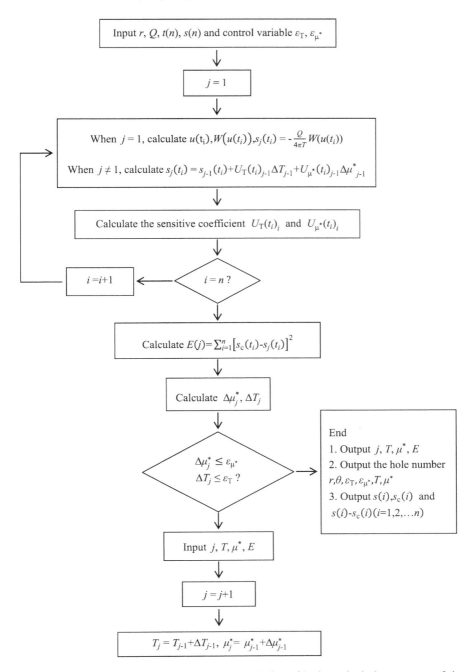

Fig. 2.47 The program block diagram of the calculation of hydrogeological parameters of the aquifer by sensitivity analysis

When using the error standard of the least square method, the identification of parameters can be formulated as the optimization problem as follows:

Objective function:

$$\min E(k_1^*, k_2^*, \ldots, k_n^*) = \sum_{i=1}^{J} \sum_{j=1}^{N} \omega_{ij} \left[H_j(t_i) - H_j^0(t_i) \right]^2 \quad (2.103)$$

Constraints:

$$\alpha_i \leq k_i \leq \beta_i \quad (i = 1, 2, \ldots, n) \quad (2.104)$$

where E is the objective function; $k_1^*, k_2^*, \ldots, k_n^*$ are a group of optimal parameters; J is the total number of observation periods; N is the total number of observation points; $H_j(t_i)$ is the calculated water table of node j at t_i; $H_j^0(t_i)$ is the measured water table of node j at t_i; ω_{ij} is the weight factor. The higher the precision is, the greater ω_{ij} is; k_i is the ith parameter of any group; α_i is the lower limit of the ith parameter; β_i is the higher limit of the ith parameter.

There are many methods of solving the optimization problems. The workload of the commonly used trial method is huge because it lacks a convergence criteria in every repeated trial calculation. The parameter in this process is blind and waste of time especially when there are a large number of unknown parameters. Hydrogeological parameter optimization based on optimization method can overcome the shortcomings of the trial method. There are many optimization methods and one of the direct unconstrained optimization methods—the stepped up simplex method will be introduced here.

2.7.3.1 The Basic Principle of the Advanced Simplex Method

The basic the principle of the advanced simplex method is: calculate the objective function value E, respectively at $n + 1$ simplex vertexes in E^n and compare them to determine the worst point, the second worst point and good points. Judge the approximate trend of function variation from the size relationship of the points. Choose reflection, extension, compression and so on to structure new simplex under different circumstances until the function value of simplex vertexes reach to the required minimum value. Then that set of parameter values are the optimal parameter values. The simplex in E^n means a polyhedron with $n + 1$ vertexes. If the edge-length equals to each other, the simplex is called the regular simplex. The following point is given in n—dimension space:

$$K^0 = (K_1^0, K_2^0, \ldots, K_n^0)^T$$

2.7 Other Methods for Hydrogeological Parameters Calculation

where K^0 is a vertex of a regular simplex with the edge-length of α. Set

$$p = \frac{\sqrt{n+1} + n - 1}{n\sqrt{2}} \alpha \qquad (2.105)$$

$$q = \frac{\sqrt{n+1} - 1}{n\sqrt{2}} \alpha \qquad (2.106)$$

The rest n vertexes:

$$K^i = (K_1^i, K_2^i, \ldots, K_n^i)^{\mathrm{T}} \quad (i = 1, 2, \ldots, n) \qquad (2.107)$$

Construct as follows:

$$K_j^i = \left(K_j^0 + q\right) (j \neq i) \qquad (2.108)$$

$$K_i^j = \left(K_i^0 + p\right) \qquad (2.109)$$

Namely:

$$\left.\begin{array}{l} K^1 = (K_1^0 + p, K_2^0 + q, \ldots, K_n^0 + q)^{\mathrm{T}} \\ K^2 = (K_1^0 + q, K_2^0 + q, \ldots, K_n^0 + q)^{\mathrm{T}} \\ \cdots \\ K^n = (K_1^0 + q, K_2^0 + q, \ldots, K_n^0 + p)^{\mathrm{T}} \end{array}\right\} \qquad (2.110)$$

then, $K_1^0, K_2^0, \ldots, K_n^0$ constitute a regular simplex with the edge-length of α.

2.7.3.2 Iterative Steps of the Advanced Simplex Method

1. The initial simplex is constructed with the given initial point K^0. The vertexes are assumed as $K^1, K^2, \ldots, K^n$. The permissible error $\varepsilon > 0$, then calculate:

$$E_i = f(K^2), \ (i = 1, 2, \ldots, n).$$

Set:

$$E_l = f(K^l) = \min\{f(K^0), f(K^1), \ldots, f(K^n)\} \qquad (2.111)$$

$$E_h = f(K^h) = \min\{f(K^0), f(K^1), \ldots, f(K^n)\} \qquad (2.112)$$

where K^l and K^h are called the best and worst points of the simplex.

If the worst point K^h is removed, the rest n vertexes $K^0, K^1, \ldots, K^{h-1}, K^{h+1}, \ldots, K^n$ constitute the simplex of $(n-1)$-dimension space. Its center is:

$$K^f = \frac{1}{n}\left(\sum_{j=0}^{n} K^j - K^h\right) \tag{2.113}$$

2. Reflection: K^h is reflected to K^r in the center of K^f.

$$K^r = K^f + \alpha(K^f - K^h) \tag{2.114}$$

Among which $\alpha > 0$ is reflection coefficient and $\alpha = 1$ generally.

Because K^h is the worst point and through reflection there will be:

$$E_r < E_h \tag{2.115}$$

then point K^r better than K^h can be obtained, just like shown in Fig. 2.48.

3. Extension: after reflection not only does Eq. (2.115) hold, but there is a further step:

$$E_r < E_l \tag{2.116}$$

which indicates that K^r is better than K^l. Thus, the reflection direction is an effective direction of reducing the function value. So the simplex is being extended in this direction. Set:

$$E^e = K^f + (K^r - K^f)$$

Among which $\gamma > 1$ is extension coefficient and generally $\gamma = 2$. If:

$$E_e < E_h \tag{2.117}$$

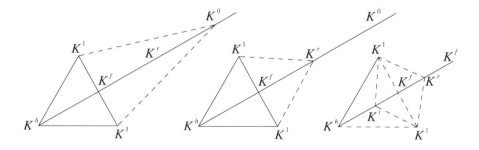

Fig. 2.48 Schematic diagram of iterative steps of the advanced simplex method

2.7 Other Methods for Hydrogeological Parameters Calculation

then K^h is replaced by K^e and the rest n vertexes are unchanged. A new simplex is constructed just like shown in Fig. 2.48(1). Turn to step 6.

If Eq. (2.111) holds and Eq. (2.117) does not, K^h is replaced by K^r to construct a new simplex, just like shown in Fig. 2.28(2). Turn to step 6.

4. Compression: if Eq. (2.116) does not hold, namely the reflection point K^r is not better than the best point K^l of the original simplex, there are two circumstances:

(1) When $j \neq h$, set $E_r \leq E_j$, which means the reflection point K^r is not worse than all the rest vertexes except the worst point K^h. Then K^h is still replaced by K^r to construct a new simplex. Turn to step 6.

(2) If for every K^h, there is:

$$E_r > E_j$$

then the reflection produces a new bad point. The simplex is being compressed in this direction. There are two cases as shown in Fig. 2.48(3).

In the first case, if

$$E_r > E_h \tag{2.118}$$

namely the reflection point is worse than the worst point of the original simplex, K^r is abandoned. Compress vector $K^h - K^f$:

$$K^e = K^f + \beta(K^h - K^f)$$

among which $0 < \beta < 1$ is compression coefficient and generally $\beta = 0.5$.

In the second case, if Eq. (2.118) does not hold, compress vector $K^r - K^f$;

$$K^c = K^f + \beta(K^r - K^f)$$

Discrimination of whether the compression point K^c is worse than the worst point of the original simplex K^h is necessary, namely whether the following equation holds

$$E_c > E_h$$

If it holds, the compression point K^c is abandoned and turn to step 5. If not, K^h is replaced by K^c to construct a new simplex and turn to step 6.

5. Decreasing the edge-length: the best point K^l of the origin simplex remains unchanged and the rest vertexes are compressed to K^l for a half distance, namely:

$$K^i = \frac{1}{2}(K^i + K^l), \quad i = 0, 1, 2, \ldots, n$$

A new simplex is obtained and the edge-length is half of the edge-length of the origin simplex. Turn to step 6.

6. Discrimination

$$\left\{\frac{1}{n+1}\sum_{i=0}^{n}(E_i - E_l)^2\right\}^{1/2} \leq \varepsilon$$

whether or not the inequality holds. If it holds, then stop calculation and $K^* = K^l$. If not, return to step 1.

2.7.3.3 Application

When fitting the optimal parameters by the advanced simplex method and the finite element program, the advanced simplex method is the main program. After determining the optimizing direction and a set of parameters, the subroutine needs to be called, as Fig. 2.49 shows. Whether the parameters optimized by the iterative steps of the advanced simplex method coincidence the required upper and lower limit range. If they coincide, finite element subroutine is called to calculate the water table of the nodes and the function value E. Return to the main program after consummation. Compare the size of the function value of every vertex of the simplex to determine the next lookup direction until the optimal vertex is found, namely the optimal parameters.

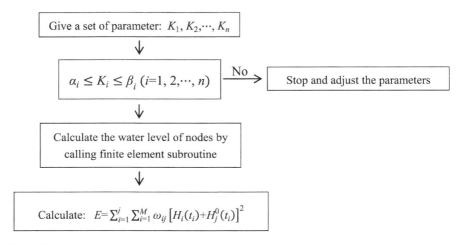

Fig. 2.49 Subdiagram of parameter adjusting

2.7 Other Methods for Hydrogeological Parameters Calculation

When the aquifer parameters of the calculation area are divided into several zones, the parameters of each zone can be called, respectively. In the end, the parameters of the whole zone will be called comprehensively.

2.8 Case Study

1. Object: multiple-hole unsteady and steady pumping tests are conducted in situ to know about the site requirement, and to figure out the experimental method and information collection of steady and unsteady flow pumping tests, additionally to calculate the hydrogeological parameters by various theories.
2. The plane arrangement of dewatering wells is present as below.

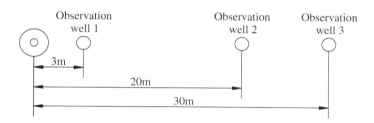

3. The requirement of pumping test site

Once the pumping test starts, any tester should keep on duty without any absence. The water table and water discharge should be measured and recorded timely.

Each team should take turns 15 min earlier to leave enough time for preparation. During the time for overlap, two teams should measure the water table together at the same location. Adjustment for the measuring tape also should be taken.

Before pumping, natural static water table should be measured (the accuracy is best for mm).

Starting pumping, the duration for measuring should be 1′, 2′, 3′, 4′, 6, 8′ 10′, 15′, 20′, 25′, 30′ 40′, 50′, 60′, 90′, 120′…(afterward measure once in each 30′); measuring: water table and water discharge in main well, water table in observation wells. (means min)

After pumping is stopped, the water table should be measured during recovery duration as: 20″, 40″, 1′, 2′, 3′, 4′, 6′, 8′, 10′, 15′, 20′, 25′, 30′, 40′, 50′, 60′, 90′, 120′ …(afterward duration is the same with above), measuring: the water table of main well and observation wells.

Each team collects and analyzes the data, including:

(1) Main well: Q-t curve, s-t curve; Observation wells: s-t curve;
(2) All wells: s-lg$t$ curve;
(3) All wells: lg$s$-lg$t$ curve;

In case of emergency occurrence, timely report should be informed to instruction teacher. The pump should be stopped if necessary. Or the power is cut down by accident; the water table in recovery duration should be measured right away.

The recorded data could not be changed if not necessary.

4. Make the report the experimental summary, including:

   (1) objects and requirement of pumping tests;
   (2) pumping method and procedure;
   (3) the main results of pumping tests;
   (4) treatment of abnormal circumstance during experiments and quality assessment and conclusions.

5. Draw the comprehensive resultant curves by pumping tests.
6. Use steady and unsteady flow method (fitting curve method, linear graphic method, water table recovery method) to calculate the hydrogeological parameters of aquifer.
7. Attach resultant table of pumping tests.

Summary of aquifer parameter results

Calculation method		Data	Aquifer parameter			
			K (m/day)	T (m^2/day)	a (m^2/day)	μ^*
Steady flow	Single well					
	Multiple wells					
Unsteady flow	Fitting curve method	Observation well 2				
		Observation well 3				
	Linear graphic method	Observation well 2				
		Observation well 3				
	Water table recovery method	Observation well 2				
		Observation well 3				
	Average					
	Range					
	Recommendation value					

2.9 Exercises

1. What parameters reflect the hydrological properties of aquifer?
2. What is the main task of pumping test?
3. What is the steady flow pumping test? And how about unsteady flow pumping test?
4. What is the difference between fully penetrated well pumping test and partial penetrated well pumping test?
5. How to collect and analyze the data of pumping test?
6. What is the object of water pressure test? How to collect and analyze the data?
7. How to measure the groundwater table, flow direction and flow velocity?
8. According to steady pumping test, what parameters can be obtained? What methods can be used to calculate hydraulic conductivity?
9. How to calculate the influence radius by steady pumping test?
10. According to unsteady pumping test, what parameters can be obtained?
11. How to estimate the coefficient of transmissibility and well loss constant from multi-water discharge test?

Chapter 3
Groundwater Engineering Problem and Prevention

During the survey, design and construction of underground engineering, groundwater is always the very crucial issue. It directly affects the properties and behaviors of rock and soils as a part of component of them, as well as some kind of underground construction environment, it also has great influence on the stability and durability of engineering projects. In design, full account must be taken on the various roles of geotechnical and underground engineering. All kinds of potential environmental geological problems which may rise due to groundwater during construction should be also paid much attention to take some appropriate preventive measures.

3.1 Adverse Actions of Groundwater

3.1.1 Suffosion

Suffosion is a kind of undermining phenomenon through removal of sediment by mechanical and corrosional action of groundwater flow. Usually, it is described the process of removal and transport of small soil particles through pores resulted in underlying caves or voids.

3.1.2 Pore-Water Pressure

In saturated soils, any small tiny variation of stress will change the pore pressure conditions. It always influences the strength, deformation of soils and stability of engineering projects, such as slope stability problem, deep foundation pit excavation in high-rise building projects, etc.

3.1.3 Seepage Flow

When water flows horizontally through an aquifer, the flow undergoes a reduction of pressure head because of friction. Thus the pressure on the upstream side of a small element is larger than on the downstream side. The water then exerts a net force on the aquifer element. The net force in the flow direction is the seepage force.

If there is an upward vertical flow, the head loss due to friction as the water flows into the pores results in an increase in the hydrostatic pressure. This in turn results in a decrease of the intergranular pressure. A point can be reached when the upward seepage force is large enough to carry the weight of the sand grains so that the sand or silt behaves like a liquid. It has no strength to support any weight on it. This condition is known as quicksand.

3.1.4 Uplift Effect of Groundwater

Groundwater has hydrostatic water pressure on the rock or soil mass below water level, which is resulted in buoyancy. From Archimedes' principle, the upward buoyant force exerted on a body immersed in a fluid is equal to the weight of the fluid the body displaces, i.e., when the water in soil pore spaces or fractures and voids in rock has hydraulic connection with groundwater, the buoyant force is buoyancy of rock mass or soil particle volume, which is the weight of displaced fluid.

3.2 Suffosion

3.2.1 Types of Suffosion

3.2.1.1 Mechanical Suffosion

Under the action of seepage force, particles of soil or rock mass are removed away from pore voids or caves by groundwater.

3.2.1.2 Chemical Suffosion

Groundwater dissolves the soluble substances in soil, breaks the cementing and weakens binding force among soil particles and thus makes soil structure loose. Generally, mechanical and chemical suffosion occur simultaneously. The chemical action takes away the soluble materials through groundwater flow leaching, which provides the circulation condition for mechanical suffosion. Suffosion actions reduce the strength of foundation soils and even form underground caves, thus results in surface subsidence and adversely affect the stability of buildings. Suffosion is greatly linked to karst terrain development and loess area and is accompanied by widespread collapse.

3.2.2 Conditions of Suffosion

It is involved in two aspects, i.e., first soil composition and second the sufficient hydrodynamic conditions. The specific details are depicted as follows:

(1) When the coefficient of uniformity ($C_u = \frac{d_{60}}{d_{10}}$, d_{60}, d_{10} are the particle-size diameters corresponding to 60 and 10 %, respectively) is large, resulting in high potential in suffosion, and specifically as $d_{60}/d_{10} > 10$.
(2) Soils with interface layers: when the permeability ratio $K_1/K_2 > 2$, suffosion is greatly conducive to occur in the interface.
(3) The hydraulic gradient. When it approaches to 5 ($I > 5$), the groundwater flow in turbulent condition facilitates suffosion. As a matter of fact, so large hydraulic gradient would rarely happen, hence a critical hydraulic gradient I_c is proposed out based on engineering practice:

$$I_c = (G_s - 1)(1 - n) + 0.5n \quad (3.1)$$

where G_s is the specific gravity, N/m^3; n is the porosity, expressed as a decimal.

3.2.3 Prevention of Suffosion

The measurements are mainly focused on following points:

(1) Reinforce the soils (e.g., grouting);
(2) Artificially lower the groundwater hydraulic gradient level;
(3) Set the filter layer.

Filter layer is a protecting measurement to prevent suffosion. It can be placed in the seepage exposure, especially directly at the exit point of the seepage. It is always composed of ranges of sizes of non-cohesive particles. Usually, these layers are arranged perpendicularly to the seepage lines and in sequence of increasing particle size (Fig. 3.1). It is extremely important to choose appropriate particle sizes when designing the filter to protect soils in place. Even when soils under really high hydraulic gradient ($I = 20$ or larger) can be protected well if the filter works effectively. Generally, the filter is designed into three layers, sometimes two layers

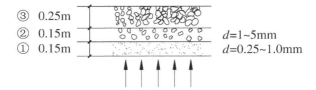

Fig. 3.1 Filter structure

as well, with thickness of 15–20 cm in each. How to control the specific thickness of each layer mainly depends on the construction conditions and the particle size used in this layer. If the filter layer could not be placed evenly or the quality could not meet the requirement, the thickness should enlarge to ensure that filter failure would not happen.

3.3 Piping and Prevention

3.3.1 Piping

The term "piping" is usually applied to a process that starts at the exit point of seepage and in which a continuous passage or pipe is developed in the soil by backward erosion, and enlarged by piping erosion. When soils around or beneath the foundation pit are loose sandy layers, if the seepage forces exerted on foundation soil are large enough and the pore spaces are large enough, water that percolates through earth dams or foundations can carry away fine soil particles. At the same time the pore spaces are also enlarged and a passage or pipe along seepage path is gradually developed through foundations or earth dams. Thus the foundation soils or earth dam soils are emptied continuously finally resulting in instability or failure. This phenomenon is namely piping. The specific process is shown in Fig. 3.2.

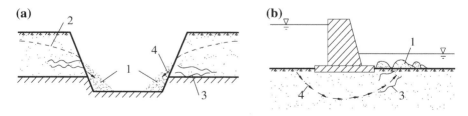

Fig. 3.2 Piping failure schematic under different conditions: **a** Slope soils; **b** Foundation soils; *Note 1* packing particle during piping; *2* groudwater level; *3* piping passage; *4* seepage direction

Scholars all over the world have studied piping extensively. Numbers of computation methodologies are figured out. A simple and practical assessment method is introduced here.

When the conditions shown as Eq. (3.2) can be met, the foundation pit is stable. The possibility of piping occurrence will be rarely small.

$$I < I_c \tag{3.2}$$

where I is the hydraulic gradient in situ, it can be calculated by:

$$I = \frac{h_w}{l} \tag{3.3}$$

I_c is the critical hydraulic gradient, in can be calculate by:

$$I_c = \frac{G_s - 1}{1 + e}; \tag{3.4}$$

h_w is the hydraulic head difference between outside and inside of retaining wall, m; l is the shortest length of flow line, m; G_s is the specific gravity, N/m³; e is void ratio.

3.3.2 Conditions of Piping

Piping generally happens in sandy soils. It is characterized by poor uniformity (gap-graded soil), i.e., some particle sizes are missing and pore void is large and well connected. It is mostly composed of low specific gravity minerals, which can be easily washed away by water percolating. In addition, good seepage exit path is also another sufficient condition for piping failure. All the above can be expressed specifically as following:

(1) The ratio of coarse and fine particles is larger than 10: $D/d > 10$;
(2) The coefficient of uniformity of the soil is larger than 10: $C_u = d_{60}/d_{10}$;
(3) The permeability ratio of two interface layers is greater than 2–3: $K_1/K_2 > 2$–3;
(4) The hydraulic gradient of seepage is greater than critical hydraulic gradient.

3.3.3 Prevention of Piping

(1) Increasing the embedded depth of retaining structure. As shown in Fig. 3.3, the length of flow lines can be extended and resultantly the hydraulic gradient is reduced. This is favorable to prevent piping.

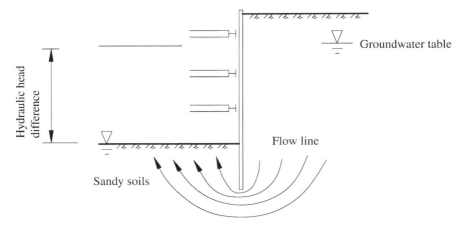

Fig. 3.3 Piping in foundation pit

(2) Artificially lowering the groundwater level and changing the groundwater seepage direction. When the foundation soils are sandy layers, and the seepage force is upward, the foundation pit underside will heave if the hydraulic gradient is greater than the critical hydraulic gradient in this condition.

To prevent this piping phenomenon (shown in Fig. 3.3), the embedded depth should be increased or artificially dewatering to lower the groundwater level before any construction.

3.3.4 Case Study

In this project, the bottom of this foundation pit is 6.5 m below the ground level, with 12 m in width and 1:1.25 slopes at both sidewalls. The foundation soils are clay layers with interbedded sand layers (Fig. 3.4). Grade 2 light wellpoint of

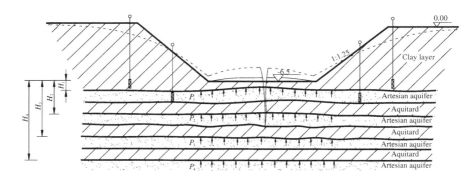

Fig. 3.4 Schematic of piping caused by foundation pit construction

dewatering is used here. When the foundation pit was dig excavated into the designate depth, the bottom heave gradually occurred. Firstly 20 cm in the center of bottom happened after 24 h. It reached up to 30 cm after another night. After 3 days, the accumulated heave is high as 1.5 m. At the beginning of heave in the pit bottom, no piping phenomenon was investigated. It happened until the heave reached a large amount finally. During the heaving, the slope and top correspondingly sunk and slid toward foundation pit.

Analysis of the reasons: the designated depth of wellpoint system was not deep enough. The water pressure in the artesian aquifer layers beneath the foundation pit was larger than self-weight of overlying layers. The clay layer in the pit bottom was uplifted and cracked, then piping happened.

Measurements: The dewatering system should be embedded into nth artesian aquifer layer, in which the water pressure in this artesian aquifer layer should be smaller than the total self-weight of overlying layers, i.e., $P_n < \sum_{i=1}^{n} H_i \cdot \gamma_i$, where H_i is the soil thickness of nth layer; γ_i is the bulk weight of nth layer.

3.4 Quicksand and Prevention

3.4.1 Quicksand

Quicksand is saturated loose sand or silty sand (including sandy silt and clayey silts as well) in the case of upward flowing water, seepage force opposes the force of gravity and suspends the soil particles. This creates liquefied soil that forms suspension and lose strength. Quicksand usually happens in uniformly graded fine or silty sands; it sometimes occurs in silts as well. The saturated sediment may appear quite solid until a sudden change in pressure or shock initiates liquefaction. All the fine particles are washed away suddenly by percolating water. Resultantly, sliding or differential subsidence occur in foundation, even worse to collapse or suspension. As shown in Fig. 3.5, quicksand happens in unexpected sudden. It is really harmful to engineering practice.

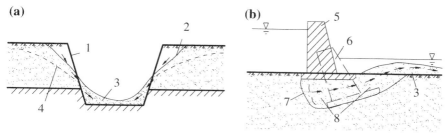

Fig. 3.5 Quicksand failure. **a** Slope soils; **b** Foundation soils, *Note 1* original slope; *2* the slope surface after quicksand occurrence; *3* quicksand packing particles; *4* groundwater level; *5* original position of structure; *6* the position after quicksand occurrence; *7* sliding surface; *8* quicksand area

3.4.2 Causes of Quicksand

(1) Large hydraulic gradient and high flow velocity make the fine particles suspense;
(2) Quicksand usually is a colloid hydrogel consisting of fine granular materials (such as sand or silt), clay and water. When saturated soil particles absorb water and swell and the density is reduced a lot. Thus it can be suspended easily by water percolating.
(3) Sand structure is destroyed by sudden vibration. In this case the vibrated force immediately increases the pore pressure of groundwater. The saturated sand loses strength and suspends away with water flow.

In practice, when a pit excavation is conducted below the groundwater level, sands with groundwater spring out frequently. This phenomenon is namely boiling sand (Fig. 3.6a). Sands spring out more serious when excavation is advancing. Quicksand brings great difficulties to construction, and also destroys the foundation strength; threaten the safety of surrounding existed buildings. This phenomenon can be explained in this simple model test. In Fig. 3.6, first open valve A to make an upward water stream in sand. When the upward hydraulic gradient was greater than 1, i.e., $I = h/l \approx 1$, the sand lose the stability. The gravel on the sand surface sunk (the sand lost the strength). Then close the valve, the sand would gain its strength again.

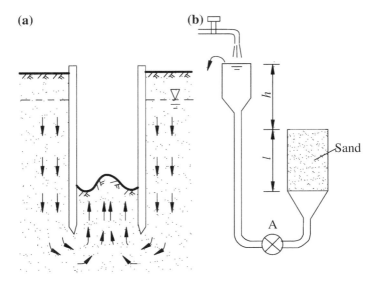

Fig. 3.6 Quicksand modeling tests

3.4 Quicksand and Prevention

3.4.3 Conditions of Quicksand

(1) Large hydraulic gradient. The seepage force exceeds the particle gravity and makes the fine suspension;
(2) The sand has large porosity. The larger, the easier to form quicksand;
(3) The sand has poor permeability. The poorer, more favorable to quicksand;
(4) Composed of more platy minerals, such as mica, chlorite, the sand is more potential to quicksand.

The influence factors and distribution of quicksand in Shanghai area is preliminarily figured out based on borehole data and geotechnical soil tests. These factors are shown as follows:

(1) The main induced external factor depends on the groundwater hydraulic head difference. With the excavation depth is increasing, the hydraulic head difference gets larger, quicksand is much easier to happen;
(2) The particle composition. In this case, the clay percentage is smaller than 10 %, while silt and sand total content is over 75 %;
(3) The coefficient of uniformity is smaller than 5. From quicksand properties data of in situ engineering projects, mostly the coefficient of uniformity is in the range of 1.6–3.2;
(4) The water content is greater than 30 %;
(5) The porosity is larger than 43 % (or void ratio is larger than 0.75);
(6) When sand soils interbedded into clay layers, the thickness of sandy soil or clayey silt should not be over 25 cm;

There are also some similar assessment standards in practical area outside China: natural porosity is greater than 43–45 % (void ratio larger than 0.75–0.80), effective particle size is smaller than 0.1 mm ($d_{10} < 0.1$ mm), and the coefficient of uniformity is smaller than 5 ($C_u < 5$), the soils have these characteristics are easier to happen quicksand.

In Shanghai area, when the water level is around 0.7 m below the ground surface, and the excavation depth is greater than 3 m, meantime the soil has properties described above; quicksand has great potential to happen. When the excavation depth is smaller than 4 m, usually sheet piles are used to excavation. When the excavation depth is over 4 m, dewatering of wellpoint system should be used.

3.4.4 Determination of Quicksand

The phenomenon of quicksand encountered in constructions:

(1) Slight—there is minor gap between sheet piles. Some sands move into the foundation pit through the gap by percolating water and make the pit much muddier;
(2) Moderate—close to the foundation pit bottom, especially nearby the sheet piles, usually there are packing fine sand particles slowly spring out. Investigated closely, it can be founded that a lot of small seepage exits in the packing sands and the water bubbles up with fine particles.
(3) Severe—if the above phenomenon happened during excavation and no measurements were taken to control and still kept excavating. In this case, the quicksand velocity would increase fast and finally formed boiling sands. At this time the sands at the pit bottom would liquefy and flow.

In Shanghai area, there are a lot of quicksand cases, such as before the People's Republic of China established, on Fujian Seven Road of Shanghai, there was a quicksand during the ditch construction. The liquefaction at the pit bottom was really serious and workers used barrels as tool to move sands and water. Another example is the pumping station construction in Shanghai Yejiazhai Road. At that time the excavation depth was very large but the embedded depth of sheet piles was not deep enough. Server quicksand happened at the lower part of the ditch. Soils at the bottom totally liquefied. All these stopped the progress of construction and made the ditch tube could not reach the designated depth and changed it as 60 cm higher. From Fig. 3.7, it is shown the quicksand circumstance during the narrow and long ditch construction. This phenomenon was also influenced by hydraulic head difference, while under the upward flow effect, closer to the edge of the sheet pile walls, more serious quicksand happened. Meantime, in practical construction land subsidence nearby sheet pile area always occurred. The ditch width was very small; the water in the surrounding soils of the ditch area flew to the ditch and concentrated at the pit bottom. Thus closer to the sheet pile walls, the flow velocity was faster. The water flux per length along the ditch can be evaluated by Eq. (3.5):

$$q = bKI, \qquad (3.5)$$

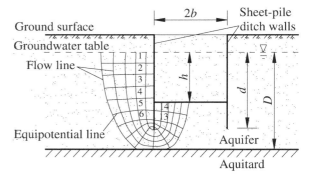

Fig. 3.7 Quicksand around the sheet pile during excavation

3.4 Quicksand and Prevention

Table 3.1 The reduction factor

d/D	0.1	0.2	0.3	0.4	0.5	0.6	0.7	0.8	0.9	1.0
Reduction factor for q (%)	5	10	15	20	25	30	40	50	65	100

where q is the water flux, m²/s; b is half of the ditch width, m; K is the hydraulic conductivity of surrounding soil, m/s; I is the hydraulic gradient.

The above Eq. (3.5) can only be used when the embedded depth of the sheet pile wall d is really small corresponding to the aquifer thickness D. If the embedded depth is very deep, the results should multiply to a reduction factor. As for the reduction factor value, it is shown in Table 3.1.

When calculating the embedded depth of sheet piles, besides the critical value obtained from Eq. (3.4), flow net methodology is still needed for calculation. And in practice, safety factor should be considered to be greater than 1, since in soils with good permeability the fine particles can be easily moved. In clay and silty clay, the seepage discharge is really small or even could not occur. Quicksand can hardly happen in these conditions. And in gravel, good permeability and large discharge amount, the seepage path is very long. Thus quicksand rarely happens as well. Hereby quicksand mostly happens in fine or silty sands with poor permeability. The fine sands or silts can easily lose strength with high seepage force exerted on; and moved by percolating water to ditch pit. Therefore, when excavation in this kind of soil, effective measurements should be taken during construction to avoid quicksand happening.

Example:
A ditch was constructed by sheet piles (Fig. 3.7). The specific gravity of the aquifer sand G_s is 2.8; the void ratio is 0.8; and $\gamma_w = 1$ g/cm³, $h = 21$ m. There are 14 flow paths in the flow net ($n = 14$). The length can be selected as 1 along the ditch. Is this project reach quick condition?

Solution:
From Eq. (3.4), $I_c = \frac{G_s-1}{1+e} = \frac{2.8-1}{1+0.8} = 1$,
From drawing flow net, $I = \frac{h}{n \times L} = \frac{21}{14 \times 1} = 1.5 > 1$,
Hence, quicksand will happen in natural condition.

3.4.5 Quicksand in Foundation Pit

Figure 3.8 presents the schematic of quicksand calculation. Due to the water level difference h' existed in the ditch around the foundation pit, a seepage flow runs down through the soils outside the sheet piles, and when it flows over the end of the sheet piles, the water advances up reaching the bottom of the pit, which is collected into the well by the ditch. Finally, all the water is pumped away. Hence, the soils beneath the pit are saturated by water and the effective unit weight γ' should be used in calculation. When the value of unit seepage force or hydrodynamic pressure G_D

Fig. 3.8 The seepage during the construction of foundation pit

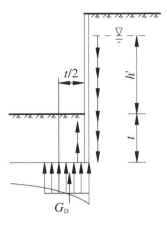

is equal to or even over the effective gravity γ', the soil particle is under a state of quicksand and is able to move free with water flow. To avoid this adverse phenomenon, the requirement as below should be met.

$$\gamma' \geq K_s G_D \quad (3.6)$$

where K_s is the safety factor, it depends on the retaining structure and soil properties; generally, $K_s = 1.5\text{--}2.0$.

According to the relevant experiment results, quicksand initially occurred within the distance of $t/2$ (t is the embedded depth of sheet pile wall) to the sheet pile wall. And the location closest to the sheet piles has the shortest seepage path and the largest seepage force can be calculated as below.

$$G_D = I \cdot \gamma_w = \frac{h'}{h' + 2t} \cdot \gamma_w$$

In conjunction with Eq. (3.6), the above condition can be changed into $\gamma' \geq K_s \frac{h'}{h'+2t} \cdot \gamma_w$.

After some transposition operation, the specific requirement for the embedded depth of sheet pile wall is calculated as Eq. (3.7).

$$t \geq \frac{K_s h' \gamma_w - \gamma' h'}{2\gamma'} \quad (3.7)$$

If the soil layers above the bottom surface of the foundation pit are coarse gravel, loose fill soil or fractured soil, the head loss in the soil layer outside the foundation pit can be neglected, so Eq. (3.7) can be simplified as

$$t \geq \frac{K_s h' \gamma_w}{2\gamma'}$$

or

$$K_s \leq \frac{2\gamma' t}{h' \gamma_w} \quad (3.8)$$

where h'/t is the hydraulic gradient in the soil outside the sheet pile wall. The increase of h'/t will result smaller K_s. When K_s less than 1, quicksand will happen. The value of h'/t as $K_s = 1$ is called limited hydraulic gradient. In the designation of embedded depth of sheet pile wall, the value of K_s should be chosen as 1.2–1.5.

3.4.6 Quicksand in the Caisson

During the construction of caisson in sands, if drainage sinking is used and the dewatering depth is not large enough or the dewatering is not effective, quicksand will easily happen in the sands beneath the caisson cutting edge under hydrodynamic pressure, shown as Fig. 3.9. Some ground subsidence and horizontal displacement maybe concomitantly occur.

During the undrained sinking of the open caisson in sands, if the water level inside the caisson is much lower than the outside, large hydrodynamic pressure is generated and results in quicksand in the bottom of the caisson. And surrounding soil movement will occur as well.

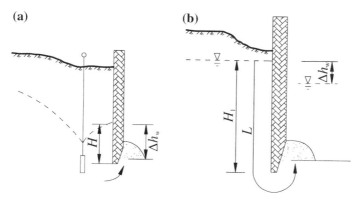

Fig. 3.9 The soil movement induced by quicksand in open caisson: **a** The depth of open caisson is not enough during drained sinking. **b** The head difference is very large during undrained sinking

The influence range can be extended (1–3) H around the caisson (H is the caisson depth). The ground subsidence amount usually depends on the soil water loss. Long-time soil water loss will induce catastrophic ground subsidence. It should be noted here, when the sands is overlying by hard clay layer, large-area collapse is probably to happen as the underlying sands is continuously removing due to quicksand.

Quicksand happening in the caisson will influence the nearby shallow foundation constructions and also the pile foundation buildings. When conducting the caisson construction in the saturated sands near some pile foundation, pile displacement and inclination may be attributed to the surrounding soil movement toward the caisson bottom during the caisson sink, rather than the soil consolidation by well dewatering. If quicksand does happen in the caisson bottom, large movement will arise in surrounding soils. Even nearby is deep foundation construction, it can be damaged greatly.

3.4.7 The Prevention and Treatment of Quicksand

As mentioned above, when the hydrodynamic pressure exceeds the buoyant (submerged or effective) unit weight or the hydraulic gradient is larger than the critical value, quicksand is probably to happen. This circumstance is usually induced by the excavation beneath ground water, the laying of underground pipes, well construction, etc. Hence, quicksand is an engineering phenomenon. It can cause large soil movement and result in ground collapse or building foundation damage. Significant difficulties can be brought into the construction, and direct influence to surrounding construction project and building stability maybe emerge. Therefore, necessary prevention and treatment should be paid attention.

In the potential quicksand area, it had better conduct in the overlying soil layer as natural soil foundation, or use pile foundation through the whole quicksand area transferring the upload into stable soil layer. In total, the excavation should avoid the quicksand area. If it could not be avoided, several treatments can be utilized as below.

(1) Artificially dewater the groundwater level to ensure it below the quicksand layer, shown in Fig. 3.10.

The prevention principle:

During the excavation, a upward seepage force is exerted on soils below the surface. As for sands, when the hydraulic gradient increases to some extent, quicksand will happen, i.e., the soil flow out of slope surface akin to a liquid state. The limited seepage hydraulic gradient inducing quicksand can also use the critical hydraulic gradient proposed by Terzaghi.

3.4 Quicksand and Prevention

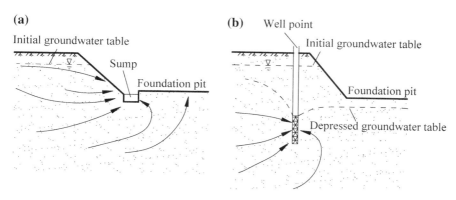

Fig. 3.10 Prevention of quicksand by well dewatering. **a** Sump **b** Well point

Fig. 3.11 The seepage force in the foundation pit

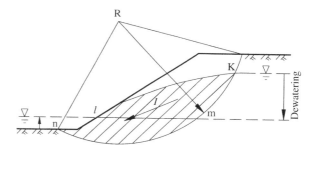

$$I_c = (1 - n)(G_s - 1) \tag{3.9}$$

where n is the porosity; G_s is the specific gravity of soil particle.

As for uniform sand soils, $I_c = 0.8$–1.2. In practice, some safety factors will be used in designation. As for the nonuniform silty sands, the critical hydraulic gradient can only be $I = 1/3$. If the hydraulic gradient exceeds the design allowable value, well dewatering should be conducted to prevent the quicksand phenomenon. The well dewatering declines the seepage head difference existing inside and outside the foundation pit, and indirectly control the hydraulic gradient within the allowable value, shown as Fig. 3.11. Simply only drainage ditch could not lower the seepage hydraulic gradient. Dewatering well cannot only decrease the hydraulic gradient within the safe value, but also change the seepage flow direction, which make the water flow moves into dewatering well pipe.

(2) Sheet pile wall can be constructed. This method has advantages in two aspects. First it can reinforce the foundation pit as retaining wall; second it prolongs the seepage path so that to decline the hydraulic gradient and slower the seepage velocity.

(3) Artificial ground freezing. This method can be used before excavation. Surrounding soils are frozen as a water-sealing wall with higher strength.

(4) Excavation in submerged condition. To avoid quicksand induced by head difference from drainage and to strengthen the stability of sands, the excavation can be performed whilst injecting (recharging) water into the foundation pit.

In addition, some other methods such as chemical reinforcement, blasting method, strengthen weighting, etc. When some local quicksand occurs during excavation, filling coarse gravel can alleviate the quicksand movement greatly.

3.5 Liquefaction of Sands and Relevant Preventions

3.5.1 Liquefaction

A number of failure of embankment, natural slopes, earth structures and foundations have been attributed to the liquefaction of sands caused by either static or seismic loading. The liquefaction phenomenon of soil deposits can be described as the reduction of shear strength due to pore pressure buildup in the soil skeleton. When some saturated loose sand (including some silt) is applied by vibration load or a static load sharply, if the pore water could not flow out in time, the contractancy of loose sands is responded in continuously increasing of pore-water pressure. Correspondingly, the effective stress σ' gradually decreases. When $\sigma' = 0$, the saturated soil substantially loses shear strength and stiffness. At the onset of initial liquefaction, loose sands will undergo unlimited deformations or flow without mobilizing significant resistance to deformation. As a result, structures supported above or within the liquefied deposit undergo significant settlement and tilting; water flows upward to the surface creating sand boils; and buried pipelines and tanks may become buoyant and float to the surface. This is usually termed as liquefaction.

The phenomenon of liquefaction is most often observed in any part of saturated loose sands. It can occur in the ground surface, or some depth underground, depending on the sand condition and vibration circumstance. Sometimes the shallow sand layers are liquefied induced by liquefaction of underlying sands. The excess pore-water pressure is dissipated by upward flowing of water. If the hydraulic gradient is so large that the upward water flow may destroy the stability of overlying sand layers and results in seepage failure. Even the failure has not shown up but the strength of overlying sand layers will be lowered severely.

Usually phenomena of sand boils, water spouts, and ground cracks appear in the areas of liquefaction. The waterspouts can reach as high as several meters and the sand concomitant accumulates as a crateriform around the spray spout in a diameter of several meters. These mostly start to happen several seconds shortly after a strong earthquake arises, and lasts decades' minutes to a few hours after the earthquake stops, or even tens of hours. However, sand boils and water spouts may not always happen in some circumstances, such as when the liquefaction sensitive sand layers are deep beneath the ground surface with very thin thickness. The upward spraying pore-water and sand particles are not sufficient to reach the ground surface. Just some sand veins are formed in the overlying layers. This kind potential liquefaction in deep soil usually will not cause tremendous amount of damage.

Liquefaction of sands induces to lose the bearing capacity and some concomitant movements. It always brings a lot of catastrophic failure and damage. Case histories of landslides or flow failures due to liquefaction are the 1937 Zeeland coast of Holland slides involving 7 million cubic meters of alluvial sands, and the 1944 Mississippi River slid near Baton Rouge containing about 4 million cubic meters of fine sands. Failures of hydraulic fill dams such as the Calaveras Dam in California in 1918, the Fort Peck Dam in Montana in 1938, and the Lower San Fernando Dam during the 1971 San Fernando Earthquake in California, just to name a few, were triggered by the liquefaction of sands. Although the importance of liquefaction of sands induced by static loading has been recognized since the work of Casagrande (1936), the subject of liquefaction of sands by seismic loading had not received a great deal of attention until 1964 when two major earthquakes shook Anchorage, Alaska, and Niigata, Japan, resulting in substantial damage and loss. The Alaska earthquake in 1964, a shock with a magnitude, M, of 9.2 on the Richter scale, destroyed or damaged more than 200 bridges and caused massive landslides. Moreover, the 7.5-magnitude earthquake of June 16, 1964, in Niigata, Japan, the extensive liquefaction of sand deposits resulted in major damage to buildings, bridges, highways and utilities. Foundations lost the bearing capacity and engineering constructions damaged severely. Over 1 meter settlement occurred in most areas. Several apartments tilted almost 80°. During the liquefaction, some groundwater spouted out from the ground cracks. Meanwhile, cars, buildings, or other objects on the ground surface sunk into underground soils. And some underground constructions damaged and were risen up to the ground surface. Some harbor port facilities were damaged a lot. It was estimated that more than 60000 buildings and houses were destroyed.

There were subsequent 229 disasters of sand liquefaction near Southwest Seaside in Netherlands in 1861–1947. The influence area was as high as 2.5 million square meters. The liquid soils in movement reach 25 million cubic meters. The original coast slope was 10°–15°, and was decreased as 3°–4° after liquefaction. A reservoir in Xinjiang, with a 3.5–7.1 m height dam, was established in 1959. In April 1961, a 9° strong earthquake occurred here and in October 1962, the second earthquake happened again. The sandy silts (just tens of centimeters in thickness) were liquefied and the dam was slid and resulted in dam foundation damage. At the 8° earthquake area in Xingtai in 1966, sands spouted out from ground cracks and hydraulic gate of dams were mostly lower down. From the Tangshan earthquake, there were four kinds of liquefaction. Flat sheeted and striped liquefaction were distributed in the view of surface (Fig. 3.12). Shallow soil and deep soil liquefactions were observed from the view of vertical profile. The sheeted liquefaction and shallow liquefaction arose in the alluvial fan areas of the river. Striped liquefaction and deep soil liquefaction mostly occurred in the downstream of the ancient river. The damage differed in these various distributions. Specific analysis on the soil distributions can make great significance to ensure more applicable designation.

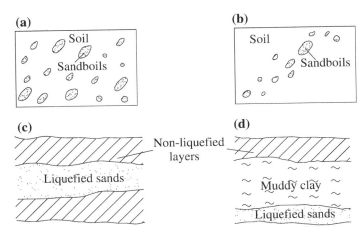

Fig. 3.12 Liquefaction properties (from Handbook of design and construction of underground engineering, 1999). **a** Sheeted liquefaction surface; **b** Striped liquefaction surface; **c** Shallow soil liquefaction; **d** deep soil liquefaction

From the statistical investigation of earthquake damage, more than half of the earthquake damages are induced by liquefaction. Taking Hatching Earthquake and Tangshan Earthquake as examples, the number of building failures due to foundation liquefaction accounted for almost 54 % of the total foundation damages. Foundation liquefaction can make the buildings tilt, collapse, or induce ground uplift, cracks, or slides of coast slope surface. Some shallow light construction (such as pipes) can be moved upward to ground surface as well. In total, all the facilities in the liquefaction area can hardly avoid the damage.

However, it is worth remarking that once liquefaction happens, the above various damages will arise but the surface movement can be alleviated. Since the liquefaction layer can effectively weaken the energy transfer of upward shear wave. At the same time the accompanied sand boils and waterspouts can consume a part of energy and results in smaller energy reaching the ground surface. Hence the vibration duration can be shortened. This is the reason that the seismic intensity in the liquefaction area is always not higher than nonliquefaction area, even smaller. Acknowledge of the advantages and adverse aspects of liquefaction in earthquake disaster is very important for improving the aseismic design level. In practice, firstly whether the foundation is sensitive to liquefaction or not should be distinguished. Then the relevant measurement can be adopted.

3.5.2 The Factors Affecting Liquefaction

From the statistical investigation of earthquake damage, more than half of the earthquake damages are induced by liquefaction. Taking Hatching Earthquake and

Tangshan Earthquake as examples, the number of building failures due to foundation liquefaction accounted for almost 54 % of the total foundation damages. Foundation liquefaction can make the buildings tilt, collapse, or induce ground uplift, cracks, or slides of coast slope surface. Some shallow light construction (such as pipes) can be moved upward to ground surface as well. In total, all the facilities in the liquefaction area can hardly avoid the damage.

However, it is worth remarking that once liquefaction happens, the above various damages will arise but the surface movement can be alleviated. Since the liquefaction layer can effectively weaken the energy transfer of upward shear wave. At the same time the accompanied sand boils and waterspouts can consume a part of energy and results in smaller energy reaching the ground surface. Hence the vibration duration can be shortened. This is the reason that the seismic intensity in the liquefaction area is always not higher than nonliquefaction area, even smaller. Acknowledge of the advantages and adverse aspects of liquefaction in earthquake disaster is very important for improving the aseismic design level. In practice, firstly whether the foundation is sensitive to liquefaction or not should be distinguished. Then the relevant measurement can be adopted.

Based on field observation and laboratory testing results, liquefaction characteristics of cohesionless soils are affected by a number of factors:

(1) Grain Size Distribution and Soil Types

The type of soil most susceptible to liquefaction is one in which the resistance to deformation is mobilized by friction between particles. If other factors such as grain shape, uniformity coefficient and relative density are equal, the frictional resistance of cohesionless soil decreases as the grain size of soils becomes smaller. Tsuchida (1970) summarized the results of sieve analyses performed on a number of alluvial and diluvial soils that were known to have liquefied or not to have liquefied during earthquakes. He proposed ranges of grain size curves separating liquefiable and nonliquefiable soils as shown in Fig. 3.13. The area within the two inner curves in the figure represents sands and silty sands, the soils with the lowest resistance to liquefaction. A soil with a gradation curve falling in the zones between the outer and inner curves is less likely to liquefy. Soils with a higher percentage of gravels tend to mobilize higher strength during shearing, and to dissipate excess pore pressures more rapidly than sands. However, there are case histories indicating that liquefaction has occurred in loose gravelly soils (Seed 1968; Ishihara 1985; Andrus et al. 1991) during severe ground shaking or when the gravel layer is confined by an impervious layer. The space between the two curves farthest to the left reflects the influence of fines in decreasing the tendency of sands to densify during seismic shearing. Fines with cohesion and cementation tend to make sand particles more difficult to liquefy or to seek denser arrangements. However, nonplastic fines such as rock flour, silt and tailing slimes may not have as much of this restraining effect. Ishihara (1985) stated that clay- or silt-size materials having a low plasticity index value will exhibit physical characteristics resembling those of cohesionless soils, and thus have a high degree of potential for liquefaction. Walker and Steward (1989), based on their extensive dynamic tests on silts, have also concluded that nonplastic and low plasticity silts, despite having their grain size distribution curves

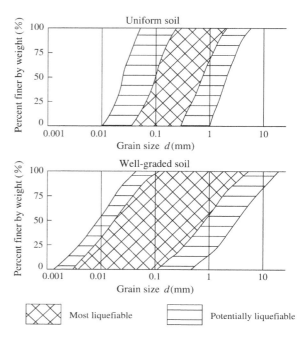

Fig. 3.13 Limits in the gradation curves separating liquefiable and nonliquefiable soils (Tsuchida 1970)

outside of Tsuchida's boundaries for soils susceptible to liquefaction, have a potential for liquefaction similar to that of sands and that increased plasticity will reduce the level of pore pressure response in silts. This reduction, however, is not significant enough to resist liquefaction for soils with plasticity indices of 5 or less.

Even though major slide movements during earthquakes have occurred in clay deposits, they are commonly considered to be nonliquefiable during earthquakes in the sense that an extensive zone of clay soil is converted into a heavy fluid condition. However, it is believed that quick clays may lose most of their strength after strong shaking and that other types of clay may lose a proportion of their strength resulting in slope failures. Frequently, landslides in clay deposits containing sand or silt lenses are initially triggered by the liquefaction of these lenses before any significant strength loss occurs in the clay. This has been supported by laboratory test results which indicate that the strain required to liquefy sands is considerably smaller than the strain required to overcome the peak strength of cohesive soils (Seed 1968; Poulos et al. 1985). There is also ample evidence to show that uniformly graded materials, generally having a uniformity coefficient smaller than five, are more susceptible to liquefaction than well-graded materials (Ross et al. 1969; Lee and Fitton 1969) and that for uniformly graded soils, fine sands tend to liquefy more easily than coarse sands, gravelly soils, silts, or clay.

(2) Relative Density

Laboratory test results and field case histories indicate that, for a given soil, initial void ratio or relative density is one of the most important factors controlling liquefaction. Liquefaction occurs principally in saturated clean sands and silty sands

Table 3.2 The relative density index D_r for possibility in liquefaction

Seismic fortification intensity	6	7	8	9
D_r	0.65	0.70	0.75	0.80–0.85

having a relative density less than 50 %. For dense sands, however, their tendency to dilate during cyclic shearing will generate negative pore-water pressures and increase their resistance to shear stress. The lower limit of relative density beyond which liquefaction will not occur is about 75 %. According to *Code for Hydropower Engineering Geological Investigation (GB50287-2006)*, it is specific that when the relative density D_r is smaller than values in Table 3.2, liquefaction probably happens during earthquake. During the Niigata earthquake of 1964 in Japan, in 7-M areas, liquefaction mostly occurred in the places with $D_r \leq 0.5$; and the sections with $D_r \geq 0.5$ can hardly be seen the liquefaction damage.

(3) Earthquake Loading Characteristics

The vulnerability of any cohesionless soil to liquefaction during an earthquake depends on the magnitude and number of cycles of stresses or strains induced in it by the earthquake shaking.

These in turn are related to the intensity, predominant frequency, and duration of ground shaking. The earthquake load is characterized in terms of the maximum acceleration. Generally when the surface maximum acceleration reaches $0.1g$ (g is the gravity acceleration, $1g = 980$ cm/s^2), liquefaction is potential to happen. Both field monitoring and experimental data indicate that liquefaction of soil under dynamic loading is related with the vibration frequency and duration. Such as the Alaska earthquake, most liquefaction occurred 90 s later after the earthquake happened. If that earthquake lasted only 45 s, it was probably that liquefaction hardly arose.

(4) Vertical Effective Stress and Overconsolidation

It is well known that an increase in the effective vertical stress increases the bearing capacity and shear strength of soil, and thereby increases the shear stress required to cause liquefaction and decreases the potential for liquefaction. From field observations it has been concluded by a number of investigators that saturated sands located deeper than 15–18 m are not likely to liquefy. These depths are in general agreement with Kishida (1969) who states that a saturated sandy soil is not liquefiable if the value of the effective overburden pressure exceeds 190 kN/m^2.

Both theory and experimental data show that for a given soil a higher overconsolidation ratio leads to higher lateral earth pressure at rest and thereby increases the shear stress ratio required to cause liquefaction. During the Xingtai Earthquake in China, there was a village in the same buried sand layer condition with other areas. Liquefaction did not happen here due to the difference with 2–3 m fill soil above. During the Niigata Earthquake in Japan, the areas with 2.75 m filling soils were all stable without liquefaction, while in other area severe liquefaction happened.

(5) Age and Origin of the Soils

Natural deposits of alluvial and fluvial origins generally have soil grains in the state of loose packing. These deposits are young, weak, and free from added strength due to cementation and aging. Youd and Hoose (1977) stated that, as a rule of thumb, alluvial deposits older than late Pleistocene (10,000–130,000 years) are unlikely to liquify except under severe earthquake loading conditions, while late Holocene deposits (1,000 years or less) are most likely to liquefy, and earlier Holocene (1,000–10,000 years) deposits are moderately liquefiable.

(6) Seismic Strain History

It has been demonstrated from laboratory test results that prior seismic strain history can significantly affect the resistance of soils to liquefaction (Finn et al. 1970; Seed et al. 1977; Singh et al. 1980). Low levels of prior seismic strain history, as a result of a series of previous shakings producing low levels of excess pore pressure, can significantly increase soil resistance to pore pressure buildup during subsequent cyclic loading. This increased resistance may result from uniform densification of the soil or from better interlocking of the particles in the original structure due to elimination of small local instabilities at the contact points without any general structural rearrangement taking place. Large strains, however, associated with large pore pressure generation and conditions of full liquefaction can develop weak zones in the soil due to uneven densification and redistribution of water content (National Research Council 1985; Whitman 1985), and thus lower the resistance of the soil to pore pressure generation during subsequent cyclic loading.

(7) Degree of Saturation

Liquefaction will not occur in dry soils. Only settlement, as a result of densification during shaking, may be of some concern. Very little is known on the liquefaction potential of partially saturated sands. Available laboratory test results (Sherif et al. 1977) show liquefaction resistance for soils increases with decreasing degree of saturation, and that sand samples with low degree of saturation can become liquefied only under severe and long duration of earthquake shaking.

(8) Thickness of Sand Layer

In order to induce extensive damage at level ground surface from liquefaction, the liquefied soil layer must be thick enough so that the resulting uplift pressure and amount of water expelled from the liquefied layer can result in ground rupture such as sand boiling and fissuring (Ishihara 1985; Dobry 1989). If the liquefied sand layer is thin and buried within a soil profile, the presence of a nonliquefiable surface layer may prevent the effects of the at-depth liquefaction from reaching the surface. Ishihara (1985) has set up a criterion to stipulate a threshold value for the thickness of a nonliquefiable surface layer to avoid ground damage due to liquefaction, as shown in Fig. 3.10. Although this figure is believed to be speculative and should not be used for design purposes, it provides initial guidance in this matter for sites having a buried liquefiable sand layer with a standard penetration resistance of less than 10 blows per 0.3 m. It should also be noted that even though the thickness of a nonliquefiable surface layer exceeds the threshold thickness shown in the figure, the ground surface may still experience some settlement which may be undesirable for certain settlement-sensitive structures. Like all of the empirical curves shown in this report, this figure, based on just three case histories, may need to be modified as more data become available.

3.5.3 Evaluation of Liquefaction Potential

To date, after 30 years of intensive research on this subject, much progress has been made in understanding the liquefaction phenomena of cohesionless soils under seismic loading. A variety of methods for evaluating the liquefaction potential of soils have been proposed. As mentioned above, the factors affecting sands or silts are various. Various procedures for evaluating the liquefaction potential of saturated soil deposits have been proposed in the past 20 years. These procedures, requiring various degrees of laboratory and/or in situ testing, may be classified into two categories: first aspect is empirical correlations between in situ characteristics and observed performance. Soil liquefaction characteristics determined by field performance have been correlated with a variety of soil parameters such as Standard Penetration Test (SPT) Resistance, Cone Penetration Resistance, Shear Wave Velocity and Resistivity and Capacitance of Soil. Second is threshold shear strain concept compared with the laboratory testing value. There exists for a given cohesionless soil a threshold shear strain, typically 0.01 %. If the peak shear strain induced by an earthquake does not exceed this strain, the shaking will not cause a buildup of excess pore pressure regardless of the number of loading cycles, and, therefore, liquefaction cannot occur. For the laboratory condition limitation for undisturbed saturated sands or silts, the former method is more applicable in practical engineering. And it has been adopted into relevant specific code. And during the in situ testing soil parameters, Standard Penetration Test (SPT) Resistance is used widely in many countries, such as China, Japan, and the United States.

The determination of vibrated liquefaction of soil may be carried out in two stages: preliminary determination and redetermination. During preliminary determination, soil stratum, which will not be excluded; and for soil stratum, which is likely to be liquefied according to preliminary determination, redetermination shall be performed to determine the liquefaction potential.

1. According to the experience, the preliminary determination can be carried out as follows:

(1) If a stratum belongs to the age of Quaternary late Pleistocene (Q_3) or earlier, it may be determined as nonliquefied soils for intensity 7–9.

(2) If the percentage of grain content with particle size bigger than 5 mm is equivalent to or larger than 70 %, it may be determined to be nonliquefaction. If the percentage of grain content with particle size larger than 5 mm is smaller than 70 %, and no other integral discriminative method is available for use, its liquefaction performance may be determined according to the portion of grain with particle size smaller than 5 mm. For soil with particle size smaller than 5 mm, if the

mass percentage of grain content is larger than 30 %, and the mass percentage of grain contents with particle size smaller than 0.005 mm, corresponding to aseismatic fortification intensity VII, VIII and IX, are not smaller than 10, 13 and 16 % respectively, it may be determined as nonliquefaction.

Note: The clay particles contain shall be determined by use of sodium heametaphosphate as the dispersant. When other methods are be used, it shall be correspond conversed according to relative provisions.

(3) For buildings with natural subsoil, the consequences of liquefaction need not be considered when the thickness of the overlying nonliquefied soils and the elevation of groundwater table comply with one of following conditions:

$$d_u > d_0 + d_b - 2$$
$$d_w > d_0 + d_b - 3$$
$$d_u + d_w > 1.5d_0 + 2d_b - 4.5$$

where d_w is the elevation of groundwater table (in m), for which the mean annual highest elevation during the reference period should be used, or the annual highest elevation in recent years may also be used; d_u is the thickness of the overlying nonliquefiable layer (in m), in which the thickness of mud and silt seams should be deducted; d_b is the foundation depth (in m), when it is less than 2 m, shall equal 2 m; d_0 is the reference depth of liquefaction soil (in m), it may be taken as the values presented in Fig. 3.14 (according to the earthquake disaster survey data some criteria are drawn in Fig. 3.14 including some safety factors).

2. When the sequence discriminated liquefaction need be considered base on the primary discrimination, the standard penetration tests shall be performed, in which the discriminated depth shall be taken as 15 m underground, but shall be taken as

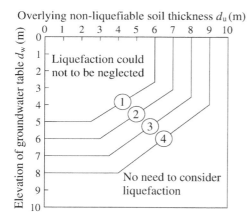

Fig. 3.14 The preliminary discrimination criteria for liquefaction according to two thicknesses (from Handbook of design and construction of underground engineering 1999)

3.5 Liquefaction of Sands and Relevant Preventions

20 m underground for the pile foundation or for the foundation buried depth greater than 5 m.

When the measured value of standard penetration resistance (in blow-number, and bar-length-modification is not included) is less than the critical value of that, the saturated soil shall be discriminated as liquefied soil; and other methods, if already proved successful, may also be used.

Within the depth of 15 m underground, the critical value of standard penetration resistance (in blow-number) for liquefaction discrimination may be calculated according to the following equation:

$$N_{cr} = N_0 \beta [\ln(0.6 d_s + 1.5) - 0.1 d_w] \sqrt{\frac{3}{\rho_c}} \quad (3.10)$$

Within the depth of 15–20 m underground, the critical value of standard penetration resistance (in blow-number) for liquefaction discrimination may be calculated according to the following equation:

$$N_{cr} = N_0 [2.4 - 0.1 d_w] \sqrt{\frac{3}{\rho_c}} \quad (3.11)$$

where N_{cr} is the critical value of standard penetration resistance (in blow-number) for liquefaction faction discrimination; N_0 is the reference value of standard penetration resistance (in blow-number) for liquefaction discrimination, it shall be taken from Table 3.3; d_s is the depth of standard penetration resistance for saturated soil (in m); ρ_c is the percentage of clay particle content; when it is less than 3 % or when the soil is sand, the value shall equal 3 %.

3. For the subsoil with liquefied soil layers, the level and thickness of soil layer shall be explored and the liquefaction index shall be calculated by the following equations, and then the liquefaction grades shall be comprehensively classified according to Table 3.4.

$$I_{le} = \sum_{i=1}^{n} \left(1 - \frac{N_i}{N_{cri}}\right) d_i w_i \quad (3.12)$$

Table 3.3 Reference value of standard penetration resistance

Design seismic group	Aseismatic fortification intensity		
	7	8	9
Group 1st	6 (8)	10 (13)	16
Group 2nd or 3rd	8 (10)	12 (15)	18

Note Values in the brackets are used for the design basic acceleration of ground motion is 0.15g and 0.30g

Table 3.4 Grade of liquefaction

Grade of liquefaction	Light	Moderate	Serious
Liquefaction index for discrimination depth is 15 m	$0 < I_{le} \leq 5$	$5 < I_{le} \leq 15$	$I_{le} > 15$
Liquefaction index for discrimination depth is 20 m	$0 < I_{le} \leq 6$	$6 < I_{le} \leq 18$	$I_{le} > 18$

where I_{le} is the liquefaction index: $I_{le} = \sum_{i=1}^{n}\left(1 - \frac{N_i}{N_{cri}}\right) d_i w_i n$; n is the total number of standard penetration test point in each bore within the discriminated depth under the ground surface; N_i, N_{cri} are measured value and critical value of standard penetration resistance (in blow-number) at ith point respectively, when the measured value is greater than the critical value, shall take as equal critical value; d_i is the thickness of soil layer (in m) at ith point, it may be taken as half of the difference in depth between the upper and lower neighboring standard penetration test points, but the upper point level shall not be less than elevation of groundwater table, and the lower point level not greater than the liquefaction depth; w_i is the weighted function value of the ith soil layer (in m^{-1}), which is considered the effect of the layer portion and level of the unit soil layer thickness. For discrimination depth is 15 m underground, such value is equal 10 when the depth of the midpoint of the layer is less than 5 m, is zero when it equals 15 m, and linear interpolation when it is between 5 and 15 m. For discrimination depth is 20 m underground, such value is equal 10 when the depth of the midpoint of the layer is less than 5 m, is zero when it equals 20 m, and linear interpolation when it is between 5 and 20 m.

4. Moreover, another method is recommended in *Code for Investigation of Geotechnical Engineering (GB50021-2001)* by Cone penetration tests to discriminate sand liquefaction. It was proposed by Ministry of Railway Institute of Science and Technology, and also suggested in international professional conference. This method is mainly based on the 125 series of testing information in different intensity area during Tangshan Earthquake. It is suitable for saturated sands and silts. The criterion is that, when the calculated specific penetration resistance or tip resistance is smaller than the critical specific penetration resistance or tip resistance, it is regarded as liquefaction.

A critical specific penetration resistance to discriminate the liquefaction of saturated sands was carried out as Eq. (3.13).

$$\begin{aligned} p_{scr} &= p_{s0}\alpha_w\alpha_u\alpha_p \\ q_{ccr} &= q_{c0}\alpha_w\alpha_u\alpha_p \\ \alpha_w &= 1 - 0.065(d_w - 2) \\ \alpha_u &= 1 - 0.05(d_u - 2) \end{aligned} \quad (3.13)$$

where p_{scr}, q_{ccr} is the critical specific penetration resistance or tip resistance of saturated sands, MPa; p_{s0}, q_{c0} is the base value of specific penetration resistance or tip resistance under conditions of 2 m groundwater table ($d_w = 2$ m) and 2 m overlying nonliquefiable soil ($d_u = 2$ m) (shown as Table 3.5); α_w is the groundwater table correction coefficient, it can be 1.13 when there is water and always has

3.5 Liquefaction of Sands and Relevant Preventions

Table 3.5 The base value of p_{s0}, q_{c0}

Aseismatic fortification intensity	7	8	9
p_{s0} (MPa)	5.0–6.0	11.5–13.0	18.0–20.0
q_{c0} (MPa)	4.6–5.5	10.5–11.8	16.4–18.2

Table 3.6 The correction coefficient of soil properties

Soil type	Sands	Silt	
Friction resistance ratio R_f	$R_f \leq 0.4$	$0.4 < R_f \leq 0.9$	$R_f > 0.9$
α_p	1.00	0.60	0.45

hydraulic connection all over a year; α_u is the correction coefficient of overlying nonliquefiable soil thickness, it can be 1.0 for deep foundation pit; d_w is the buried depth of groundwater table; d_u is the overlying nonliquefiable soil thickness, m; α_p is the correction coefficient of cone penetration friction resistance ratio, shown as Table 3.6.

One more method is the shear wave velocity discrimination for liquefaction according to the *Code for Investigation of Geotechnical Engineering (GB50021-2001)*.

When the shear wave velocity of soil stratum is larger than the upper limit one calculated by Eq. (3.14) or Eq. (3.15), it may be determined as nonliquefaction.

Sand:

$$v_{scr} = \sqrt{K_c(d_s + 0.01 d_s^2)} \quad (3.14)$$

Silt:

$$v_{scr} = \sqrt{K_c(d_s - 0.0133 d_s^2)} \quad (3.15)$$

where v_{scr} is the critical value of shear wave velocity of saturated sand or silt, m/s; K_c is the empirical coefficient, it is 92, 130, 184 in saturated sands and 42, 60 and 84 in saturated silts for intensity 7, 8, and 9 respectively; d_s is the depth of measuring point for shear wave velocity in sand or silt, m.

According to the specific code, any single method results should comprehensively analyzed with other method when the discrimination could not be determined easily.

There is another method called maximum pore-water pressure discrimination, proposed by the Institute of Science in water resource and hydropower of China in the Fifth International Conference on soil mechanics and foundation engineering. A relevant paper conducted the research on liquefaction analysis on sand foundation and sandy slope. This paper suggested that the triaxial testing apparatus on shaking table (vertical vibration) can be used for liquefaction study in the

Fig. 3.15 The pore-water pressure under different stress conditions

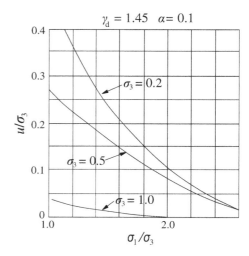

laboratory. Maximum and minimum principal stresses σ_1 and σ_3 were applied as confining pressure and vertical pressure in the triaxial test on the shaking table. The vertical pressure were employed to vibrate during $\sigma_1 \pm \Delta\sigma_1$, where $\Delta\sigma_1 = \sigma_1 \frac{\alpha}{g}$, α was the vibration acceleration. The maximum pore-water pressure u (undrained) was measured during the loading of σ_3 and $\sigma_1 \pm \Delta\sigma_1$. By virtue of this, the dynamic stability of sand foundation can be checking based on the method in soil mechanics.

Figure 3.15 is the maximum pore-water pressure field measured data under different dynamic loading in a muddy fine sands (D_{50} = 0.06 mm, coefficient of uniformity $\mu_u = 1.4$). From the figure, the smaller the confining pressure σ_3 is, the maximum pore-water pressure generated is larger. u/σ_3 increased as the stress ratio σ_1/σ_3 declined.

5. Simplified stress comparison method

Basically, this method is to compare the shear stress generated during vibration loading to the critical shear stress inducing liquefaction (i.e., the shear strength under a certain dynamic loading) and hereby to discriminate the range of liquefaction. For this purpose, several problems should be figured out. First the shear stress values in different depths under the vibration loading, by practical experience or theoretical computation. Second the critical shear stress for liquefaction under different stress conditions, analyzing in situ for liquefied area and nonliquefied area, or through laboratory tests, such as dynamic triaxial tests and reciprocated simple shear tests. These two aspects are much complicated. Here a simplified method proposed by Seed is introduced.

(1) The simplified calculation for the shear stress generated during earthquakes:

$$\tau_{av} \approx 0.65 \frac{\gamma h}{g} \alpha_{max} \cdot \gamma_d \tag{3.16}$$

3.5 Liquefaction of Sands and Relevant Preventions

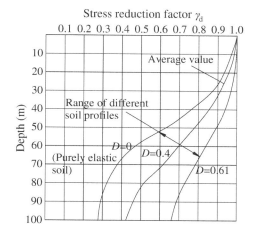

Fig. 3.16 Stress reduction factor in relation to depth (Seed and Idriss 1971)

where τ_{av} is the average peak shear stress; γ is the unit weight of the soils above the studying depth; h is the buried depth of purpose soils; α_{max} is the maximum horizontal ground acceleration during earthquakes; γ_d is the reduction factor of dynamic stress, whose value is smaller than 1, depending on the soil type and buried depth shown as Fig. 3.16.

From the above figure, it can be seen that in the upper 9.00–12.00 m, the variation of γ_d marginally changed. The average value in the dash curve can be used. When in the depth of:

$$3\,\text{m},\ \gamma_d = 0.98$$
$$6\,\text{m},\ \gamma_d = 0.95$$
$$9\,\text{m},\ \gamma_d = 0.92$$
$$12\,\text{m},\ \gamma_d = 0.85$$

Nevertheless the deviation brought in is generally less than 5 %. According to Eq. (3.16), if the maximum acceleration generated during the earthquake can be known, as well as unit weight of soil, then the average shear stress under different depth during the earthquake can be calculated.

(2) The simplified calculation for critical shear stress inducing liquefaction

The liquefied shear stress under reciprocated vibration loading can be determined by analyzing the stress condition of liquefaction during earthquakes. It can also be figured out by specific laboratory tests.

According to the previous research data, the relation of the liquefied shear stress ratio in situ and measured in the laboratory presented in Eq. (3.17).

$$\left(\frac{\tau_d}{\sigma'_0}\right) = \left(\frac{\Delta\sigma_1}{2\sigma_3}\right)_{50} \cdot C_r \frac{D_r}{50} \qquad (3.17)$$

Fig. 3.17 The relation of modification factor with the relative density

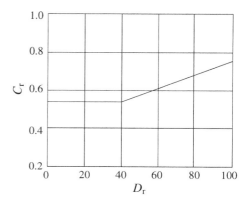

where τ_d is the liquefied shear stress on the horizontal surface; σ_0' is the initial stress; $\sigma_1 + \Delta\sigma_1$ is the vertical stress under cyclic loading; σ_3 is the initial consolidation stress, i.e., confining pressure. C_r is the modification factor from laboratory data to the in situ value, shown as Fig. 3.17; $\left(\frac{\Delta\sigma_1}{2\sigma_3}\right)_{50}$ is the liquefied stress ratio under triaxial tests. The relative density of the sand was controlled in 50 % during the tests.

Figure 3.19 was the result of the stress ratios under different particle size (represented by d_{50}) and the same relative density of 50 %. Even though these two curves obtained from two different researchers, the results were much consistent with each other. Hence we can use Fig. 3.18 to get the rough linear relationship between liquefied shear stress ratio and relative density of sand, combined with Eq. (3.15), $\left(\frac{\tau_d}{\sigma_0'}\right)$ can be calculated.

(3) Comparing the value calculated from Eq. (3.16) τ_{av} and Eq. (3.17) τ_d, the area of $\tau_d < \tau_{av}$ is the range of liquefaction, presented in Fig. 3.19.

6. The critical acceleration method

This method is based on the laboratory dynamic triaxial tests. During the tests, the sample was saturated and drained consolidation under confining pressure σ_3

Fig. 3.18 The relation between shear stress ration and grain size

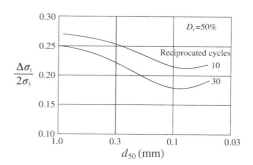

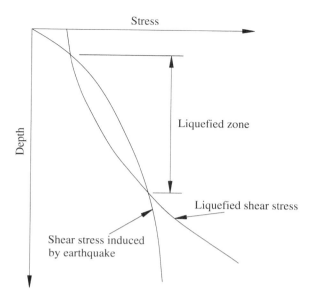

Fig. 3.19 The relation of liquefaction discrimination

was performed, then the drainage valves were closed for the undrained vibration loading. Gradually increasing the vibration acceleration, a critical acceleration to make sand liquefaction can be found under a certain confining pressure σ_3. Then the sand sample were remolded to measure the critical acceleration under different confining pressures. A relation of $\sigma_3 \sim a_c$ can be deduced out. In addition, the critical acceleration a_c under different σ_3, σ_1 can be figured out during the tests as well. Then the curve of $(\sigma_1 - \sigma_3) \sim a_c$ was acquired.

To discriminate the potential of liquefaction, it can be derived by calculation method or experimental method. The maximum and minimum primary stress of σ_3, σ_1 are both figured out in each calculation point of sand foundation under the designed loading; according to the $(\sigma_1 - \sigma_3) \sim a_c$ curve, the a_c value of each point can be determined. All the same a_c values are connected to draw isolines in each a_c value. During a practical earthquake, the acceleration is a'_c. All the areas of $a_c \leq a'_c$ are the potential liquefaction places.

The above six methods, except maximum pore-water pressure method and the critical acceleration method are time consuming and contain a lot of work based on laboratory dynamic triaxial tests, are really convenient and simply to be carried out.

3.5.4 Anti-Liquefaction Measurement

Numerous case histories on earthquake activities have documented that liquefaction of cohesionless soils is one of the major causes for structure damage and human casualties. However, one can ensure that liquefaction in loose cohesionless soils cannot be triggered if the effective stress of the soil during shaking is always greater than zero. The development of initial liquefaction in dense sands is often of no practical significance, since subsequent straining will decrease the amount of pore pressure generated. Hence, on one hand, if the potentially liquefiable soil layer is located at the ground surface and is not thicker than 3.5 m, the most economical solution may be removal and replacement with properly compacted nonliquefiable soils. However, for liquefaction-prone soil layers located deeper than 3.5 m from the ground surface, ground reinforcement techniques such as dynamic compaction, vibroflotation, stone columns and grouting may be the optimal solution. Or using of piling to bypass the potentially liquefiable zones. This is the brute force and cost-expensive solution. Piling would need to be designed for the unsupported length equivalent to the liquefied depth and for potential negative skin friction from clay layers overlying liquefiable zones. On the other hand, from the view of superstructure without soil improvements, increasing the overall stiffness and balance-symmetric ability in the superstructure (avoiding to employ differential settlement-sensitive structure) or strengthening the integrity and rigidity of the foundation (such as raft foundation, box foundation, or cross-strip foundation) can effectively improve the ability of balancing the differential settlement in buildings and then mitigate the consequences of foundation liquefaction damage.

From the investigation of earthquake disasters, the circumstances differed a lot whether the liquefied layers located directly underlying the foundation or interbedded by a nonliquefied layer. The consequences of latter were mitigated greatly. Therefore, if there is a nonliquefied layer close to the ground surface; and the building upper loads are not so large, shallow foundation should be applied to best utilize this nonliquefiable layer as the bearing layer. Similarly, raising the ground surface elevation to increase the overlying pressure by filling soil is also an effective measurement.

In total, a rational anti-liquefaction measure is really important, in which safety and cost should be both cared. Comprehensively considering the foundation liquefaction grade and the specific superstructure configuration, the option can be determined by the seismic code or previous practical experience.

3.6 Pore-Water Pressure Problems

3.6.1 *The Influence of Pore-Water Pressure on Shear Strength*

In an undrained triaxial test on a saturated foundation clay, each increase of the cell pressure will lead to increase of the pore-water pressure. According to effective stress theory and Skempton's formula, this can be described in Eq. (3.18).

3.6 Pore-Water Pressure Problems

$$\begin{cases} \sigma = \sigma' + u \\ \Delta u = B[\Delta\sigma_3 + A(\Delta\sigma_1 - \Delta\sigma_3)] \end{cases} \quad (3.18)$$

where σ is the total stress, kPa; σ' is the effective stress, kPa; $\Delta\sigma_1$, $\Delta\sigma_3$ are the maximum and minimum primary stress increment, respectively, kPa; A, B are coefficients of pore-water pressure called as Skempton's coefficients: B is related to the saturation of soil, in which complete saturation refers to $B = 1$; complete dry condition $B = 0$. The values of B observed in tests are usually somewhat smaller than 1. A is related to the stress history of soils. Higher overconsolidation ratio results in smaller A value. The coefficient A various values, usually between 0 and 0.5 are found, but sometimes even negative values have been obtained.

Hence, when changes occur, positive or negative pore-water pressure can either be generated. Since the effective stress on the soil particle skeleton, equals to the difference between the total pore-water pressure and pore-water pressure, the undrained pore-water pressure variation only affects the effective normal stress applied on the soil skeleton, while has no influence on the shear stress.

$$\tau = \frac{\sigma'_1 - \sigma'_3}{2}\sin 2\alpha = \frac{(\sigma_1 - u) - (\sigma_3 - u)}{2}\sin 2\alpha = \frac{\sigma_1 - \sigma_3}{2}\sin 2\alpha \quad (3.19)$$

Under a certain total stress condition, the positive pore-water pressure will weaken the shear strength of soils. From Fig. 3.20, if the initial stress condition is represented by the Mohr circle A, and a positive pore-water pressure is generated during the undrained triaxial test, resulting the left movement in the Mohr circles and closer to the strength envelop. When the Mohr circle is tangent to the strength envelop curve, such as B, the soil strength failure happens. On the contrary if the negative pore-water pressure is generated the Mohr circle moves to right and results in safer circumstance. In practical engineering, acknowledge of the variation of pore-water pressure can really make great sense.

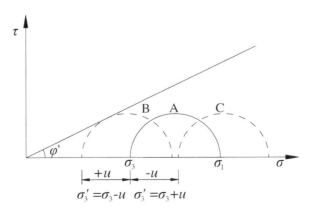

Fig. 3.20 The influence of pore-water pressure on shear strength

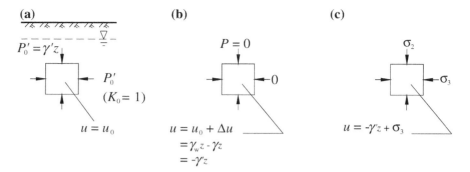

Fig. 3.21 The undrained stress variation during sampling. **a** In-situ condition. **b** After sampling. **c** Tri-axial testing

Figure 3.21 presents an example of undrained stress variation process during sampling. In Fig. 3.21a shows the undisturbed stress condition in situ. Assuming the coefficient of the lateral pressure at rest $K_0 = 1$, the field consolidation pressure is $P'_0 = \gamma' \cdot z$; and the initial pore-water pressure is $u_0 = \gamma_w \cdot z$; total stress is $P = \gamma \cdot z$. If the sampling technology is advanced enough to hardly bring disturbance for the soil, the stresses originally applied on the sample are all released. The variation of total stress is $-P = -\gamma \cdot z$. It transfers into pore-water pressure $\Delta u = -\gamma \cdot z$ under undrained condition. Then the whole pore-water pressure is $u = u_0 + \Delta u = \gamma_w \cdot z - \gamma \cdot z = -\gamma' \cdot z$, resulting the effective stress on the soil particle skeleton as:

$$\sigma' = \sigma - u = 0 - (-\gamma' \cdot z) = \gamma' \cdot z = P'_0$$

This calculation indicates that, the stress is released after soil sampling, but the effective stress on soil particle skeleton has not changed. Figure 3.21b, c present the stress conditions deploying in laboratory triaxial tests.

The above analysis makes significant sense on the excavation engineering. The excavation can be considered as a negative load, which will result in decreasing total stresses, and therefore decreasing pore pressures immediately after the excavation. Due to consolidation, however, the pore-water pressures later will gradually increase, and they will ultimately be reduced to their original value, as determined by the hydrologic conditions. Thus the effective stresses will be reduced in the consolidation process, so that the shear strength of the soil is reduced. This means that in the course of time the risk of a sliding failure may increase. A trench may be stable for a short time, especially because of the increased strength due to the negative pore pressures created by the excavation, so during the excavation construction, the soil in the bottom of foundation pit should be protected as soon as possible and lay the cushion and pour the lining plate in short time.

3.6.2 Instantaneous and Long-Term Stability in Foundation Pit in Saturated Clay

During the stability analysis in foundation pit excavation, the shear strength of soil should be considered under the influences of loading mode and time. Analyzing the relative variation of stress and strength is the first step in the stability study and also the most important part. By virtue of this, all variety of stages in the whole foundation pit project can be under well consideration and control.

Figure 3.22 shows an embankment project in the saturated soft clay foundation. The stress condition of point a is fully depicted in Fig. 3.23a, b. The shear stress increases with the filling load rising. It approaches the highest at the onset of completion. And the initial pore-water pressure equals to the static pore-water pressure $\gamma_w \cdot h_0$. Due to the poor permeability, undrained condition can be logically assumed, i.e., the excess pore-water pressure could not dissipate during the filling load and the pore-water pressure ramps as the filling height rises. Shown as Fig. 3.23b, the coefficient A of pore-water pressure is arbitrary and pore-water pressure always positive value only otherwise large negative A exists. On the completion of embankment, the shear strength is consistent with the undrained shear strength at the beginning of construction (Fig. 3.23c).

After the completion of soil filling in embankment, i.e., at the time of t_1, the total stress keeps in a constant but the excess pore-water pressure dissipates gradually and reach zero at the full consolidated time of t_2. Consolidation makes the pore-water pressure decline, void ratio decrease and effective stress and shear strength augment both. Provided pore-water pressure value, the shear strength at any time can be evaluated according to the effective stress indices c' and φ'. Hence, the stability analysis on the completion should utilize total stress and undrained strength methods. And the long-term stability should apply effective stress and effective indices analysis. From Fig. 3.23d can be easily seen that, after the completion of filling, the foundation gets through a most adverse circumstance. Over this stage, the safety degree is increasing with time.

Figure 3.24 presents the excavation in saturated soft clay. The stress condition of point a is shown as the figure (Fig. 3.25). Excavation releases the overlying pressure, resulting in decrease of pore-water pressure and occurrence of negative

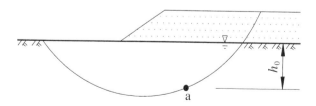

Fig. 3.22 The filling embankment on the soft foundation

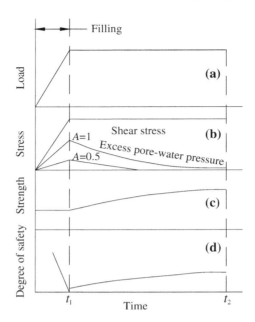

Fig. 3.23 The stability conditions during filling

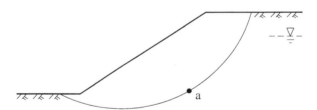

Fig. 3.24 The excavation on the soft foundation

pore-water pressure. If the pore-water pressure coefficient B equals to 1, the variation of pore-water pressure is following Eq. (3.20).

$$\Delta u = \Delta\sigma_3 + A(\Delta\sigma_1 - \Delta\sigma_3) \quad (3.20)$$

During the slope excavation, the minor primary stress declines more than the major primary stress. Hence the variation of the minor primary stress $\Delta\sigma_3$ is negative; the excess pore-water pressure Δu is negative at most circumstance. At the onset of completion of excavation, the shear strength of point a reaches the highest value; and because of the negative excess pore-water pressure, the shear strength is still equal to the initial shear strength before excavation. With the expansion of the soft clay after the excavation unloading, the negative pore-water pressure dissipates

3.6 Pore-Water Pressure Problems

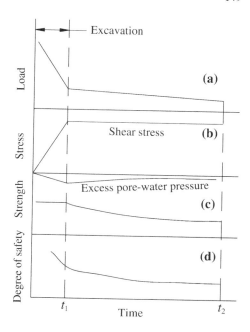

Fig. 3.25 The stability conditions of excavation **a** Load; **b** Stress; **c** Strength; **d** Degree of safety

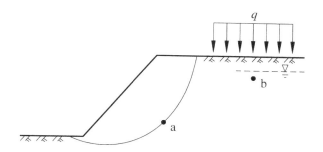

Fig. 3.26 Surcharge conditions

gradually; and the shear strength decreases accordingly. During a long-time duration, the negative pore-water pressure dissipates to zero resulting in lowest shear strength. Therefore, it is not hard to understand, excavation is opposite to filling circumstance. The stability after the completion is better than the long-term stability. The safety degree decreases with time.

Figure 3.26 shows the surcharge influence on foundation pit stability. The excess pore-water pressure induced by large-area surcharge on the slope top, such as heavy buildings or piling, etc., constructions, radiantly dissipates to drainage exit. The water flows from b to a, which increase the pore-water pressure at point a.

The stability conditions are depicted in Fig. 3.27. Assuming the surcharge load has some distance from the slope surface, the stress conditions is not conspicuously influenced on the circle sliding surface. And the shear stress keeps the same

Fig. 3.27 The stability conditions under variable loading **a** Shear stress; **b** Δu; **c** Strength; **d** Degree of safety

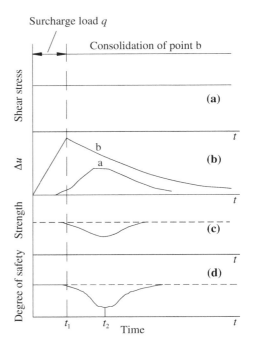

(Fig. 3.27a). The pore-water pressure at point b increases by surcharge loading. As the water radiantly drains down away to the drainage exit, the excess pore-water pressure gradually ramps to highest value at point a. The augment of pore-water pressure makes the shear strength and safety degree both decreases at point a. It can be seen that, at a certain time t_2, the safety degree has a minimum value. In this circumstance, the potential dangerous is greatest. Hence, even if there are enough instantaneous stability and long-term stability, the slope is still has possibility to failure.

According to the above analysis, the stability of foundation pit is related to the loading mode, pore-water pressure, effective stress and soil strength. Some empirical experiences are summarized in Table 3.7 as reference.

3.7 Seepage

3.7.1 The Stability of Foundation Pit with Retaining Wall Under Seepage Condition

During the excavation in saturated soft clay, supporting structures need to be conducted. Sheet piles, underground diaphragm wall, cement mixing piles, or some other bored piles are usually utilized to seal the groundwater during construction.

3.7 Seepage

Table 3.7 The measurements for improving the stability of foundation

Loading	Variation in stability	Measurement
Filling (loading)	Poorest stability on the completion of filling and then increase with time	Control the loading rate to have enough time for the dissipation of excess pore-water pressure; sand well can be used in foundation
Excavation (unloading)	Highest stability on the completion of excavation and then decrease with time	Protect the soil at pit bottom to avoid disturbance. Place cushions as soon as excavating to the design elevation
Large-area surcharge (overloading)	Most dangerous condition occurs after a certain duration after the completion of construction	Reasonably arrange surcharging area. Avoid piling, blasting activities nearby the slope top

Due to the high groundwater level, groundwater flow lines and equipotential lines are focused around the supporting structure, shown as Fig. 3.28. Hence the seepage failure can easily happen at the bottom of foundation pit. So the embedded depth should be designed appropriately to resist the seepage failure and enough safety degree for the seepage stability.

Figure 3.28 shows a foundation pit with supporting structure. The planar seepage calculation is shown as Fig. 3.29. 3-3' and 7-7' are assumed to be the water level equipotential lines, by which the foundation pit is divided into two parts I and II. Part I has the same seepage mode of the entrance and exit in foundation seepage calculation of gate dam. Part II is equivalent to the half part of $2S_2$ length flat floor seepage condition (Fig. 3.30). According to the fluid mechanics, the drag coefficients of these two conditions are presented in Fig. 3.31, in which, ξ_1 is the drag coefficient of Part I, determined by parameter $\frac{S_1}{T_1}$ and the $\frac{T_2}{b} = 0$ curve; ξ_2 is the drag coefficient of Part II, determined by parameters $\frac{S_2}{T_2}$ and $\frac{T_2}{b}$. Hereby the seepage capacity from one single side of sheet piles is:

$$q = Kh \frac{1}{\xi_1 + \xi_2} \qquad (3.21)$$

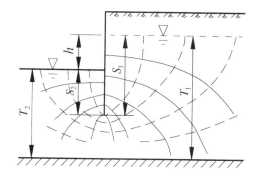

Fig. 3.28 Groundwater flow lines and equipotential lines

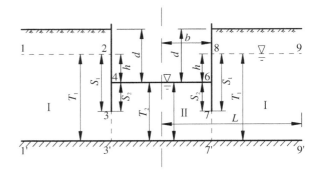

Fig. 3.29 Planar seepage calculation schematic

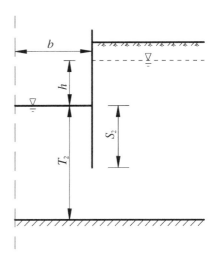

Fig. 3.30 Schematic in Part II (from Handbook of excavation engineering 1997)

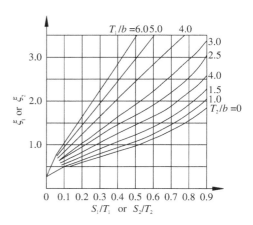

Fig. 3.31 Drag coefficients

3.7 Seepage

The water head at the end of supporting structure of point 3 or 7 is:

$$h_F = h \frac{\xi_2}{\xi_1 + \xi_2} \quad (3.22)$$

Then the hydraulic gradient of the exit of foundation pit bottom (point 3 or 7) is:

$$I_F = \frac{h_F}{S_2} \quad (3.23)$$

The critical hydraulic gradient for seepage stability is $I_c = 1 = I_F$; then the embedded depth of supporting structure should be:

$$S_2 \geq h_F \quad (3.24)$$

For the three-dimensional seepage calculation, it can be modified by the planar calculation results.

For the circular foundation pit,

$$q = 0.8Kh \frac{1}{\xi_1 + \xi_2} \quad (3.25)$$

$$h_F = 1.3h \frac{\xi_2}{\xi_1 + \xi_2} \quad (3.26)$$

where q is the seepage flux over unit length sheet pile, m³/day. Thus the total seepage flux in circular foundation pit is $Q = 2\pi R q$, where R is the radius of the foundation pit.

For square foundation pit,

$$q = 0.75Kh \frac{1}{\xi_1 + \xi_2} \quad (3.27)$$

$$h'_F = 1.3h \frac{\xi_2}{\xi_1 + \xi_2} \quad (3.28)$$

$$h''_F = 1.7h \frac{\xi_2}{\xi_1 + \xi_2} \quad (3.29)$$

where q is the seepage flux over unit length sheet pile, m³/day. Thus the total seepage flux in circular foundation pit is $Q = 8lq$ (m³/day), where l is the half length of the foundation pit side. h'_F and h''_F are the water head of center point and corner point of a foundation pit side, respectively.

Calculation value indicates the water head has highest value in the corner point. Thus the seepage instability can easily happen in the corner point. Hereby the embedded depth should be designed deeper in the corner than in the center positions.

As for the foundation pit in other geometries, such as triangular foundation pit, the water head in the corner point of short side can be calculated the same as square foundation pit, while for the head water in the center point of longer side, when the length–width ratio is close to or over 2, it can be calculated by planar seepage, without modification; As for polygon foundation pit, it can be equivalent to a circular foundation pit for the calculation.

3.7.2 The Stability of Slope Under Seepage Condition

During the excavation without well dewatering, seepage flow exists in the slope surface. The dynamic hydraulic action brings in adverse influence for the slope stability. Figure 3.32 describes the circumstance of seepage curve flowing through the slope surface. The groundwater flows downward to generate a hydrodynamic force, promoting the soil to slide down. The hydrodynamic force can be calculated by flow net analysis. In practice, it can be simply determined by mean hydraulic gradient.

In Fig. 3.32, point A and B are the intersection points of seepage line and sliding surface. Hence the mean hydraulic gradient is the slope of line AB. Hereby the total hydrodynamic force T of the sliding soil above the seepage line is:

$$T = \gamma_w I A, \qquad (3.30)$$

where γ_w is the water unit weight, kN/m^3; I is the horizontal hydraulic gradient over the applying area; H is the water head difference between point A and B, m; L is the horizontal distance between point A and B, m; A is the sliding area of the soil above the seepage line, m^2.

The seepage force T is conducted on the soil downward, resulting in a sliding moment of $T \cdot e$, where e is the distance of seepage force to the sliding center O. The point of action T can be assumed in the centroid of area A; and the direction is

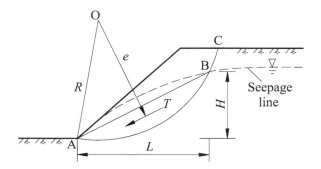

Fig. 3.32 The influence of seepage on stability

3.7 Seepage

parallel to line AB. Thus the stability calculation formula can be modified as Eq. (3.31).

$$F_s = M_{\text{slide-resistant}} / (M_{\text{slide}} + Te) \tag{3.31}$$

where $M_{\text{slide-resistant}}$ is the slide-resistant moment; M_{slide} is the slide moment; Te is the seepage-slide moment.

3.8 Piping and Soil Displacement in Foundation Pit Bottom

The water exists between two stable aquitards bearing static water pressure is called the confined pressure water. It is formed closely related to the geological development and plays an important role in the underground environmental geological problems.

3.8.1 Piping in the Foundation Pit

When confined water layer exists under the foundation pit, excavation decreases the thickness of overlying aquitard to some extent; the water head of confined water may break or destroy the pit bottom and results in piping. There are several different piping behaviors as below:

(1) Cracking of pit bottom; mesh or branch fissures occur and water pouring out with fine particles.
(2) Quicksand in the pit bottom; slope instability and the entire foundation suspending flow.
(3) Boiling sands; water accumulates in the pit and disturbs the foundation.

Some conditions inducing the occurrence of piping in foundation pit during excavation are presented in Fig. 3.33. From the equilibrium condition of aquitard

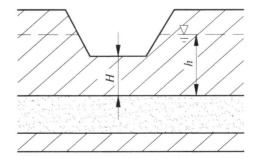

Fig. 3.33 The minimum aquitard thickness beneath the foundation pit

thickness beneath the pit bottom during excavation and the confined water pressure, there is some requirement on the minimum thickness H.

$$H = \frac{\gamma_w}{\gamma'} \cdot h \qquad (3.32)$$

When

$$H > \frac{\gamma_w}{\gamma'} \cdot h \qquad (3.33)$$

piping can hardly happen;
When

$$H < \frac{\gamma_w}{\gamma'} \cdot h \qquad (3.34)$$

piping may happen, where H is the thickness of aquitard beneath the pit bottom after excavation, m; γ' is the effective (buoyant) unit weight of relevant soil, kN/m^3; γ_w is the unit weight of water, kN/m^3; h is the water head difference between confined water pressure and the elevation of aquitard baseline, m.

When $H < \frac{\gamma_w}{\gamma'} \cdot h$, measurement should be taken to avoid piping. Relief well is a good way to decrease the confined water head of the foundation pit bottom. During the dewatering process of relief well, the pore-water pressure in the soil should be monitored in real time. Shown as Fig. 3.34, the pore-water pressure of point C at the roof of the confined aquifer should be smaller than 70 % of the total stress. When the excavation surface is very narrow, this condition can be marginally flexible, since the shear strength of soil has some resistance to the bottom heave.

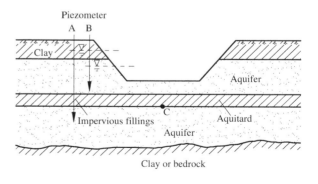

Fig. 3.34 The confined water pressure circumstance in foundation pit excavation

3.8.2 Soil Displacement

When the open caisson sinks close to the design depth, the thickness of the aquitard beneath is not large enough; it is probable cracked by the confined water pressure in the underlying sand layers (Fig. 3.35). The consequence is that large amount of sand boils rush into the caisson; the caisson sinks suddenly and substantial large-area ground subsidence occurs surrounding the caisson. When the caisson sinks undrained; and the water depth in the caisson is not enough; the plain concrete in the bottom is insufficient to balance the confined water pressure beneath; it can also induce the bottom floor of the caisson is cracked and punched by confined water pressure. The reason of the above problem is mainly contributed by the lack of enough borehole geological information. The engineering geological and hydrological conditions within the areas in 1.3 times of excavation depth are not well known before excavation. The stability of the finite-thickness aquitard overlying the confined water layer (Fig. 3.36) can be determined by following formula, by assuming the confined water head is stable at the elevation of ±0.00.

$$c \cdot u \cdot (mH) + F \cdot \gamma_s \cdot (mH) \geq F \cdot \gamma_w \cdot H_w$$

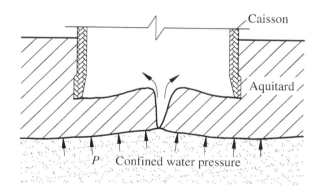

Fig. 3.35 The soil displacement induced by confined water pressure in the pit bottom

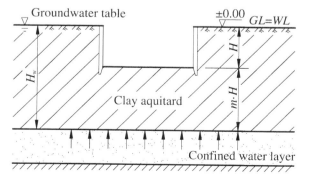

Fig. 3.36 The open caisson above a certain thickness of aquitard

Since the tension resistance of soil material is really poor, herein the cohesive effect c is ignored; the equilibrium condition can be simplified.

$$F \cdot \gamma_s \cdot (mH) \geq F \cdot \gamma_w \cdot H_w$$

$$\gamma_s \cdot (mH) \geq \gamma_w \cdot H_w$$

Because, $H_w = H + mH = H(1+m)$, thus

$$\gamma_s \cdot (mH) \geq \gamma_w \cdot H \cdot (1+m) \tag{3.35}$$

From Eq. (3.35), we can get m by Eq. (3.36):

$$m \geq \frac{\gamma_w}{\gamma_s - \gamma_w} \tag{3.36}$$

where F is the bottom area of open caisson, m^2; γ'_s is the unit weight of the below aquitard layer, kN/m^3; u is the perimeter of inner wall of cutting edge, m; γ_w is the unit weight of water, kN/m^3; H_w is the confined water head of the underlying sand layer below the aquitard, m; H is the depth of the open caisson, m.

If $\gamma_w = 10$ kN/m^3, $\gamma_s = 18$ kN/m^3, the equilibrium condition could not be broken when Eq. (3.36) can be achieved.

$$m \geq \frac{10}{18 - 10} = 1.25$$

3.8.3 The Foundation Pit Bottom Stability Encountering Confined Water Pressure

If the thickness of the aquitard layer is not enough beneath the pit bottom, and at the same time, the overlying soil weight could not balance the underlying confined water pressure, the pit bottom may heave and failure can occur. Shown as Fig. 3.37,

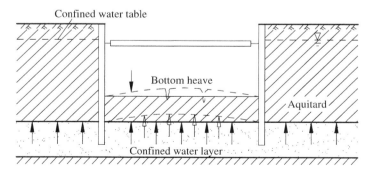

Fig. 3.37 The pit bottom heave induced by confined water pressure

3.8 Piping and Soil Displacement in Foundation Pit Bottom

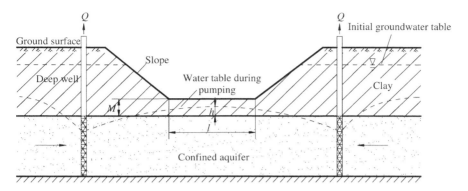

Fig. 3.38 Deep well dewatering

when designing the underground diaphragm wall before construction, the confined water pressure circumstance should be checked; and the stability analysis of the bottom heave can be examined as follows.

Firstly the balance of overlying soil weight and the underlying confined water pressure should be considered. The safety coefficient can be chosen as 1.1–1.3. When this condition could be met, the additional friction force of supporting structure can also be taken into account for the balance, as for the small-scale foundation pit with spatial effect or narrow strip pit. The friction coefficient can be determined according to the specific project by experiments. The earth pressure applied on the supporting structure can use the positive earth pressure for a safe consideration. In addition, the safety factor is taken as 1.2. Hereby if the balance still could not be satisfied, some measurements should be taken to prevent the instability of foundation pit. There are usually two methods:

(1) Underground diaphragm wall to cut off the hydraulic connection of aquifer;
(2) Lowing the confined water pressure by deep well dewatering.

When the thickness of clay layer beneath the pit bottom could not bear the upward confined water pressure, deep well dewatering is usually used to decrease the confined water head to ensure the stability of the pit bottom (Fig. 3.38). Under this circumstance, the stability condition is:

$$\gamma_w \cdot h \leq M \cdot \gamma \tag{3.37}$$

where M is the thickness of clay layer beneath the pit bottom, m; γ is the unit weight of clay layer beneath the pit bottom, kN/m³; γ_w is the unit weight of water, kN/m³; h is the confined water head after dewatering, m.

3.8.4 The Measurements of Foundation Pit Piping

3.8.4.1 Range of Reinforcement

When the foundation pit encounters piping problems and the dewatering could not be easily used, the soil improvement can be utilized. After the deep geological survey and the calculation analysis on surrounding soil displacement, some rational reinforcement can be pre-conducted on the weak places as for the foundation pit. The required locations and range should be within the following conditions:

1. The clay layer with high thixotropic and rheological properties and the liquid index over 1.
2. Confined aquifer exists below the pit bottom and has large potential to crack the aquitard beneath the pit bottom.
3. Transitional layer of clay aquitard interbedded with confined aquifer exists between the confined aquifer and the pit bottom.
4. Some special external deviator loading conditions on foundation pit:

    (1) Great difference between the surrounding pit surface and groundwater level;
    (2) Some loose soil or cavity exists outside the retaining wall;
    (3) High surcharge loading outside the retaining wall in foundation façade;
    (4) The soil hardness varies a lot from inside to outside of the foundation pit;
    (5) Addition pressure arise due to the adjacent site piling or grouting.

5. Abundant sandy layer with large thickness or water storage body such as abandoned basement pipelines exists.
6. Abundant groundwater with great flow motive connectivity to gravel layer or old building waste layer exists.
7. Some settlement-sensitive construction facilities such as high-rise tower, flammable pipes, underground railway and tunnel exist around the outside of foundation pit.

As for the above adverse circumstances, specific engineering geological and hydrological and the construction conditions should be considered in detail to predict the soil displacement surrounding the foundation pit. After the carefully optimized structure designation of retaining wall, supporting system and excavation technology, if the surrounding soil displacement is still over the allowable deformation amount, some rational soil reinforcement should be considered at some weak stability locations. For the place where the failure potential is really high, the safety factor should be increased accordingly. And grouting in real-time tracking during the excavation can be used to reliably control the differential settlement of protected objects. As for the place where piping and soil erosion may happen, some reliable soil improvement is much more important. The reinforcement place, location, range, and the properties indices after reinforcement should be calculated specifically. Some requirement to check the reinforcement effects needs to be proposed. The reinforcement methods can take the following representative method as reference.

3.8.4.2 Pit Bottom Soil Improvement to Resist the Confined Water Pressure

Piping or bottom heave is the most dangerous problem in foundation pit excavation. When the pit bottom foundation soil could not balance the underlying confined water pressure, some reliable soil improvement should be taken. There are usually three traditional methods as below:

(1) Chemical grouting or high pressure triple jet grouting method. Before excavation, the bottom of underground diaphragm wall is sealed by grouting and connects to the reinforced aquitard layer as a whole mass body to get higher the weight of the overlying soil above the pit bottom. Then it can well balance the underlying confined water pressure (Figs. 3.39 and 3.40). The calculation is seen as Eq. (3.39):

$$h \cdot \gamma_{cp} \geq H \cdot \gamma_w \qquad (3.39)$$

where h is the height between the pit bottom to the reinforced soil baseline; γ_{cp} is the mean unit weight above the reinforced soil baseline; $H\gamma_w$ is the confined water pressure.

In the Phase I project of Shanghai combined sewage treatment, the strip deep foundation of Peng-yue-pu Pumping Station is adjacent to some multistory residents' buildings. The total length is 160 m; width is 5.8 m and depth is 15 m. The clay aquitard beneath the pit bottom is only 5 m. It could not bear the underlying 16 t/m²confined water pressure. Then the above recommended method was utilized and the project was completed safely. The excavation in foundation pit of subway

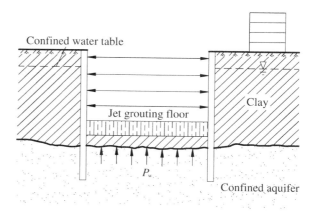

Fig. 3.39 Soil reinforced by jet grouting to resist the confined water pressure

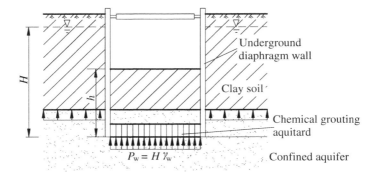

Fig. 3.40 Soil reinforced by chemical grouting to resist the confined water pressure

tunnel in Rotterdam, Netherlands was also applied this method to solve the confined water pressure problem.

(2) Deep well dewatering is conducted inside or outside the foundation pit, and at the same time, recharging is also applied in soil layers of adjacent buildings to control the surrounding settlement. When the foundation pit locates at some open area, recharging is no need (Fig. 3.41).

(3) Sealing curtain is deployed outside the foundation pit. In the loose sand, gravel or high permeability layers under groundwater level, some sealing curtain should be made by mixing piles, jet grouting piles, cement or chemical grouting piles, around the sheet pile retaining wall or outside the poor-sealing wall, to prevent soil erosion and piping at the bottom edge of retaining wall. The imbedded depth of the sealing curtain should meet the requirement of resisting piping (Fig. 3.42).

(4) Pre-consolidation method by dewatering inside the pit. In the high-density urban building area, some dewatering measurements can be taken in the sandy soil

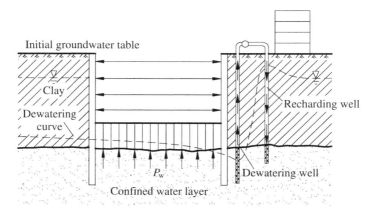

Fig. 3.41 Stabilizing the pit bottom by well dewatering

3.8 Piping and Soil Displacement in Foundation Pit Bottom

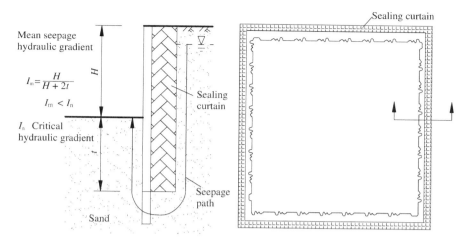

Fig. 3.42 The sealing curtain of retaining wall in foundation pit

or soft clay imbedded with thin sandy layer inside the foundation pit with good sealing curtain wall. Rational wellpoints' arrangement can drain the water in the soils between ground surface and some depth below pit bottom. The pre-dewatering before the pit excavation, can facilitate the soil drainage and consolidation to easy excavation, most importantly resulting in the increase of strength and stiffness, and also decreasing the rheology, to meet the stability and deformation requirement. The time of pre-consolidation is determined by the dewatering depth and the permeability of soils. In the sand imbedded muddy layer of Shanghai, the horizontal coefficient of permeability is about 10^{-4} cm/s. The vertical is smaller than 10^{-6} cm/s. When the dewatering depth is 17–18 m in this layer; and the excavation

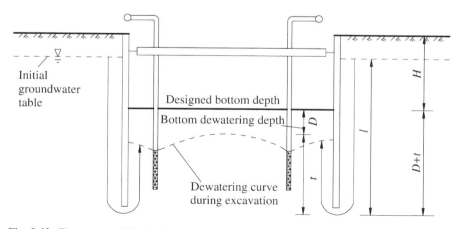

Fig. 3.43 The pre-consolidation by dewatering in foundation pit

duration is 30 d; the pre-consolidation time should be larger than 28 d. In practice, it indicates that the strength of the sand imbedded soft clay layer is augmented by 30 % through the dewatering consolidation. It works more effectively in sandy soils. For better reinforcement effects, the dewatering depth should be checked and rationally determined (Fig. 3.43).

3.9 Exercises

1. Which are the adverse effects of groundwater?
2. What conditions may induce suffosion? How to prevent it?
3. What conditions may induce piping? How to prevent it?
4. What conditions may induce quicksand? How to prevent it?
5. What is the sand liquefaction? What factors may influence it? How to prevent it?
6. What is the mechanism of pore-water pressure influencing on soil strength?
7. What properties of instantaneous stability and long-term stability for saturated clay foundation pit?
8. How groundwater seepage influences the stability of foundation pit or slope?
9. What are the behaviors of piping in foundation pit? How to prevent it?

Chapter 4
Construction Drainage

4.1 Summary

In the process called open pumping, water can be allowed to flow into the excavation as it is advanced. The water is collected in ditches and sumps and pumped away. Open pumping from sumps and ditches is usually the least expensive method from the standpoint of direct dewatering cost. Under favorable conditions, it is a satisfactory procedure. But if conditions are not conducive, attempts to handle the water by open pumping can result in delays, cost overruns, and occasionally catastrophic failure. The key is to identify those conditions that are or are not favorable for open pumping, and to recognize which conditions predominate in a given job situation. Generally, the main sump is placed in the middle of the excavation, with the result that the entire subgrade was turned into a quagmire because the water had to travel across subgrade to enter the sump. The condition is exacerbated by the presence of stratified or fine-grained soils at or near excavation subgrade that inhibit vertical drainage. This method suits in dense sands, coarse sands, graded sands, hard fissured rock and clay with surface runoff drainage. But in loose sand, soft soil or rock, problems of slope stability and boiling of the bottom must be anticipated. Tables 4.1 and 4.2 tabulate the conditions that, in authors' experience, may affect whether open pumping is viable on a given project.

Table 4.1 Conditions favorable to open pumping

Condition	Explanation
Soil characteristics	
Dense, well-graded granular soils, especially those with some degree of cementation or cohesive binder	Such soils are low in hydraulic conductivity and seepage is likely to be how to moderate in volume. Slopes can bleed reasonable quantities of water without becoming unstable. Lateral seepage and boils in the bottom of an excavation will often clear in a short time, avoiding the transport of excessive fines from soils so that foundation properties are not impaired
Stiff clay with no more than a few lenses of sand, which are not connected to a significant water source	Only small quantities of water can be expected from the sand lenses, and it should diminish quickly to a negligible value. No water is expected from the clay
Hard fissured rock	If the rock is hard, even moderate-to-large quantities of water can be controlled by open pumping. As in typical quarry operations (for soft rock and rock with blocked fissures, see Table 4.2)
Hydraulic characteristics	
Low to moderate dewatering head	These characteristics indicate that groundwater seepage will be low, minimizing problems with slope stability and subgrade deterioration, and facilitating the construction and maintenance of sumps and ditches
Remote source of recharge	
Low to moderate hydraulic conductivity	
Minor storage depletion	
Excavation methods	
Dragline, clamshell, and backhoe (if operated from ground surface or elevated bench above excavation subgrade)	These methods do not depend on traction within the excavation, and the unavoidable temporarily wet condition due to open pumping does not hamper progress
Excavation support	
Relatively flat slopes	Flat slopes, appropriate to the soils involved, can support moderate seepage without becoming unstable
Steel sheeting, slurry diaphragm walls or other cutoff structures	These methods cut off lateral flow, and assuming there are no problems at the subgrade, open pumping is satisfactory
Miscellaneous	
Open, unobstructed site	If there are no existing structures nearby, so that minor slides are only a nuisance, some degree of risk can be taken
Large excavation	In a large excavation the time necessary to move the earth is sometimes such that the slow process of lowing water with sumps and ditches does not seriously affect the schedule

(continued)

4.1 Summary

Table 4.1 (continued)

Condition	Explanation
Light foundation loads	When the structure being built puts little or no load on the foundation soils (for example, a sewage pump station) slight disturbance of the subsoil may not be harmful

Table 4.2 Conditions unfavorable to open pumping (predrainage or cutoff usually advisable)

Condition	Explanation
Soil characteristics	
Loose, uniform granular soils without plastic fines	Suck soils have moderate-to-high hydraulic conductivity and are very sensitive to seepage pressure. Slope instability and loss of strength at subgrade are likely when open pumping
Cohesive less silts, and soft clays or cohesive silts with moisture contents near or above the liquid limit	Such soils are inherently unstable. And slight seepage pressures in permeable lenses can trigger massive slides
Soft rock; rock with large fissures filled with granular soft soils, erodible materials or soluble precipitates, sandstone with uncemented sand layers	If substantial quantities of water are open pumped, soft rock may erode. Soft materials in the fissures of hard rock may be leached out. Uncemented sand layers can wash away. The quantity of water may progressively increase, and massive blocks of rock may shift
Hydrology characteristics	
Moderate to high dewatering head	These characteristics indicate the potential for high water quantities. Even well-graded gravels can become quick if the seepage gradient is high enough. Problems with construction and maintenance of ditches and sumps are aggravated
Proximate source of recharge	
Moderate to high hydraulic conductivity	
Large quantities of storage water	If the aquifer to be dewatered is high in hydraulic conductivity and porosity, large quantities of water from aquifer storage must be expected during the early phase of lowering the water table. This higher flow can greatly aggravate problems with open pumping. With predrainage, pumping can be started some weeks or months before excavation, the pumping rate will decrease and the problem can be mitigated
Artesian pressure below subgrade	Open pumping cannot cope with pressure from below subgrade since, if water reaches the excavation, damage from heave or piping has already occurred. Predrainage with relief well is advisable

(continued)

Table 4.2 (continued)

Condition	Explanation
Excavation methods	
Scrapers, loaders and trucks	These methods require good traction for efficient operation. Unavoidable temporarily wet conditions due to open pumping can seriously hamper progress. If horizontal drains and sumps can be prepared well in advance with drainage or backhoe, mass excavation with scrapers may be feasible
Excavation support	
Steep slopes	Steep slopes are sensitive to erosion and sloughing from seepage, and can also suffer rotary slides unless the water table is lowered sufficiently in advance of excavation
Soldier beams and lagging	Excavating a vertical face to place lagging boards is costly and sometimes dangerous under lateral flow conditions
Miscellaneous	
Adjacent structures	When existing structures would be endangered by slides or loss of fines from the slopes, open pumping cannot be tolerated
Small excavation	In small excavation, delays due to open pumping can seriously delay the work
Heavy foundation loads	When the structure being built bears heavily on the subsoils, even minor disturbance must be avoided
Excavating to clay or rock subgrade	Conditions will improve with extended pumping time. Extra pumping time is usually not available when open pumping

4.2 Open Pumping Methods

4.2.1 Open Ditches and Sump Pumps

4.2.1.1 Stage Excavation Drainage

Shown as Fig. 4.1, the final sump must be deep enough so that when it is pumped out the entire excavation will be drained. This is an obvious point but surprisingly it is often violated. Digging the sump down that extra several feet, or meters, is difficult and sometimes risky; there is a tendency to give up too soon. If necessary, a temporary sump at a shallower level should be constructed and pumped long enough to improve conditions so that the final sump can be safely constructed to the proper depth. Generally, the ditches are stratified dug at one/two sides or in the middle of foundation pit. And the sumps are placed at each 20.00–30.00 m distance

4.2 Open Pumping Methods

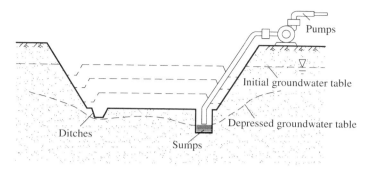

Fig. 4.1 Stage excavation drainage

for the water to be collected and pumped out. The depth of ditches and sumps can be deepened as the excavation advances. The bottom of ditches should be always kept 0.30–0.60 m lower than the pit bottom elevation. Usually in small excavations, depth of ditches can be 0.30–0.6 m with the width of 0.40 m and slope ratio of 1:1–1:1.5. And small slope of 0.2–0.5 % can be set in the ditches bottom for the drainage. The sectional area of sumps should be 0.60 × 0.60–0.80 × 0.80 m. And the bottom elevation should be kept 0.40–1.00 m lower than the ditches. The sump walls can be reinforced by bamboo cages and wood plates. The pumping should be continuously conducted until the backfill is completed.

4.2.1.2 Double Well-Point Drainage

Shown as in Fig. 4.2, the cement concrete pipes with diameter around 80–100 cm are driven into the earth section by section. The water table outside or in the bottom of foundation pit is lowered by a centrifugal pump. Usually, single well is sufficient for construction requirement. Double system is just for the very deep drawdown.

Fig. 4.2 Double well-point drainage

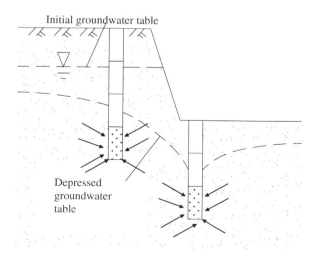

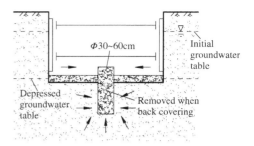

Fig. 4.3 Main central sump pumps

The last section of the well is the filter, which is drilled as quincunx holes in 15–20 cm space for better inflowing of water. The diameter of the quincunx hole is set large out and small in, which is filled by sackcloth. Sand filter material is employed in the filter to block the soil particle flowing through with water.

4.2.1.3 Main Central Sump Pumps

Shown as Fig. 4.3, in the condition that there are no sheet piles surrounding, or slope excavation and no drilled-in supporting, could not meet the construction requirement; some failures would happen, such as slope collapse. Thus, a main seepage well is established for the sump-pump system in the center of foundation pit. This system can be set during the whole construction period. Until the foundation is completed, it is sealed to prevent water seepage.

4.2.1.4 Range of Application

The above three methods are generally applicable for the water drainage in the common foundation, medium area group foundations, or building foundation pit. Easily constructed, simple equipments, low costs, they are mostly used.

4.2.2 Multilayer Open Pumping from Ditches and Sumps

4.2.2.1 Method

Shown as Fig. 4.4, along the slope of foundation pit, 2–3 ditches and sumps system is set to collect the groundwater and to block the water out of excavation area. The distribution and specific sizes of ditches and sumps are almost the same with those in the above common ditches and sumps. It should be paid attentions that to prevent the water in the upper ditches flowing down to the lower ditches. If so it is probable that the slope of foundation pit may collapse by the water seepage.

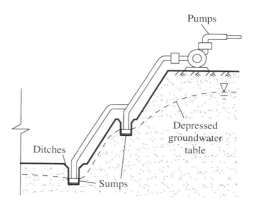

Fig. 4.4 Multilayer open pumping from ditches and sumps

4.2.2.2 Range for Application

This method is used in the very deep foundation pit project, in which the initial groundwater table is relatively high and multilayer permeable soils. Establishing multilayer ditches and sumps can effectively prevent the slope collapse when the groundwater in the upper layers scours the underlying layers. The single pumping head and slope height can be shortened but the excavation area and earthwork volume are both increased.

4.2.3 Deep Ditches Pumping

4.2.3.1 Method

In appropriate locations or upstream of groundwater in the construction site, a longitudinal deep ditch is dug as a main collector, in which the groundwater flows away or is pumped out (Fig. 4.5). Sub-ditches are connected to the main ditch and equipped all round to induce the water directions. The main ditch should be deepest and the depth is lower than the bottom of the foundation pit 1.00–2.00 m. Sub-ditches must be set to be shallower than the main ditch by 0.50–0.70 m. At the locations through the foundation, blind ditches should be set by gravels and sands. Before foundation pit backfilling, they are blocked by clays to prevent the groundwater flowing in the ditches to cause the failure of the subgrade. The deep ditches can also be set in the permanent drainage places in or surrounding the buildings.

4.2.3.2 Range for Application

This method is suitable for dewatering of large-area deep basement, caisson foundation, and group foundations.

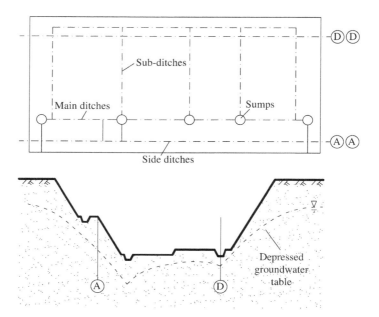

Fig. 4.5 Deep ditches pumping

4.2.4 Combined Pumping

4.2.4.1 Method (Fig. 4.6)

Based on deep ditches pumping, combined the multilayer ditches and sumps pumping, or light well-point dewatering in the upper soil layers, this combined pumping method is employed to drain large amount of underground water.

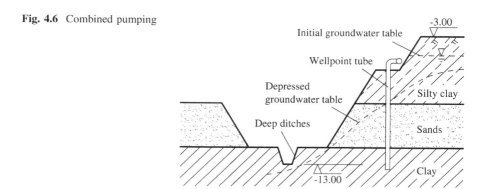

Fig. 4.6 Combined pumping

4.2 Open Pumping Methods

4.2.4.2 Range for Application

This method is used in the very uniform soil condition and deep foundation pit, or large amount of water discharge in large-area foundation excavation. The effectiveness is very good by this method but the cost is relatively high.

4.2.5 Dewatering by Infrastructure

4.2.5.1 Method (Fig. 4.7)

In this method, the deep foundation of the plant is constructed firstly, which is set to be the total water collecting site; or the surrounding drainage and sewer system is built previously, so that open sump pumps or blind seepage ditches are established in one/two sides along the foundation pit to induce the water into the main drainage and sewer system.

4.2.5.2 Range for Application

It is specially employed in group foundation dewatering of the large scale infrastructure construction (such as underground garage, oil depot).

4.2.6 Open Pumping in Sheet Pile Supporting System

Shown as Fig. 4.8, when sheet piles are constructed for the support of foundation pit excavation, small scale side ditches are set in the foundation pit edge beside the sheet piles, which is also called collecting ditches. Groundwater flows into the ditches and is pumped away immediately. Gravels and sands are filled in the ditches as filter. The depth of the ditch depends on the water amount. Generally, it is 0.60–1.00 m. Sometimes it can be set outside the foundation pit just beside the outer edge for convenient manipulation.

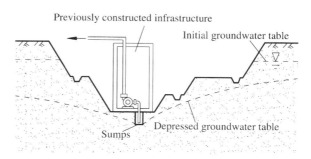

Fig. 4.7 Dewatering by infrastructure

Fig. 4.8 Open pumping in sheet pile supporting system

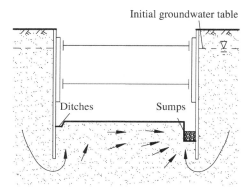

4.3 Calculation on Open Pumping Amount

4.3.1 Formulas

In industrial and civil engineering construction, very high groundwater table is usually encountered, which is much adverse to the excavation of foundation pit. Thus it is necessary to take some dewatering measures to depress the groundwater table. The dewatering mode and size can be hardly unified. Generally, some simplification is employed to estimate the rate of groundwater flow.

4.3.1.1 Long and Narrow Foundation Pit

Long and narrow foundation pit is defined as the ratio of foundation pit length B to the width C is larger than 10:

$$\frac{B}{C} > 10 \qquad (4.1)$$

When groundwater flows into a long and narrow foundation pit, it can be regarded that the groundwater laterally infiltrates in from two sides. According to Dupuit's equation:

Unconfined aquifer:

$$Q = KB \frac{H_0^2 - H_w^2}{R} \qquad (4.2)$$

More specifically, the flow rate of groundwater in two ends along width should be considered. So the calculation mode is divided into two parts (Fig. 4.9). The lateral flow rate can be estimated just by Eq. (4.2). As for the two ends, each can be approximate as a half of well with radius of $C/2$, which is sum up as an entire dewatering well. Thus,

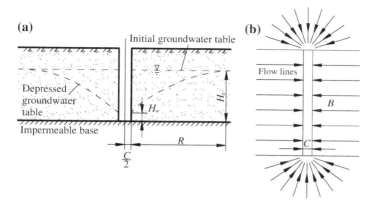

Fig. 4.9 Long and narrow foundation pit. **a** Cross sectional profile. **b** Plane view

Unconfined aquifer:

$$Q = KB\frac{H_0^2 - H_w^2}{R} + \frac{\pi \cdot K(H_0^2 - H_w^2)}{\ln R - \ln \frac{C}{2}} \qquad (4.3)$$

Confined aquifer:

$$Q = 2KBM\frac{H_0 - H_w}{R} + \frac{2\pi \cdot K \cdot M(H_0 - H_w)}{\ln R - \ln \frac{C}{2}} \qquad (4.4)$$

where C is the width of the foundation pit, m; B is the length of the foundation pit, m; Q it the flow rate, m³/d; K is the hydraulic conductivity, m/d; H_0 is the initial groundwater table, m; H_w is the water table in the well, m; R is influence radius, m; M is the thickness of confined aquifer, m.

When the lateral recharge conditions in two ends of the foundation pit are different, the calculations should be correspondingly various. Then the total flow rate must be summation of the two parts, i.e., $Q = Q_1 + Q_2$. This circumstance mostly occurs in unconfined aquifer, shown as Fig. 4.10.

Unconfined aquifer:

$$Q_1 = KB\frac{H_1^2 - H_{w1}^2}{2 \cdot l_1} \qquad (4.5)$$

$$Q_2 = KB\frac{H_2^2 - H_{w2}^2}{2 \cdot l_2} \qquad (4.6)$$

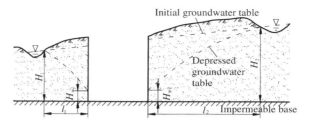

Fig. 4.10 Different lateral recharge boundaries to the foundation pit

where l_1, l_2 are the distances from the recharge boundaries to the foundation pit, m; H_{w1} and H_{w2} are the water tables on the lateral walls of the foundation pit. Others are same as previous equations.

In the case of two paralleling fully drainage channels (Fig. 4.11), the calculation can be considered as the combination of Channel I and Channel II.

Channel I:

$$Q_I = KB\frac{H_1^2 - H_w^2}{2l_1} \quad (4.7)$$

Channel II:

$$Q_{II} = KB\frac{H_2^2 - H_w^2}{2l_2} \quad (4.8)$$

where B is the length of the foundation pit. H_w is much smaller than the thickness of aquifer, then it can be neglected, so the calculation can be largely simplified.

$$\frac{B}{C} < 10 \quad (4.9)$$

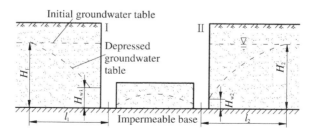

Fig. 4.11 Two paralleling fully penetrated drainage channels

Table 4.3 The value of η

C/B	0	0.2	0.4	0.6	0.8	1.0
η	1.0	1.12	1.16	1.18	1.18	1.18

The flow rate can be estimated as large well method regardless of the shape is rectangle, square, or some others. The reference radius of the hypothesized large well can be calculated as follows.

In the case of square foundation pit, it is

$$R_0 = \eta \frac{C+B}{4} \tag{4.10}$$

where the value of η can be selected from Table 4.3 based on the ratio of width over length of the foundation pit.

In the case of irregular shape foundation pit, the reference radius can be estimated by Eq. (4.11).

$$R_0 = \sqrt{\frac{F}{\pi}} \tag{4.11}$$

where F is the area of the foundation pit, m^2; R_0 is the reference radius in the calculation of large well method, m.

4.3.1.2 The Fully Penetrated Large Well Method in Horizontal Impermeable Base

In the case of the foundation pit fully penetrating an unconfined aquifer, shown as Fig. 4.12a, the calculation formula is as follows as Eq. (4.12).

$$Q = \frac{\pi \cdot K(H_0^2 - H_w^2)}{\ln \frac{R+R_0}{R_0}} \tag{4.12}$$

In the case of the foundation pit fully penetrating a confined aquifer, shown as Fig. 4.12b, the dewatering of groundwater must depress the water table down into the confined aquifer. The groundwater farer than the distance of a is the confined

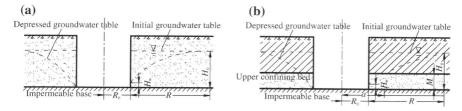

Fig. 4.12 A foundation pit fully penetrating an aquifer. **a** Unconfined aquifer, **b** Confined aquifer

groundwater, while it is the free-surface flow within the range of a. According to the principle of continuity, under the condition of steady flow, it has the relationship of $Q_{\text{unconfined}} = Q_{\text{confined}}$.

$$Q_{\text{unconfined}} = \frac{\pi \cdot K(M^2 - H_w^2)}{\ln \dfrac{a}{R_0}} \tag{4.13}$$

$$Q_{\text{confined}} = \frac{2\pi \cdot K \cdot M(H_0 - M)}{\ln \dfrac{R + R_0}{a}} \tag{4.14}$$

In conjunction with Eq. (4.13) and (4.14), eliminating $\ln a$, it has:

$$Q = \frac{\pi \cdot K(2MH_0 - M^2 - H_w^2)}{\ln \dfrac{R + R_0}{R_0}} \tag{4.15}$$

Assuming $H_w = 0$, $s = H_0$ when dewatering for the foundation pit, the flow rate can be deduced as Eq. (4.16).

$$Q = \frac{\pi \cdot K \cdot M(2s - M)}{\ln \dfrac{R + R_0}{R_0}} \tag{4.16}$$

where s is the groundwater drawdown, m; others in the equation are the same as above.

In the case of the foundation pit partially penetrating the unconfined aquifer, shown as Fig. 4.13a, the flow rate per unit width can be estimated as Eq. (4.17).

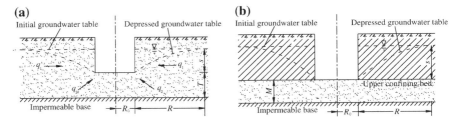

Fig. 4.13 A foundation pit partially penetrating an aquifer. **a** Unconfined aquifer, **b** Confined aquifer

4.3 Calculation on Open Pumping Amount

$$q = q_1 + q_2 = \frac{\pi \cdot K \cdot s^2}{\ln\frac{R+R_0}{R_0}} + \frac{2\pi K s R_0}{\frac{\pi}{2} + 2\text{arcsh}\frac{R_0}{T+\sqrt{T^2+R_0^2}} + 0.515\frac{R_0}{T}\ln\frac{R+R_0}{4T}} \quad (4.17)$$

where q_1 is the flow rate per unit width from lateral seepage of the foundation pit, m²/d; q_2 is the flow rate per unit width from the foundation pit bottom, m²d; T is the thickness from the impermeable base to the bottom of foundation pit, m; arcsh is the inverse hyperbolic cosine function.

So the entire flow rate of the foundation pit in this circumstance is:

$$Q = q \cdot B = B(q_1 + q_2) \quad (4.18)$$

where B is the width of the foundation pit, m.

In the case of the foundation pit partially penetrating the confined aquifer, shown as Fig. 4.13b, the foundation pit bottom just penetrates the upper confining bed, it has

$$Q = \frac{2\pi K s R_0}{\frac{\pi}{2} + 2\text{arcsh}\frac{R_0}{M+\sqrt{M^2+R_0^2}} + 0.515\frac{R_0}{M}\ln\frac{R+R_0}{4M}} \quad (4.19)$$

4.3.2 Empirical Method

If the project scale is not large, under the moderate groundwater head, an empirical method of unit area seepage amount can be employed. Table 4.4 provides the empirical values of seepage amount under different conditions.

Table 4.4 Seepage amount on unit area in foundation pit

Soil condition	Seepage amount per area (m³/d)	Soil condition	Seepage amount per area (m³/d)
Fine sands	0.16	Coarse sands	0.30–3.0
Medium sands	0.24	Fissured rock	0.15–0.25

Note 1. If the construction is in the cofferdam, the seepage from the cofferdam should be taken into consideration. Specifically, the value in the table should be multiplied by a factor of 1.1–1.3
2. The number of pumps should be consider a certain safe factor based on the estimation value in this table

Table 4.5 The section of the drainage ditch

Graphical schematic	Area of foundation pit (m²)	Section symbol	Silt clay			Clay		
			Depth beneath groundwater level (m)					
			4	4–8	8–12	4	4–8	8–12
	<1000	a	0.5	0.7	0.9	0.4	0.5	0.6
		b	0.5	0.7	0.9	0.4	0.5	0.6
		c	0.3	0.3	0.3	0.2	0.3	0.3
	5000–10,000	a	0.8	1.0	1.2	0.5	0.7	0.9
		b	0.8	1.0	1.2	0.5	0.7	0.9
		c	0.3	0.4	0.4	0.3	0.3	0.3
	>10,000	a	1.0	1.2	1.5	0.6	0.8	1.0
		b	1.0	1.5	1.5	0.6	0.8	1.0
		c	0.4	0.4	0.5	0.3	0.3	0.4

4.4 The Common Section of the Ditches in Foundation Pit

The section of the drainage ditch is generally as Table 4.5.

4.5 The Calculation of the Power of Pumps in Requirement

The power in requirement can be calculated by Eq. (4.20).

$$N = \frac{K_s \cdot Q \cdot H}{102 \cdot \eta_1 \cdot \eta_2} \tag{4.20}$$

where H is the total water head, including pumping head, suction head, and head loss generated by various resistance; K_s is the safe factor, generally $K_s = 2$; η_1 is the pump efficiency, 0.4–0.5; η_2 is the dynamic mechanical efficiency, 0.75–0.85.

To ensure the successful construction, there are always emergency pumps in preparation in case of the accident mechanical failure.

4.6 The Performance of Common Pumps

The performance of common pumps is presented in Table 4.6.

Table 4.6 Performance of general pumps

Type		Flow rate (m³/h)	Total pumping head (m)	Suction head (m)	Motor power (kW)	Weight (kg)	
B	BA					B	BA
1.5B17	1.5BA-6	6–14	20.3–14.0	6.6–6.0	1.7	17	30
2B31	2BA-6	10–30	34.5–24.0	8.7–5.7	4.5	37	35
2B19	2BA-9	11–25	34.5–24.0	8.0–6.0	2.8	19	36
3B33	3BA-9	30–5	35.5–28.8	7.0–3.0	7.0	40	50
3B19	3BA-13	32.4–52.2	21.5–15.6	6.5–5.0	4.5	23	41
4B20	4BA-18	65–110	22.6–17.1	5	10.0	51.6	50

Note 2B19 represents the inlet diameter is 2 in. (50 mm); the total pumping head is 19 m by a single pump

4.7 Case Study

Calculate the hydraulic parameter of aquifer by sensitivity analysis method

1. Compile the sensitivity analysis method program by any available software, adding instruction by block diagram;
2. Use pumping test data to calculate the parameter of aquifer by the above designed program. The pumping test data is shown in the following table.

Pumping test data

Radius of pumping well r (mm)	20			Discharge Q (m³/d)	2592		
Time t (min)		Drawdown s (m)	Time t (min)	Drawdown s (m)		Time t (min)	Drawdown s (m)
1		0.160	300	0.566		930	0.617
2		0.228	330	0.569		960	0.617
3		0.285	360	0.575		990	0.619
4		0.293	390	0.580		1020	0.622
6		0.321	420	0.583		1050	0.624
8		0.341	450	0.585		1080	0.626
10		0.370	480	0.591		1110	0.627
15		0.387	510	0.595		1140	0.627
20		0.410	540	0.596		1170	0.625
25		0.422	570	0.597		1200	0.624
30		0.443	600	0.598		1230	0.625
40		0.454	630	0.598		1260	0.623
50		0.471	660	0.600		1290	0.624

(continued)

(continued)

Radius of pumping well r (mm)	20		Discharge Q (m³/d)	2592		
Time t (min)	Drawdown s (m)	Time t (min)	Drawdown s (m)	Time t (min)	Drawdown s (m)	
60	0.484	690	0.602	1320	0.624	
90	0.515	720	0.603	1350	0.625	
120	0.531	750	0.605	1380	0.625	
150	0.541	780	0.608	1410	0.626	
180	0.547	810	0.610	1440	0.629	
210	0.556	840	0.610	1470	0.629	
240	0.560	870	0.613	1500	0.631	
270	0.563	900	0.615	1530	0.632	

4.8 Exercises

1. What kinds of open pumping method are commonly used? What are application conditions?
2. How to estimate the open pumping water discharge in foundation pit?

Chapter 5
Wellpoint Dewatering in Engineering Groundwater

With the development of social economy, the improvement of modern industrialization and urbanization, and the increase of population, the shortage of urban ground space becomes more and more serious. To take full advantages of limited land, it has been paid attention on the high-level space and underground space. In recent years, the emergence of a large number of high-rise buildings, and underground projects such as the subway, underground commercial street, underground power plants, and pumping stations are well developed.

In the construction of high-rise buildings and underground projects, the deep excavation accounts for a large percentage, which has became a preferred method in construction. However, engineering accidents, which is caused by quicksand, piping, the instability of the pit bottom, or the collapse of the pit wall, have happened almost every year, resulting in inestimable casualties and loss in economy. Such accidents can be prevented by dewatering the groundwater table in advance of excavation. The wellpoint system has been in general use in construction dewatering, which has became the most versatile of pre-drainage methods, being effective in all types of soils. Dewatering wells are set around the foundation pit, deeper than the bottom. When dewatering begins, the water level goes down and forms the cone of depression. The water table should be 0.50–1.00 m lower than pit bottom to keep the soil dry during excavation.

Wellpoint dewatering technology has been developed over a hundred years of history. In the early days, only some simple ditches and sumps were set during excavation. Later, the filter wells appeared, and pump was used for water drainage. Practice shows that when the effective diameter d_{10} is less than 0.10 mm, the time required for dewatering sharply increases; when the d_{10} is less than 0.05 mm, this simple approach cannot achieve the purpose of dewatering. Later, it was found that a certain vacuum degree around the tube can break through this limit, thus the vacuum wellpoint, also known as light wellpoint, occurred in 1925–1930. Then in 1930s, electroosmosis wellpoint also had been used in dewatering. With the increasing dewatering depth, the multistage wellpoint, ejector wellpoint, and deep wellpoint have been developed.

Table 5.1 Application for different types of wellpoint

Wellpoint types	Hydraulic conductivity (m/day)	Drawdown (m)
One-stage light wellpoint	0.10–80.00	3.00–6.00
Two-stage light wellpoint	0.10–80.00	6.00–9.00
Electroosmosis wellpoint	<0.10	5.00–6.00
Tube wellpoint	20.00–200.00	3.00–5.00
Ejector wellpoint	0.10–50.00	8.00–20.00
Deep well pump point	10.00–80.00	>15.00

In the excavation and construction of deep foundation pit, dewatering with wellpoints to reduce phreatic or confined underground water table has become a necessary engineering measure. Wellpoint dewatering has a significant effect on avoiding quicksand, piping, and pit bottom heave, keeping dry construction environment, and improving soil strength and pit slope stability. Thus, in engineering practices, it has been widely used.

Wellpoint dewatering in general includes light wellpoint, ejector wellpoint, tube wellpoint, electroosmosis wellpoint, deep well pump, and so on. The soil permeability, drawdown in requirement, equipment condition, and engineering characteristics should be considered to make choice, shown in Table 5.1. Figure 5.1 visibly presents various dewatering methods for different types of soils.

In soft soil area, the most commonly used is light wellpoint, followed by the ejector wellpoint. Electroosmosis wellpoint also has been used in some practical projects.

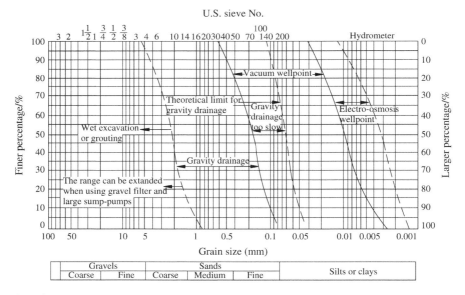

Fig. 5.1 Different dewatering methods depending on grain size distribution (From Leonards 1962)

Table 5.2 Soil permeability versus dewatering methods (Shanghai area) (From Si 1957)

Soil types	Permeability (m/day)	Effective diameter (mm)	Dewatering methods	Notes
Clay	0.001	<0.003	Electroosmosis wellpoint	Usually with open drainage
Silty clay	0.001–0.05			
Clayey silt	0.05–0.10			Electroosmosis wellpoint can be used in deep foundation pit
Sandy silt	0.10–0.50	0.003–0.025	Light wellpoint, ejector wellpoint	In Shanghai area, these methods are mostly used in those soil layers
Silty sand	0.50–1.00			
Fine sand	1.00–5.00	0.10–0.25	Common wellpoint, ejector wellpoint	
Medium sand	5.00–20.00	0.25–0.50		
Coarse sand	20.00–50.00	0.50–1.00		
Gravel	≥ 50.00		Multistage wellpoint, deep well pump point	It sometimes needs underwater excavation

Field dewatering tests have been conducted in Shanghai area and the appropriate dewatering methods for different kinds of soil have been summarized, shown in Tables 5.2 and 5.3.

Table 5.3 Excavation depths versus dewatering methods (From Si 1957)

Excavation depths (m)	Soil types			
	Silty clay, sandy silt, silty sand	Fine sand, medium sand	Coarse sand, Gravel	Large gravel, coarse pebbles (with sand)
<5	Single-stage wellpoint	Single stage light wellpoint	Wellpoint, open drainage, dewatering with pump	
5–12	Multistage wellpoint, Ejector wellpoint		Multistage wellpoint	
12–20		Ejector wellpoint		
>20		Tube wellpoint, deep well pump point		

5.1 Light Wellpoint Dewatering

5.1.1 Range of Application

Light wellpoint is set around or along the side of the foundation pit. The wellpoint pipes have small diameter and are penetrated into aquifers, which are deeper than foundation bottom. The top of the wellpoints connects the header pipe, through which the water is pumped out by vacuum, and then the water table can be depressed until it is below the pit bottom. This method can be applied in the soil layer with the hydraulic conductivity of 0.10–80.00 m/day, especially with large amount of fine sand and silty sand. It can prevent quicksand; increase slope stability; make it convenient for construction, and reduce the earth pressure that act on temporary supports.

Light wellpoint can be divided into two types: mechanical vacuum pump wellpoint and water jet pump wellpoint. The main difference between these two kinds of light wellpoints is the mechanism of the creation on vacuum.

5.1.2 Major Equipment

Light wellpoint system consists of wellpoint pipe, connection pipe, header pipe, pump devices, and other components, as shown in Fig. 5.2.

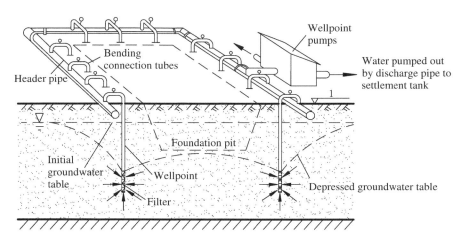

Fig. 5.2 Schematic of light wellpoint

5.1.2.1 Wellpoint Pipe

The wellpoint pipe is a steel pipe with a diameter of 38–55 mm and a length of 5.00–7.00 m. A filter is installed at the lower end of the pipe, and its structure is shown in Fig. 5.3. The filter has the same diameter with the pipe, and the length commonly is 1.00–1.70 m. On the wall of filter, there are drilled holes with 12–18 mm diameter that are arranged in quincunxes. The filter wall is covered by two filter screens. The inner one is fine-mesh brass wire gauze or raw silk gauze with 30–50 holes per centimeter. The outer one is coarse-mesh iron wire fabric or nylon wire fabric with 8–10 holes/cm. To avoid the filter pore blockage, iron wire is twined around the filter inside the inner filter screen and a thick wire protection mesh has also been set outside the outer filter screen. A cone-shaped cast iron head is installed on the lower end of the filter. The upper end of the wellpoint pipe is connected with the header pipe by connection pipe.

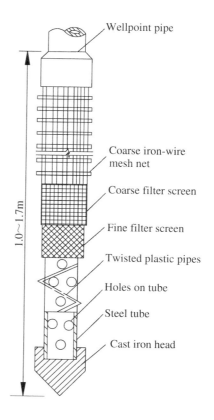

Fig. 5.3 Structure of the filter

5.1.2.2 Connection Pipe and Header Pipe

The connection pipe is composed of rubber hose, plastic hyaline pipe, or steel pipe, with a diameter of 38–55 mm. A valve should be installed for every connection pipe to overhaul the pipe. The header pipe usually consists of several steel tubes, each with a diameter of 100–127 mm and a length of 4.00 m. It is spaced by every 0.80–1.60 m to connect a wellpoint.

5.1.2.3 Dewatering Devices

The dewatering devices usually include one vacuum pump, two centrifugal pumps (one for spare), and one gas–water separator. Their working principles are shown in Fig. 5.4. The technical features of wellpoint system equipment are shown in Tables 5.4 and 5.5.

The vacuum produced by a mechanical vacuum pump in the collecting tank makes the groundwater get in through the filter, wellpoint pipe, header pipe, filtration chamber, and other parts. The pressure is relative low in the tank. When the float chamber rises to a certain height, pump will start working to pull water outside the tank.

Water jet pump light wellpoint equipment is relatively simple, only two centrifugal pumps and ejector are needed, and their working principles are shown in Fig. 5.5a. Figure 5.5b shows the working principles of water ejector. As the water flows through the nozzle, a sudden increase in flow velocity generated vacuum

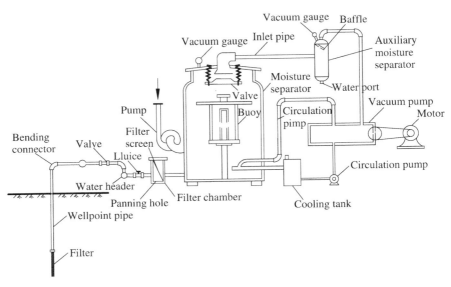

Fig. 5.4 The mechanical vacuum pump in the wellpoint system

5.1 Light Wellpoint Dewatering

Table 5.4 Technical features of dewatering devices

Items	V5 vacuum pump wellpoint	S-1 Ejector pump wellpoint
Drawdown (m)	6	8
Wellpoint pipe: diameter (mm) × length (mm)	50 × 6000	50 × 6000
Quantity	70	75
Header pipe: diameter (mm) × length (mm)	125 × 100000	100 × 100000
Space between connection pipes (m)	0.8	0.8
Vacuum degree	750	750
Ancillary electrical equipment	One model-V5 vacuum pump One model-B or model-BA centrifugal pump	Two model-3LV-9 centrifugal pump
Rated power (kW)	11.5	15
Size: length (mm) × width (mm) × height (mm)	2400 × 1400 × 2000	2300 × 1000 × 1350
Weight (kg)	1800	800

around the wellpoint to suck out groundwater. The pressure in water tank is 1 atm. Such wellpoints are well developed in the 1970s. They have the advantages of low power consumption, large drawdown compared with mechanical vacuum pump, and small influence range due to the steep cone of depression. The technical features of water jet pump wellpoint devices are shown in Table 5.6.

5.1.3 Wellpoint Arrangement

Arrangement of wellpoint system should be based on the shape and size of the foundation pit, the soil properties, the groundwater table and flow direction, and requirement of drawdown.

5.1.3.1 Plane Layout

When the pit or ditch width is less than 6 m and the drawdown is no more than 5 m, single-row linear wellpoints are available. The wellpoints should be arranged in the upstream side of groundwater flow, and not exceed the width of the pit. If the width is greater than 6 m or for very poor soils, then double-row linear wellpoints are needed. If the pit area is very large, wellpoints should be annularly arranged, or in U-shape in convenience for the transportation of excavator and dump trucks. The distance between the wellpoint pipe and the pit wall should be 0.7–1.0 m in order to

Table 5.5 Technical features of model-Shanghai wellpoint system equipment

Items drawdown		Unit	Notes 5.5–6.0 (m)
Centrifugal pump	Model		Model-B or Model-AB
	Working rate	m³/b	20
	Lift	m	25
	Pumping height	m	7
	Diameter of suction port	mm	50
	Electromotor power	kW	2.8
	Electromotor rotate speed	r/min	2900
Reciprocating-type vacuum pump	Model		V5 Type (W6 Type)
	Unit water yield	m³/min	4.4
	vacuum degree (mercury column height)	mm	747
	Electromotor power	kW	5.5
	Electromotor rotate speed	r/min	1450
Specifications of wellpoint connecting pipe and header pipe	Size (length × width × height)	mm	2600 × 1300 × 1600
	Weight	kg	1500
	Number of filter pipe		100
	Diameter of header pipe	mm	127
	Section length	m	1.6–4
	Number of sections		25
	Interval of connecting point	m	0.8
	Number of elbow pipe		100
	Number of punching pipe		1

prevent local gas leak. The wellpoint pipes should be usually spaced in 0.8–1.6 m, determined by preliminary calculation or practical experience. To take full advantages of the capacity of dewatering pump, the header pipe should be as close as possible to the groundwater table, and drawn along the flow direction with 0.25–0.5 % upslope degree. In determining the number of the wellpoints, it should be considered arranging more pipes in each corner of the pit. The general plane layout of wellpoint system is presented in Table 5.7.

5.1.3.2 Elevation Layout

The drawdown of the light wellpoint near the well wall generally can be 6–7 m. Required burial depth of wellpoint pipe (not including filter) can be calculated as follows:

5.1 Light Wellpoint Dewatering

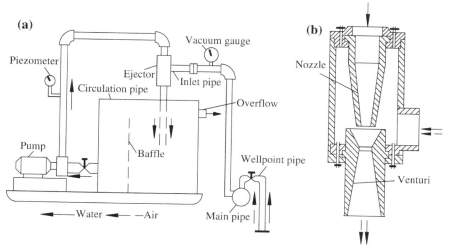

Fig. 5.5 Water jet pump wellpoint. **a** Header map. **b** Profile of ejector

Table 5.6 Technical features of water jet pump wellpoint devices

Items	Model		
	QJD-60	QJD-90	JS-45
Working depth (m)	9.5	9.6	10.26
Discharge rate (m³/h)	60	90	45
Working pressure (N/mm²)	≥0.25	≥0.25	≥0.25
Rated power (kW)	7.5	7.5	7.5

$$H \geq H_1 + h + IL$$

where H is the distance between the top of the wellpoint pipe to the bottom of the pit, m; h is the distance from the water table to the pit bottom center, usually 0.5–1.0 m; I is the hydraulic gradient, for annular arrangement is 1/10 and for single-row linear wellpoint is 1/4–1/5; L is the horizontal distance from the wellpoint pipe to the pit center, m. In addition, the wellpoint pipe generally should be above the ground about 0.2 m.

If the calculated depth (H) is less than 6 m, one-stage wellpoint is suitable. If the H is slightly larger than 6 m, the burial surface can be lowered to meet the dewatering requirement. However, providing the one-stage wellpoint cannot achieve the dewatering requirement, then two-stage wellpoint should be considered. The elevation layout of wellpoint system is shown as Table 5.8.

Table 5.7 Plane layout of wellpoint system

Type	Plane layout	Note and instruction
Single-row linear wellpoint -partially denser	(diagram: wellpoints spaced along one side of foundation pit, with $L_a/20$ spacing at ends, L_a in middle; width <6.0 m; distance 0.8–1.0 m)	1. Width of the pit is less than 6 m; drawdown is no greater than 6 m 2. At the two ends of the pit's width, the wellpoint can be denser
Single-row linear wellpoint -extending	(diagram: wellpoints extended 10.0–15.0 m beyond each end of foundation pit of length L_a)	3. Or extended by 10–15 m along two ends (this is better for section construction in long distance work)
Single-row linear wellpoint -cornering	(diagram: wellpoints along one long side and wrapping around one corner; width <6.0 m)	4. Or wellpoints are added in the corners (shown as figure in the left), which is much favorable for the upstream water
Double-row linear wellpoint system	(diagram: wellpoints on both sides of foundation pit; width >6.0 m; spacing 1.0–1.5 m)	5. Width of the pit is greater than 6 m 6. In mucky clay, even if the width of a foundation pit is not greater than 6 m, this wellpoint layout is necessary
Semiannular wellpoint system	(diagram: wellpoints around deep foundation and partially around shallow foundation; width B; extension $B/2$)	7. In some special circumstance, semiannular arrangement can only be employed, shown as the figure left. More wellpoints should be extended at the nonclosed end for about length of $B/2$
Annular wellpoint system	(diagram: wellpoints fully surrounding foundation pit with pump and valve opposite; width <40.0 m)	8. Width of the pit is mostly less than 40 m (generally 30–40 m). A valve should be set opposite to the pump group, to shunt the water flow to avoid turbulence. Or directly disconnect the pipe opposite to the pump 9. Make the wellpoint denser in the four corners, almost by 1/5 length part

(continued)

5.1 Light Wellpoint Dewatering

Table 5.7 (continued)

Type	Plane layout	Note and instruction
Annular wellpoint system		10. Width of the pit is greater than 40 m. Additional internal wellpoints inside the pit are necessary for considering the geological conditions 11. In case the total length of annular length is over 100–120 m, two-stage pump group should be employed using valves or sluices
Octangle ring wellpoint system		12. In construction of circular caisson, octangle header pipe is set by 45° corner joints. With upper excavation, the surface elevation is depressed after excavation; then pump and header pipe are equipped
Attentions	1. Try best to make most constructions or buildings into the wellpoint system, so that the main project can be successfully in progress 2. Try best to reduce the area of wellpoint system. Header pipe is probably equipped on the periphery of the foundation pit. All the wellpoints are set toward the foundation side 3. The header pipe is equipped paralleling to the contour of the foundation pit, but try best to avoid tortuous and complicated pavement, only along line or polyline for easy installation 4. The width of header pipe terrace generally should be 1–1.5 m. The plane layout should consider the drainage outlet. The discharged groundwater should be drained as far as possible 5. L_a is the main wellpoint calculation section length	

5.1.4 Wellpoint Construction Processes

The construction of light wellpoint can be roughly divided into the following processes: preparation, installation, usage, and demolition.

Preparation include wellpoint equipment, power, water source, and other necessary materials, the excavation of drainage ditch, the elevation observation of nearby buildings, and the settlement measurement of nearby buildings.

Wellpoint installation program includes: placing the header pipe, burying wellpoint pipes, connecting the header pipe and the wellpoint pipes with connection pipe, and installing dewatering devices.

The wellpoint pipe is generally installed through flushing water. This process is divided into two processes of punching and burying, shown in Fig. 5.6.

Table 5.8 Elevation layout of wellpoint system

Type	Layout schematic	Note and instruction
Single-row linear wellpoint		1. According to the requirement of drawdown, the length of wellpoint pipe and buried depth can be determined (generally 6–7 m, not including the filter) 2. The drawdown curve of single-row wellpoint system can be arranged as hydraulic gradient of $i = \frac{1}{3} - \frac{1}{5}$. In initial stage of dewatering, the slope of drawdown curve is very steep, gradually into stable of 1/10 for best
Double-row or annular wellpoint system		3. The slope of drawdown curve is generally considered as hydraulic gradient of 1/10. For the safe value of drawdown in requirement, it depends on specific project. And Δs is usually no less than 0.5 m. If possible 1.0 is better 4. Try best to make full usage of effective drawdown, to lower the elevation of header pipe
Two-stage wellpoint system		5. When one-stage wellpoint could not meet the drawdown requirement, it should try best to equip auxiliary or temporary special drainage method (shown as Table 5.7) 6. If necessary, two-stage wellpoint system can be installed as the left figure. First-stage wellpoint can be equipped first to drainage the groundwater, then the second-stage wellpoint can be set in the bottom
Concrete well combined with one-stage deeper wellpoint system		7. The groundwater is depressed by 80 mm inner diameter reinforcement concrete pipe, and then foundation pit excavation is conducted. Until the designed elevation, the wellpoint system is set to continue to dewatering the groundwater. After the caisson is installed in the predesigned elevation, the spouting construction can be conducted in dry condition

(continued)

5.1 Light Wellpoint Dewatering

Table 5.8 (continued)

Type	Layout schematic	Note and instruction
Attentions	1. The elevation of the head pipe is best to set close to the groundwater table, or slightly 20 mm higher 2. The elevation of pump is better to keep identical with the header pipe. To avoid the surface runoff water into the foundation pit, cofferdam usually should be set surrounding the pit 3. Whether linear or annular wellpoint system, all the well pipes in a certain system should be the same length. Try best to make the elevation of each filter of wellpoint be the same, preventing large elevation difference resulting dewatering effect 4. Pump system and head pipe should be installed reliable and safe terrace. Generally before equipping the pump, sleeper must be set, otherwise the site should be flattened by ramming	

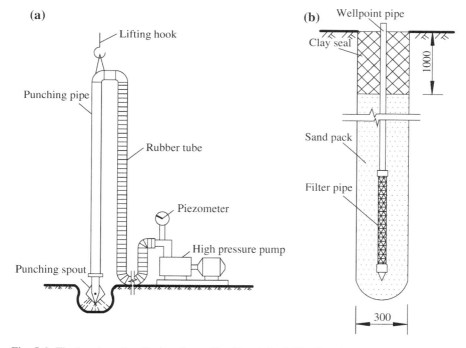

Fig. 5.6 The burying of wellpoint pipe. **a** Punching hole. **b** Pipe burying (unit: mm)

First, lift a punching pipe with 50–70 mm diameter and penetrate it on the location of well pipe. Then turn on the high-pressure water pump and loosen the soil. The punching pipe should be vertically penetrated during working and swung in all direction, in order to quickly loosen the soil. Press the pipe while punching. The diameter of punched hole is usually 300 mm to ensure there is enough space

Table 5.9 Required water pressures of different soil layers

Soil type	Punching pressure (MPa)	Soil type	Punching pressure (MPa)
Loose fine sand or clay in great plastic state	0.25–0.45	Clay in plastic state	0.60–0.75
Silty clay	0.25–0.50	Gravel with clay	0.85–0.90
Compacted humus	0.50	Clay in less plastic state or silty clay	0.75–1.25
Compacted fine sand	0.50	Coarse sand	0.80–1.15
Loose medium sand	0.45–0.55	Medium gravel	1.00–1.25
Loess	0.60–0.65	Hard clay	1.25–1.50
Compacted medium sand	0.60–0.70	Compacted coarse gravel	1.35–1.50

Notes
a. The most reliable punching pressure is obtained from in situ trail punching. The values shown in this table is for choosing suitable pump and air compressor
b. The minimum distance for domestic light wellpoint between two well centers is 80 cm, which requires punching points to not be too close to each other to prevent two holes connection (generally the diameter of punched hole is 30 cm). The distance between two adjacent well centers in light wellpoint system is 0.80–1.60 m for better
c. The punched hole should be 50 cm deeper than the end of filter pipe. When reaching that depth, reduce the water pressure rapidly. Then pull out the punching pipe and at the same time bury in the well pipe, and fill the sand filter immediately

for sand filter. The depth should be 0.50 m deeper than the end of the filter pipe. The punching water pressures in different soil layers are shown in Table 5.9.

When punching is completed, pull out the punching pipe and put in the wellpoint pipe immediately, and rapidly fill the space between the hole and well pipe with sand to prevent collapse. The quality of sand filter filling is the key to the success of dewatering. Usually choose clean coarse sand and fill it uniformly. The sand fill should be 1.00–1.50 m above the top of filter in order to ensure water flows smoothly. After sand filling, the top of this gap should be sealed with clay to prevent gas leaking.

Notes for wellpoint usage:

(1) After completion on installation of dewatering system, trail pumping test should be carried out to check whether there is gas leakage. The dewatering should be continuous, if not, the filter can be easily blocked and the fine particles will flow away, which may result in settlement and cracking of surrounding buildings. The normal drained water should be continuous and clean.
(2) Two batteries should be prepared in order to keep the wellpoint working continuously. The degree of vacuum is the criterion for whether the wellpoint system works well. It should be investigated frequently, and the readings of pressure should be not lower than 400–500 mm height of mercury column. The insufficiency of vacuum degree is usually caused by gas leaking of the

pipeline, which should be repaired in time. If the pipe is blocked, check it by listening to the water flowing sounds, touching the pipe wall and feeling the vibration, feeling the temperature of pumped water. If the pipeline is heavily blocked, the well pipe should be washed by high-pressure water or reburied one by one.

(3) The wellpoint can be removed only after the underground structure is completed and the pit is backfilled. Usually, chain block and crane are used to remove the wellpoint pipes. The holes left should be filled with sand or soils. If the foundation needs to be antiseepage, the hole below subsurface 2.00 m should be filled with clay.

5.1.5 Parameter Calculation

The purpose of light wellpoint calculation is to obtain the dewatering discharge, so that the wellpoint number and space, and the suitable dewatering devices can be determined.

Influenced by many uncertain factors, such as hydrogeological conditions and dewatering devices conditions, the results of calculated parameters cannot be very accurate. However, if the hydrogeological condition data is carefully analyzed and the formulas are chosen appropriately, the error can be limited and the results can meet the engineering requirement. For the area with abundant engineering experience, the wellpoints can be arranged according to practical data, maybe without calculation. But for multistage wellpoint system, aquifers with large hydraulic conductivity or nonstandard wellpoint system, careful and comprehensive calculation is necessary.

Before calculation, the following information should be collected first:
1. Necessary hydrogeological data
(1) The properties of aquifer, including unconfined or confined layers.
(2) The thickness of aquifer.
(3) The coefficient of permeability and influence radius of aquifer.
(4) The recharge conditions of aquifer, and the flow direction and hydraulic gradient of groundwater.
(5) The burial depth of groundwater, the water table, and variation information.
(6) The properties of wellpoint system, whether fully penetrated well or partially penetrated well.

2. Dewatering requirements
(1) The layout and range of the engineering project, and the distributions and structures of the surrounding buildings.
(2) The depth of foundation and drawdown in requirement.
(3) The permitted settlement amount and range resulting from dewatering.

5.1.5.1 Calculation for Dewatering Discharge of Single Wellpoint

1. Calculation formulas

The water discharge of wellpoint system is calculated based on well theory. The wells can be divided as fully penetrated well and partially penetrated well according to whether it reaches the impermeable base. On the other hand, the wells can also be divided as confined well and unconfined well according to groundwater surface pressure. The water discharge of single wellpoint can be calculated by the formulas in Table 5.10.

Before water discharge calculation, the coefficient of permeability and the radius of influence should be determined first.

2. Determine the hydraulic conductivity K

For the deep foundation pit excavation project in soft soil area, the test result of hydraulic conductivity must be included in the geotechnical investigation report, which can be directly used in calculation. The hydraulic conductivity can also be calculated by the data of in situ dewatering test.

(1) Calculate K based on the consolidation coefficient.

The consolidation coefficient of soil is determined by consolidation test.

$$C_V = \frac{K \cdot (1+e)}{a_V \cdot \gamma_w}$$

So

$$K = \frac{C_V \cdot a_V \cdot \gamma_w}{1+e} \tag{5.1}$$

where C_V is the consolidation coefficient of soil, cm²/s; a_V is the compression coefficient of soil, cm²/s; γ_w is the unit weight of water, kN/m³; e is the void ratio of soil.

(2) Calculate K according to effective grain size of soil:

$$K = C \cdot (d_{10})^2 \tag{5.2}$$

where d_{10} is the effective grain size of soil, mm; C is the coefficient determined by laboratory tests and local project experience.

(3) Calculate K based on the laboratory permeability tests.

The laboratory permeability test, as shown in Fig. 5.7, can be used to estimate the hydraulic conductivity of soil or rock sample.

(1) Constant head permeability test (left)

$$K = \frac{V \cdot L}{h \cdot A \cdot t} \tag{5.3}$$

where V is the volume of water flowing through the soil sample during a period of time t, m³; L is the length that water flows through, m; h is the water head difference, m; A is the cross-sectional area of soil sample, m².

5.1 Light Wellpoint Dewatering

Table 5.10 Calculation formulas for water discharge of single wellpoint

Groundwater types	Well types	Calculation formulas	Schematic diagram	Notes
Unconfined	Fully penetrated	$Q = 1.366 \dfrac{K(H^2 - h^2)}{\lg R - \lg r}$		H—Thickness of aquifer s—Drawdown in well h—Water level in well K—Hydraulic conductivity R—Radius of influence r—Radius of well
	Partially penetrated	$Q = 1.366 \dfrac{K(H_0^2 - h_0^2)}{\lg R - \lg r}$		H_0—Effective depth h_0—Distance from the water table in well to the effective region

(continued)

Table 5.10 (continued)

Groundwater types	Well types	Calculation formulas	Schematic diagram	Notes
Confined	Fully penetrated	$Q = 2.73 \dfrac{KM(H-h)}{\lg R - \lg r}$		H—Confined water head, starting from the bottom of aquifer M—Thickness of aquifer
	Partially penetrated	$Q = \dfrac{2.73 K s L}{\lg(1.32 L) - \lg r}$		L—The length of filter's working part

5.1 Light Wellpoint Dewatering

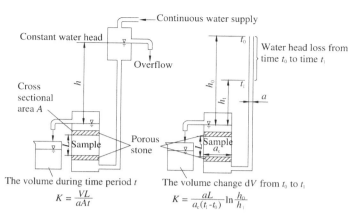

Fig. 5.7 Laboratory permeability test

(2) Falling head permeability test (right)

$$K = \frac{a \cdot L}{a_c \cdot (t_1 - t_0)} \ln \frac{h_0}{h} \tag{5.4}$$

where a_c is the cross-sectional area of soil sample, m²; a is the cross-sectional area of piezometer tube, m²; h_0 is the initial water head, m; h is the water head after a period of time $(t_1 - t_0)$, m. Other symbols have the same meaning with the former.

The disadvantage of this method is that the soil sample may be disturbed, which means the particle orientation and pore structure may change. These will influent the accuracy of hydraulic conductivity. In addition, the mud drilling method is a factor for soil sample disturbance. The combination of laboratory permeability tests and field dewatering tests is a better way, in which an empirical relationship can reduce project costs.

(4) Calculate K according to dewatering test data.

Before dewatering tests, according to local hydrogeological characteristics, such as geological structure, thickness, and properties of the aquifer and flow direction of groundwater, dewatering wells including main wells and several observation wells in typical region are constructed to form a test net. The observation wells should be arranged parallel or perpendicular to the flow direction of groundwater, as shown in Fig. 5.8. The distances between dewatering well and observation wells can consult Table 5.11.

The diameter of main well should not be less than 200–250 mm for convenience in burying the dewatering well pipe. The diameter of observation well should not be less than 50–75 mm. The main well and observation wells should all be equipped with filter pipe. The dewatering should be continuously conducted to reach a stable drawdown curve, after which another 6–8 h continues before termination. The water table should be still observed to find out the water recovering circumstances

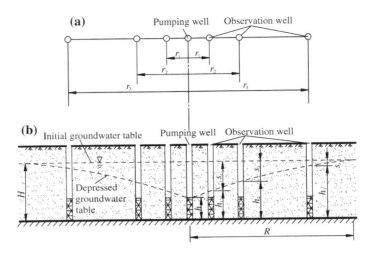

Fig. 5.8 Schematic of dewatering test. **a** Plane view. **b** Cross sectional profile

Table 5.11 Distance between main well (MW) and observation wells (OW)

Soil type	Linear distance (m)			MW-farthest OW (m)	
	MW-OW1	OW1-OW2	OW2-OW3	Minimum	Maximum
Silty clay	2–3	3–5	5–8	10	16
Sand	3–5	5–8	8–12	16	25
Gravel	5–10	10–15	15–20	30	45

until the water tables are totally recovered. At last, the profile curve of depression cone can be obtained.

For unconfined fully penetrated well, the hydraulic conductivity of soil can be calculated by

$$K = \frac{Q(\ln r_2 - \ln r_1)}{\pi(h_2^2 - h_1^2)} \quad (5.5)$$

where Q is the water discharge of the main well; h_1 and h_2 are the drawdowns in two different observation wells; and r_1 and r_2 are their distances to main well.

For confined fully penetrated well, the hydraulic conductivity of soil can be calculated by

$$K = \frac{Q(\ln r_2 - \ln r_1)}{2\pi M(s_1 - s_2)} \quad (5.6)$$

(5) Reference value of K

The reference values of hydraulic conductivity of different types of soil are shown in Table 5.12.

5.1 Light Wellpoint Dewatering

Table 5.12 Reference values of hydraulic conductivity for different types of soil

Soil type	K (m/day)	Soil type	K (m/day)
Clay	<0.005	Medium sand	5.00–20.00
Silty clay	0.005–0.10	Uniform medium sand	35.00–50.00
Clayed silt	0.10–0.50	Coarse sand	20.00–50.00
Loess	0.25–0.50	Gravel	50–100
Silty sand	0.50–1.00	Pebble	100.00–500.00
Fine sand	1.00–5.00	Pure pebble	500.00–1000.00

(6) Influence factors for hydraulic conductivity

According to above formulas, the hydraulic conductivity is proportional to dewatering discharge Q, which also determines the employment of model of devices. The accuracy of hydraulic conductivity will determine success or failure of the project indirectly.

(1) Pay particular attention to check whether there are thin silty or sand layers in sediment, whether there are clay interlayers in aquifer. In dewatering test, precipitation, influence of nearby drainage wells, flow direction and stratigraphic structure should also be considered.
(2) The hydraulic conductivity determined by laboratory test should select most representative samples. If there is thin silty sand layer, horizontal permeability test should be carried out as well. The undisturbed sand sample is difficult to take, for which hydraulic conductivity cannot be very accurate.
(3) The environmental temperature and the salt content will also influence the test result.

3. Determine the radius of influence R

The most reliable method to determine the radius of influence is dewatering test. According to dewatering test data, draw s-lg$r$ curve or (H^2-h^2)-lg$r$ curve, and then connect the water level of each observation well by a smooth curve and extend it to intersect with or tangent to initial water table. The r value of the point of intersection or tangency is the radius of influence. The radius of influence can also be backcalculated by the test data of water discharge Q and drawdown s.

The radius of influence can also be determined by comparison of experience value and empirical calculation value. According to soil properties, there are experience values for influence radius, as shown in Table 5.13. Meantime some empirical formulas are provided as below.

(1) For unconfined well И.П.Кусакин formula is

$$R = 1.95s\sqrt{H \cdot K} \qquad (5.7)$$

Table 5.13 Empirical value for radius of influence

Soil type	Silty sand	Fine sand	Medium sand	Coarse sand	Extremely coarse sand	Little gravel	Medium gravle	Large gravel
Grain size (mm)	0.05–0.1	0.1–0.25	0.25–0.5	0.5–1.0	1.0–2.0	2.0–3.0	3.5–5.0	5.0–10.0
Proportion (%)	<70	>70	>50	>50	>50			
R (m)	25–50	50–100	100–200	200–400	400–500	500–600	600–1500	1500–3000

(2) For confined well W.Sihardt formula is

$$R = 10s\sqrt{K} \qquad (5.8)$$

where s is the distance from initial water table to the dynamic water table in dewatering well, m; H is the thickness of aquifer, m; K is the hydraulic conductivity of soil, m/s.

5.1.5.2 Calculation for the Water Discharge of Wellpoint System

The interference of depression cone of single wellpoint will make the water discharge of single well less than the calculated value. However, the total drawdown is larger than that caused by single wellpoint. This is a favorable situation for the water drainage in dewatering.

1. Unconfined fully penetrated circular wellpoint system (Fig. 5.9)

The calculation formula for total water discharge is

$$Q = \frac{\pi K(2H - s')s'}{\ln R' - \ln r'} \qquad (5.9)$$

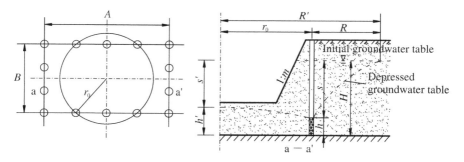

Fig. 5.9 Calculation sketch of unconfined fully penetrated circular wellpoint system

5.1 Light Wellpoint Dewatering

The discharge of single wellpoint in the system is

$$Q' = \frac{\pi K(2H-s)s}{n \ln R' - \ln(rnr_0^{n-1})} \qquad (5.10)$$

If the pit shape and the wellpoint system arrangement are irregular, it can be calculated as

$$Q' = \frac{\pi K(2H-s)s}{n \ln R - \ln(r_1 r_2 \cdots r_n)} \qquad (5.11)$$

where r_0 is the reference radius of the well group, m; $R' = R + r_0$ is the reference influence radius of the well group, m; s' is the drawdown in the pit center, m; Q' is the water discharge of any single wellpoint, m³/day; r is the radius of a wellpoint, m; n is the number of wells; s is the drawdown in a certain well, m; r is the distance from any point in the pit to the wellpoint pipe (in m), when calculating the water discharge of single wellpoint, it is the radius of filter pipe. Other symbols are shown in Fig. 5.10.

Providing the wellpoint system is arranged in a rectangle, in order to simplify the calculation, the discharge amount can also be calculated by Eq. (5.10), but the r_0 represents the reference radius of the wellpoint system, which can be calculated by following equations according to the ratio of the length A and the width B:

when $A/B < 2 \sim 3$,

$$r_0 = \sqrt{\frac{F}{\pi}} \qquad (5.12)$$

when $A/B > 2 \sim 3$ or pit in an irregular shape,

$$r_0 = \frac{P}{2\pi} \qquad (5.13)$$

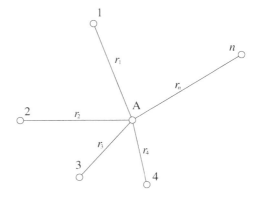

Fig. 5.10 Interference of wellpoint group

where F is pit area surrounded by wellpoint system, m^2; P is the perimeter of the irregular pit, m.

2. Unconfined partially penetrated wellpoint system (Fig. 5.11)

In order to simplify the calculation, the water discharge can also be calculated by Eqs. (5.9) and (5.10), in which H should be replaced by effective depth H_0, which can be obtained from Table 5.14. In case the calculated effective depth is larger than aquifer's thickness H, H is also selected for calculation.

3. Confined fully penetrated wellpoint system (Fig. 5.12)

The total water discharge of wellpoint system is calculated by

$$Q = \frac{2\pi K M s'}{\ln R' - \ln r_0} \quad (5.14)$$

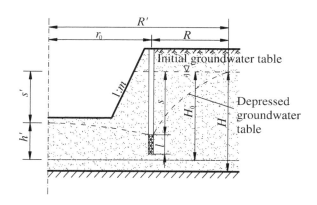

Fig. 5.11 Calculation schematic of unconfined partially penetrated wellpoint system

Table 5.14 Values of effective depth

$\frac{s}{s+l}$	0.2	0.3	0.5	0.8
H_0	1.3 $(s+l)$	1.5 $(s+l)$	1.7 $(s+l)$	1.85 $(s+l)$

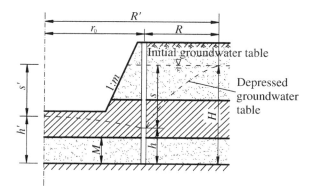

Fig. 5.12 Calculation schematic of confined fully penetrated wellpoint system

5.1 Light Wellpoint Dewatering

The water discharge of single wellpoint in the system is

$$Q' = \frac{2\pi KMs}{n \ln R' - \ln\left(rnr_0^{n-1}\right)} \quad (5.15)$$

The water discharge of single wellpoint in irregular arranged in the system is

$$Q' = \frac{2\pi KMs}{n \ln R - \ln(r_1 r_2 \cdots r_n)} \quad (5.16)$$

where M is the thickness of confined aquifer, m. The means of other symbols are shown in Fig. 5.12.

5.1.5.3 The Burial Depth of Wellpoint Pipe

The calculation schematic of the wellpoint pipe burial depth is shown in Fig. 5.13.

$$H = h_1 + h_2 + \Delta h + I \cdot L_1 + l \quad (5.17)$$

where H is the burial depth of wellpoint pipe, m; h_1 is the distance between initial water table and pit bottom, m; h_2 is the distance from initial water to the top of the wellpoint pipe, m; Δh is the safety distance between decreased water table and pit bottom, m; I is the hydraulic gradient, commonly 1/10; L_1 is the horizontal distance from the well pipe center to the pit center, m; l is the length of filter pipe, m.

5.1.5.4 The Number of Wellpoints and Their Interval

The flow capacity of single wellpoint is

$$q = 65\pi dl \cdot \sqrt[3]{K} \; (\mathrm{m}^3/\mathrm{day}) \quad (5.18)$$

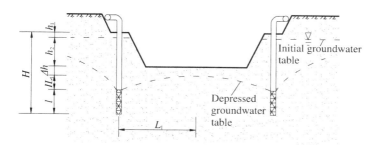

Fig. 5.13 Burial depth calculation schematic of wellpoint pipe

where d is the diameter of filter pipe, m; l is the length of filter pipe, m; K is the hydraulic conductivity, m/day.

The minimum wellpoint number is

$$n = 1.1 \frac{Q}{q} \qquad (5.19)$$

The maximum space between wellpoints is

$$D = \frac{L}{n} (\text{m}) \qquad (5.20)$$

where L is the length of header pipe, m; 1.1 is the safety factor of wellpoints for reserve.

The calculated space should be larger than $15d$, which should also meet the standard space of the adapter of header pipe (0.80 m, 1.20 m, 1.60 m, etc.).

After determining the number of wellpoints and the spacing, checking calculation must be done to find if the dewatering requirements are met or not.

5.1.5.5 The Water Head in the Pit with Fully Penetrated Circular Wellpoint System

(1) For unconfined aquifer
The water head at any location in the pit is

$$h' = \sqrt{H^2 - \frac{Q}{\pi K} \left[n \ln R' - \ln \left(rn r_0^{n-1} \right) \right]} \qquad (5.21)$$

The water head in the pit center is

$$h'_0 = \sqrt{H^2 - \frac{Q}{\pi K} \left[\ln R' - \ln r_0 \right]} \qquad (5.22)$$

(2) For confined aquifer
The water head at any location in the pit is

$$h' = H - \frac{Q}{2\pi KM} \left[n \ln R' - \ln \left(rn r_0^{n-1} \right) \right] \qquad (5.23)$$

The water head in the pit center is

$$h'_0 = H - \frac{Q}{2\pi KM} \left[n \ln R' - \ln r_0 \right] \qquad (5.24)$$

5.1.5.6 The Water Head in the Pit with Fully Penetrated Irregular Wellpoint System

(1) For unconfined aquifer
The water head at any location in the pit is

$$h' = \sqrt{H^2 - \frac{Q}{\pi K}\left[\ln R' - \frac{1}{n}(r_1 r_2 \cdots r_n)\right]} \qquad (5.25)$$

(2) For confined aquifer
The water head at any location in the pit is

$$h' = H - \frac{Q}{2\pi KM}\left[\ln R' - \frac{1}{n}\ln(r_1 r_2 \cdots r_n)\right] \qquad (5.26)$$

where h' is the water head at any location in the pit, m. For fully penetrated well, the calculation should be continued to the reference point of the well bottom; and for partially penetrated well, the reference point is the effective depth; r_i is the horizontal distance from any point in the pit to the wellpoint pipe (in m), when calculating the water head outside the filter, it is the radius of well.

The drawdown of the pit center is

$$s' = H - h' \qquad (5.27)$$

If the result of checking calculation cannot meet the dewatering requirements, adjust the burial depth of wellpoint pipe until the requirements are all satisfied.

5.1.6 Choice of Filter Screen and Sand Pack

5.1.6.1 The Importance of Filter Screen and Sand Pack

The choice of filter screen directly influencing the dewatering effect. In fine sand layer, if there is no sand pack, the water head will decrease heavily after flowing into the filter pipe. If the sand filled in could not be up to the standard or the holes in filter screen are too big, the fines in soils will flow away during drainage, which will lead to the decrease in foundation bear capacity, and the blocking of filter pipe, connection pipe, or header pipe.

5.1.6.2 Sand Back Filling Condition

1. No requirement for sand pack
   (1) The aquifer's hydraulic conductivity is larger than 10 m/day.
   (2) The filter screen meets the following criterion:

$$d_c \leq 2d_{50}$$

where d_c is the net distance between filter holes; d_{50} is the medium grain size of the aquifer soil.

   (3) The dewatering is in the aquifer without impermeable base.

All the above three requirements should be met; otherwise the sand pack is needed.

2. Requirements for sand pack

The coarse-filled sand should be suitable for natural soil composition, which can be represented by following criterion:

$$5d_{50} \leq D_{50} \leq 10d_{50} \tag{5.28}$$

where D_{50} is the medium grain size of the sand filling.

The sand pack for all wellpoints is better to choose an identical type. The coefficient of nonuniformity μ_u should be

$$\mu_u = \frac{D_{60}}{D_{10}} \leq 5 \tag{5.29}$$

where D_{60} is the diameter containing 60 percent fines, called the limited grain size of the filled sand; D_{10} is the diameter containing 10 percent fines, also called effective grain size of the sand pack.

3. The thickness of sand pack
   (1) For silty sand aquifer or sandy silt aquifer

The diameter of sand pack should be larger than 30 cm. Any void should be prevented during punching of sand pack, especially in clay layers with high compressibility. Figure 5.14 shows this phenomenon, which is caused because of negligence in construction or hurry construction. The sand pack is discontinuous in clay layer part and necking occurs. During dewatering, the groundwater above impermeable clay layer cannot be drained out. Therefore, the quality of wellpoint pipe should be ensured during construction.

   (2) For fine sand aquifer

The hydraulic conductivity of this kind of aquifer is larger than 5 m/day. The diameter of sand pack can be smaller, but no less than 20–25 cm. The sand pack void phenomenon also should be avoided.

Fig. 5.14 Sand pack void

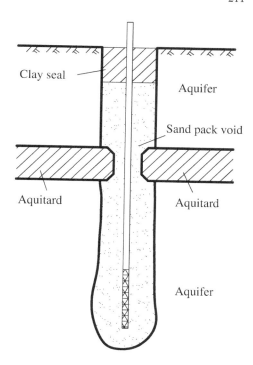

5.1.6.3 Common Filter Screen Types and Specifications

The common filter screen types mainly include quadrate knitmesh, diagonal knitmesh, and parallel knitmesh, as shown in Table 5.15. The specifications of quadrate knitmesh-type filter screen are shown in Table 5.16 and the specifications of parallel knitmesh-type filter screen are shown in Table 5.17.

5.1.6.4 The Backfilled Material Size and Wire Wrapping Interval

The backfilled material size and wire wrapping interval are shown in Table 5.18.

5.2 Ejector Wellpoint

5.2.1 Scope of Application

The ejector wellpoint is suitable for deep excavation pit, in which the dewatering depth is more than 6 m, and the site is too narrow to set multistage light wellpoint

Table 5.15 Common filter screen types

Filter screen types	Optimal hole diameter (mm)		Notes
	For uniform sand	For nonuniform sand	
Quadrate knitmesh	(2.5–3.0) d_{cp}	(3.0–4.0) d_{50}	d_{cp}—Average grain size; d_{50}—Medium grain size.
Diagonal knitmesh	(1.25–1.5) d_{cp}	(1.5–2.0) d_{50}	
Parallel knitmesh	(1.5–2.0) d_{cp}	(2.0–2.5) d_{50}	
Legend	Quadrate knitmesh		
	Diagonal knitmesh		
	Parallel knitmesh		

Table 5.16 Specifications of quadrate knitmesh type filter screen

Net number (Lines quantity in 2.5 cm²)	Lines quantity in 1 cm²	Line diameter (mm)	Hole net diameter (mm)	Weight (kg/m³)
8	3	0.50	3.13	1.10
10	4	0.50	2.32	1.34
12	5	0.45	1.86	1.38
15	6	0.40	1.41	1.32
18	7	0.35	1.14	1.22
20	8	0.35	0.99	1.36
25	10	0.30	0.76	1.19
28	11	0.25	0.69	0.96
30	12	0.25	0.63	1.03
32	13	0.23	0.59	0.93
35	14	0.20	0.55	0.77
40	16	0.16	0.47	0.73
45	18	0.15	0.43	0.55
50	20	0.15	0.35	0.63
55	22	0.14	0.33	0.59
60	24	0.14	0.29	0.65
Screen width: 1.00–5.00 m				

Table 5.17 Specifications of parallel knitmesh type filter screen

Net number (vertical/horizontal in 2.5 cm²)	Vertical/horizontal lines in 1 cm²	Line diameter (mm) Vertical	Line diameter (mm) Horizontal	Hole net diameter among horizontal lines (mm)	Weight (kg/m³)
6/40	2.5/16	0.60	0.65	0.65	6.70
6/70	2.5/28	0.70	0.40	0.34	3.80
7/70	3/28	0.60	0.40	0.34	3.75
10/75	4/30	0.55	0.37	0.32	3.56
10/90	4/35	0.45	0.30	0.27	2.69
12/90	5/36	0.45	0.30	0.27	3.00
14/100	3.5/40	0.45	0.28	0.23	2.95
16/100	6/40	0.40	0.28	0.23	2.90
16/130	6/52	0.38	0.22	0.17	2.30
18/130	7/52	0.33	0.22	0.17	2.30
18/140	7/56	0.30	0.20	0.16	2.00
20/160	8/64	0.28	0.18	0.14	2.00

Notes
1. Parallel knitmesh is better suitable for fine sands, while diagonal knitmesh for medium sands and quadrate knitmesh for coarse sands and gravels.
2. The knitmesh is made by antirust material, such as copper or bronze.
3. The inner filter screen for light wellpoint in Shanghai is cooper knitmesh with 30 holes per centimeter. The outer filter screen is iron knitmesh with 5 holes per centimeter, which can also be nylon knitmesh. The filter pipe and filter screen are separated by laddered iron wires.

Table 5.18 The backfilled material size and wire wrapping interval

Soil type	Backfilled material size (mm)	Wire wrapping interval (mm)
Fine-medium sand	2–4	0.75–1.0
Coarse gravelly sand	4–6	2.0
Gravel and cobble	8–15	3.0

system. Its dewatering depth can reach to 10–20 m. The ejector wellpoint can be applied in sand layer with the hydraulic conductivity of 3–50 m/day.

5.2.2 *Major Equipment and Working Principles*

The ejector wellpoint can be divided into two types: water ejector wellpoint and gas ejector wellpoint. The major equipment include ejector well, high-pressure water pump or high-pressure gas pump, and pipeline system, as shown in Fig. 5.15. The former is performed by pressured water, and the later is performed by pressured gas.

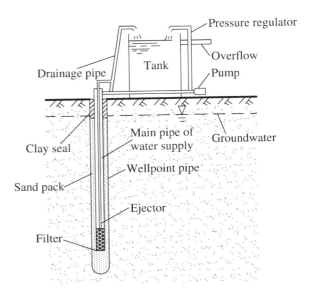

Fig. 5.15 Working principle of ejector wellpoint system

The structure of ejector wellpoint can be divided into parallel type (also known as external type) and concentric type, as shown in Fig. 5.16a, b. Their working principles are the same. The concentric ejector wellpoint includes two parts: inner pipe and outer pipe. The ejector is installed in the lower end of the inner pipe (Fig. 5.16c), and connected with filter pipe (Fig. 5.17). The ejector is composed of jet nozzle, mixing chamber, the diffusion chamber. Its structure is determined by five factors, which are the jet nozzle diameter D, the mixing chamber length L_4, the diffusion chamber taper φ, the diffusion chamber length L_5, and the distance from the top of jet nozzle to the end of diffusion chamber L_2. These five factors should match well for each other, especially for the ratio of jet nozzle diameter over the diffusion chamber diameter. If the ratio is appropriate, the wellpoint can reach maximum efficiency, whereas there will be sharp decline in efficiency. Presently, the design of ejector wellpoint system is mainly determined by the combination of empirical method and theoretical calculation.

When the ejector wellpoint system starts working, the working flow is pumped into the annular space between the inner and outer tubes by high pressure and then reaches jet nozzle. Because the cross-sectional area of water flow suddenly reduces, the flow velocity rapidly raises to maximum value, about 30.00–60.00 m/s. The water rushes into the mixing chamber, and causes a vacuum near the jet nozzle. Under vacuum suction effect, the groundwater is brought into the mixing chamber through suction tube and mixed with working flow, then flowed into diffusion chamber.

Then the water kinetic energy transforms into potential energy. The water flow gradually slows down while the water pressure raises, making the mixed water flow into the water tank. Part of the water can be reused as high-pressure working flow,

5.2 Ejector Wellpoint

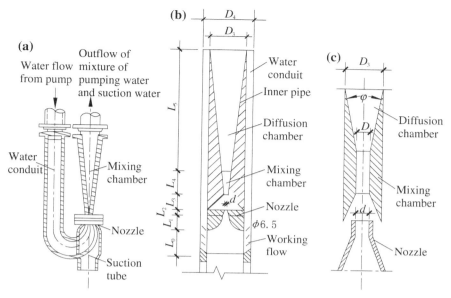

Fig. 5.16 Ejector wellpoint structure. **a** External type. **b** Concentric type (jet nozzle diameter is 6.5 mm). **c** ejector

and the rest is drained away by low-pressure pump. This cyclic operation gradually lowers the groundwater table to a demand depth.

5.2.3 Design of the Pumping Device Structure

(1) According to the outflow rate of the pit and the arrangement of the wellpoints, determine the required single well discharge Q_0 and the suction head H.

(2) According to the required suction head H, determine the working pressure P_1 by the following equation:

$$P_1 = \frac{0.1H}{\beta} \; (\text{N/mm}^2) \tag{5.30}$$

where β is the ratio of suction head over the working pressure, refer to Table 5.19.

(3) According to the single well discharge Q_0, determine the working flow Q_1 by the following equation:

$$Q_1 = \frac{Q_0}{\alpha} \; (\text{m}^3/\text{day}) \tag{5.31}$$

Fig. 5.17 Ejector wellpoint structure in detail

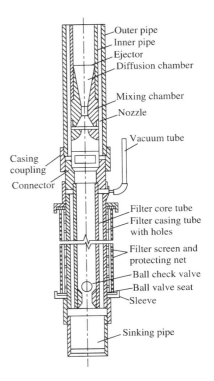

Table 5.19 Empirical parameters

Hydraulic conductivity K (m/day)	β	α	M	r
$K < 1$	0.225	0.8	1.8	4.5
$1 \leq K \leq 50$	0.25	1.0	1.0	5.0
$K > 50$	0.30	1.2	2.5	5.5

where α is the ratio of single well discharging flow and the working flow, refer to Table 5.19.

(4) According to the working flow Q_1 and the working pressure P_1, determine the diameter of jet nozzle d_1.

$$d_1 = 19\sqrt{\frac{Q_1 \times 10^{-6}}{v_1 \times 3600}} \text{ (mm)} \quad (5.32)$$

$$v_1 = \phi\sqrt{2gH} = \phi\sqrt{2gP_1 \times 10} = \phi\sqrt{20gP_1} \quad (5.33)$$

5.2 Ejector Wellpoint

where v_1 is the velocity of working flow at the exit of jet nozzle, m/s; ϕ is the velocity coefficient of jet nozzle, the approximation is 0.95; P_1 is the working pressure, N/mm^2; g is the acceleration of gravity, value as 9.8 m/s^2.

(5) According to the jet nozzle diameter d_1, determine the diameter of mixing chamber D as

$$D = M \cdot d_1 \, (\text{mm}) \tag{5.34}$$

where M is the ratio of mixing chamber diameter to jet nozzle diameter, refer to Table 5.19.

(6) According to the jet nozzle diameter d_1, determine the length of mixing chamber L_4 as

$$L_4 = r \cdot d_1 \, (\text{mm}) \tag{5.35}$$

The value of parameter r is referred to in Table 5.19.

(7) When the angle of throat is 7°–8°, the energy loss is minimum, so that diffusion chamber length can be determined as

$$L_5 = 8.5 \left(\frac{D_3}{2} - \frac{D}{2} \right) (\text{mm}) \tag{5.36}$$

where D_3 is the diameter of inner pipe, mm; D is the diameter of mixing chamber, mm.

(8) According to working flow Q_1 and the maximum allowable velocity, $v_{max} = 1.5$–2 m/s, determine the length of the inflow hole on inner pipe L_0.

$$L_0 = \frac{Q_1 \times 10^{-6}}{2a \cdot v_{max} \times 3600} \, (\text{mm}) \tag{5.37}$$

where a is the width of the inflow hole, mm.

(9) The necking length L_3 and the cylinder length L_2 of jet nozzle are determined by structural demand.

$$L_3 = 2.5 d_1 \, (\text{mm}) \tag{5.38}$$

$$L_2 = (1.0 \sim 1.5) d_1 \, (\text{mm}) \tag{5.39}$$

(10) The diameters of inner pipe D_3 and the outer pipe D_4 can be determined by trial method. Amend them by following equations:

$$D_3 = \sqrt{\frac{4Q_0 + Q_1 \times 10^{-6}}{\pi v_{max} \times 3600}} \, (\text{mm}) \tag{5.40}$$

$$D_4 = \sqrt{\frac{4Q_0 \times 10^{-6}}{\pi v_{\max} \times 3600}} \text{ (mm)} \tag{5.41}$$

The power efficiency of the high-pressure pump used in the ejector wellpoint dewatering is generally 55 kW, the flow is generally 160.00 m^3/h and the suction head is 70 m. Each pump can drive 30–40 wellpoints.

5.2.4 Layout of Ejector Wellpoint and Attention for Construction

(1) The arrangement of ejector wellpoint system is basically similar with light wellpoint.
(2) Wellpoint is generally spaced at 2.00–3.00 m, and the punching diameter is 400–600 mm. The wellpoint should be 1.00 m deeper than filter pipe bottom. To prevent ejector abrasion, casing method can be used for drilling. Use water and compressed air to eject mud, and then put the well casing when the mud content is less than 5 % in casing pipe. The wellpoint surface, approximately 0.50–1.00 m, should be sealed by clay.
(3) The pump should be run before putting the well pipe. Each well pipe should be connected with header pipe immediately after settling down. Pump out the mud by trial running, and the degree of vacuum should be measured at the same time, which should be not less than 93.3 kPa. Trial pumping should last until the pumped water becomes clear.
(4) The return header pipe should be switched on after all wellpoints are set in position. Trial dewatering should be taken out before formal experiments.
(5) Pumping device, including jet nozzle, mixing chamber, the diffusion chamber, etc., should be produced accurately.
  Different units of inflow tube and water return tube should be separated by valves. A ball valve should be set in the end of filter pipe to prevent working water backflowing.
(6) The working water should be kept clean to prevent jet nozzle and pump impeller abrasion.

5.3 Tube Wellpoint

5.3.1 Scope of Application

Tube wellpoint is suitable for aquifers composed by coarse sand and pebble, where the light wellpoint is not applicable. Those confined and unconfined aquifers usually have large hydraulic conductivity, with great water discharge and large dewatering depth, which is usually 8.00–20.00 m.

5.3.2 Major Equipment and Working Principles

5.3.2.1 Well Pipe

Well pipe includes two parts: well casing and filter, as shown in Fig. 5.18. Well casing usually is cast iron pipe, concrete pipe, or plastic pipe with a diameter of 200–350 mm. The filter can be made by twining galvanized wires out of the casing pipe with punched holes (Fig. 5.19), or covering the welded steel frame by a specific filter screen (Fig. 5.20).

5.3.2.2 Pump

If the dewatering drawdown is less than 7.00 m, centrifugal pump is enough. While it exceeds 7.00 m, submersible pump or deep well pump is necessary.

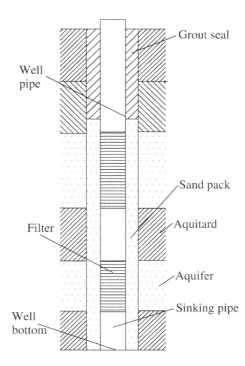

Fig. 5.18 Well pipe structure

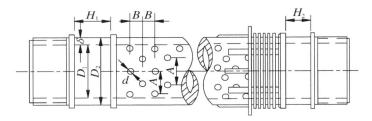

Fig. 5.19 Cast iron filter

Fig. 5.20 Tube wellpoint structure

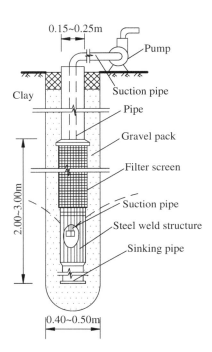

5.3.3 Construction Method

5.3.3.1 Wellpoint Arrangement

Determine the outflow rate first, then check the limited discharge of single wellpoint, and finally determine the wellpoint quantity. Uniformly arrange the wellpoint around the pit and connect them by water collecting tube.

5.3 Tube Wellpoint

5.3.3.2 Hole Creating Method

Choose percussion drill or rotary drill according to soil conditions and hole depth. If the well depth is less than 15.00 m, water pressure casing method with long auger can also be used. The diameter of drill hole is usually 500–600 mm. When the hole reaches a predetermined depth, clean out the mud in the drill hole and put the cast iron pipe or cement-gravel pipe with 300–400 mm diameter in it. In order to ensure the discharge amount and prevent fine sand flow into the pipe, filter materials should be backfilled around the well pipe. Its thickness should not be less than 100 mm. The grain diameter of backfilled material shall be 8–10 times of the aquifer's d_{50}–d_{60}. The specification is shown in Table 5.18.

5.3.3.3 Well Washing

After backfilling, the well should be washed. For cast iron pipe, piston or air compressor can be chosen. For other material pipe, wash it by air compressor until the water becomes clear.

5.4 Electroosmosis Wellpoint

5.4.1 Scope of Application

In saturated clay layer, especially for sludge or mucky clay soil, the permeability is relatively poor and the water retention capacity is strong. In these soils, the common light wellpoint dewatering and ejector wellpoint dewatering are less effective. With the corporation of electroosmosis wellpoint, water can be easily drained from impermeable soil layer.

5.4.2 Major Equipment and Working Principles

The electroosmosis wellpoints are arranged around the pit. The light wellpoints or ejector wellpoints act as cathode and the steel pipes (ϕ 50–75 mm) or steel bars (ϕ 50–75 mm) act as anode, which are inserted closer to the pit than wellpoints. The cathode and anode are connected by wires to form the access. Then apply a strong direct current on anode, as shown in Fig. 5.21. The voltage makes the negatively charged soil particles move to the anode, and the positive charged pore water moves to the cathode. The former phenomenon is called electrophoresis and the later is called electroosmosis. The combined action of electroosmosis and vacuum makes the pore water gather around the well pipe. The water will be pumped out

Fig. 5.21 Arrangement of electroosmosis wellpoints

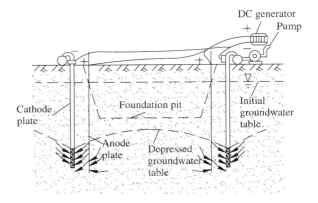

continuously and the water level will gradually drop. The soils between electrodes will become telescreen, which can prevent groundwater from flowing into the pit.

5.4.3 Key Points and Attention of Construction

(1) The electroosmosis wellpoint can be created by casing method with hydraulic giant.

(2) The anode should be vertical penetrated, and cannot contact with cathode. The anode should be 50 cm deeper than wellpoint pipe and 20–40 cm above the ground.

(3) The anode and cathode spacing is generally 0.80–1.50 m. For light wellpoint, the spacing is 0.80–1.00 m, and for ejector wellpoint it is 1.20–1.50 m. The anodes and cathodes are stagger ranged in parallel. The number of anodes and cathodes better is equal. The number of anodes can be more than the number of cathodes, if necessary.

(4) Direct current generator can be replaced by direct current welding machine. Its power can be calculated as follows:

$$P = \frac{U \cdot J \cdot F}{1000} \qquad (5.42)$$

where P is the power of welding machine, kW; U is the electroosmosis voltage, generally 45–65 V; J is the electric current density, better be 0.5–1.0 A/m^2; F is the electroosmosis area (in m^2), F = electric conduction × wellpoint perimeter.

(5) In order to prevent the electric current passing through the ground surface, which will reduce the electroosmosis effect, before power on, the electric conductors on the ground between anode and cathode should be cleaned up. If possible, coat the ground with an insulating layer of asphalt conditional. Besides, the part of

anode inside the permeable layer should be coated with two insulating layers of asphalt conditional to reduce power consumption.

(6) The electroosmosis wellpoint should be intermittently electrified. That is, outage of the electric generator 2–3 h after 24 h running, to save electric energy and prevent resistance increase.

5.5 Recharge Wellpoint

The dewatering will lead to pore water decrease and effective stress increase in clay layer, and further lead to layer compaction. This will result in differential settlement, influencing the safety of nearby buildings. To minimize the amount of ground settlement, in domestic and foreign engineering projects, dewatering wellpoints are combined with recharge wellpoints.

5.5.1 Working Principles

The working principles of recharge wellpoint are as follows. Bury a row of recharge wells in the ground between dewatering wellpoints and nearby buildings, as shown in Fig. 5.22. Keep the radius of dewatering influence no more than the range of recharge wellpoints by injecting water, and the water level will remain unchanged. The formed water purdah can prevent the loss of groundwater near the buildings.

5.5.2 Key Points and Attentions of Construction

(1) The recharging water should be clear. The recharging amount and pressure should be calculated by well theory and adjusted according to observation data.
(2) The dewatering wellpoints and recharge wellpoints should be started or stopped synchronously.
(3) The filter of recharge wellpoint should be 0.50 m higher than water table and reached to the well pipe bottom. The structure of recharge wellpoint can be the same with dewatering wellpoint, but should guarantee the quality of well hole and backfilled sand.
(4) The recharge wellpoint and dewatering wellpoint should maintain a certain distance. The length of recharge wellpoint is decided by depth of the permeable stratum.
(5) A certain number of settlement and water level observation wells should be set in nearby area, in order to adjust recharge amount in time.

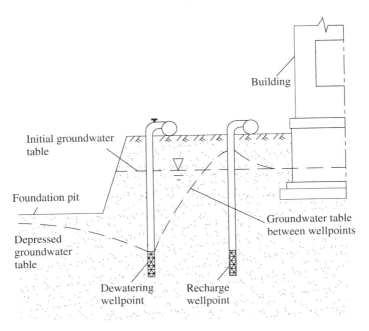

Fig. 5.22 Arrangement of recharge wellpoints

5.6 Monitoring of Wellpoint Dewatering

In an important project, a lot of monitoring equipments should be set.

5.6.1 Flow Observation

Flow observation is important, which usually can be done by flow recorder of triangular weir. If the flow is too large while the dewatering is very slow, it should be considered to change a larger rate centrifugal pump. On the contrary, replace the centrifugal pump by a small rate pump to prevent high temperature and save electric power. For example, in Shanghai, one project first adopted a centrifugal pump, whose diameter is 75 mm and power efficiency is 7.5 kW. However the flow is not very large, so it had been replaced by a small centrifugal pump, whose diameter is 75 mm and power efficiency is 3 kW. In another project, the former pump can work well.

5.6 Monitoring of Wellpoint Dewatering

5.6.2 Water Table Observation

The wellpoints also can act as observation wells. In first 3 days of dewatering, observe the water level every 4–8 h to learn the function of the entire system. After 3 days and before the water table dropping to the predetermined level, observe it 1–2 times per day. When the water table dropped to the predetermined level, the observation can be taken weekly. But in raining days, the observation time should be increased.

5.6.3 Pore Water Pressure Measurement

Pore water pressure measurement is to learn the changes of pore water pressure during dewatering, so that to estimate the foundation strength, deformation, and slope stability.

The pore water pressure is usually measured more than once time every day. If there is abnormal phenomenon, for example, pit slope cracking or nearby buildings subside largely; the observation time should be increased to more than 2 times per day.

5.6.4 Total Settlement and Layered Settlement Observation

In dewatering project, the benchmark should be set out of the influence range, in order to make the settlement observation more accurate. The settlement observation points should be set inside the influence range and close to nearby building. In multi-soil layers, if the dewatering depth is large, layered settlement observation points also need to be set, in order to measure every layer's settlement and check the calculated total settlement. The observation times are the same with pore water pressure measurement. Settlement observation should avoid direct sunlight and strong wind.

5.6.5 Earth Pressure Measurement

The earth pressure measurement includes subgrade reaction measurement and lateral earth pressure measurement. The measurement times are the same with pore water pressure measurement.

Other measurement devices, such as extensometer, concrete pressure capsule, resistance thermometer, piezometer, and high-precision inclinometer, are adopted depending on the design requirements.

5.7 Design Cases of Dewatering Projects

5.7.1 Ejector Wellpoint Case

The sectional view of an open caisson of one engineering project in Shanghai is shown in Fig. 5.23. The site elevation is +3.75 m. The foot blade of this open caisson is in mucky clay with thin silty sand layers and its elevation is −13.70 m. Water table is 2.05 m below the ground surface. The layers are identified as

(1) The first layer is isabelline silty clay between +3.75 to +2.50 m.
(2) The second layer is gray mucky clay between +2.50 to +0.60 m. Its hydraulic conductivity is 1.50×10^{-7} cm/s, thus can be considered as impermeable layer.
(3) The third layer is gray silty sand between +0.60 to −7.55 m. Its hydraulic conductivity is $(3.00–4.50) \times 10^{-4}$ cm/s, thus can be considered as confined aquifer.
(4) The fourth layer is gray clayed silt with thin silty sand layers between −7.55 to −11.55 m. Its hydraulic conductivity is 5.32×10^{-5} cm/s.
(5) The fifth layer is gray mucky clay with thin silty sand layers under −11.55 m. Its hydraulic conductivity is 1.02×10^{-7} cm/s.

According to the geological and hydrogeological conditions of the site, ejector wellpoint is adopted in order to prevent quicksand.

The wellpoints are circularly arranged according to site conditions, as shown in Figs. 5.23 and 5.24. The wellpoint is confined fully penetrating well. The hydraulic conductivity is taken the average value of third and forth layers, as 0.39 m/day. The steps of wellpoint design are as follows.

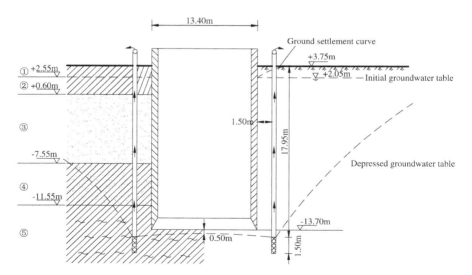

Fig. 5.23 Arrangement of ejector wellpoints

5.7 Design Cases of Dewatering Projects

Fig. 5.24 Plane arrangement of wellpoints

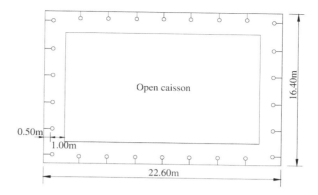

5.7.1.1 Arrangement of Wellpoints

Wellpoints are rectangularly arranged. The main tube is 1.5 m away from the open caisson. The length and width are 19.60 and 13.40 m, respectively. The hydraulic gradient I is 1/10.

(1) The length of main tube

$$[(19.60 + 1.50 \times 2) + (13.40 + 1.50 \times 2)] \times 2 = 78.00 \text{ m}$$

(2) The burial depth of wellpoint pipe

$$H = 17.45 + I \cdot L + \Delta h = 17.45 + \frac{1}{10} \times \frac{16.40}{2} + 0.5 = 18.77 \text{ m}$$

So, take the length of wellpoint pipe as 18.80 m.

(3) The length of filter pipe is 1.50 m and the diameter is 38 mm.

(4) The punched diameter of borehole is 600 mm and 1.00 m deeper than the filter bottom end, which means the depth of borehole is 18.80 + 1.50 + 1.00 = 21.30 m. The space between the walls of wellpoint pipe and borehole is filled by coarse sand as filter layer, and the part that 1.0 m under the ground is filled by clay in order to prevent gas leaking, as shown in Fig. 5.25.

5.7.1.2 Calculation of Pit Dewatering Amount

(1) The requested dewatering depth in the pit center is

$$s' = 13.70 + 2.05 + 0.50 = 16.25 \text{ m}$$

(2) The length of wellpoint pipe beneath the original water table equals to the drawdown in the wellpoint, which is

Fig. 5.25 Construction schematic of ejector wellpoint

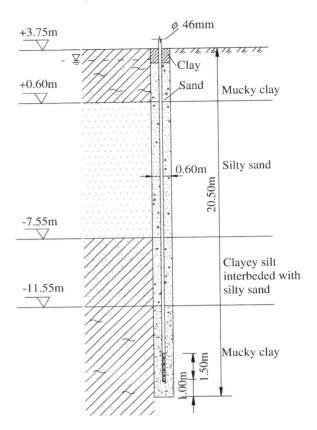

$$s = 16.25 + \frac{1}{10} \times \frac{16.40}{2} = 17.07 \text{ m}$$

(3) The confined water table is at the elevation of +2.05 m, and the elevation of third layer—silty sand layer—is +0.60 m. Thus, the confined water level is 1.45 m. The thickness of aquifer is 12.15 m.

(4) The radius of influence is

$$R = 10s\sqrt{K} = 10 \times 17.07 \times \sqrt{0.39} = 106.60 \text{ m}$$

(5) The reference radius is

$$r_0 = \sqrt{\frac{F}{\pi}} = \sqrt{\frac{22.60 \times 16.40}{3.14}} = 10.86 \text{ m}$$

(6) The total dewatering amount of the foundation pit is

$$Q = \frac{2\pi KMs}{\ln R - \ln r_0} = \frac{2 \times 3.14 \times 0.39 \times 12.15 \times 17.07}{\ln 117.46 - \ln 10.86} = 213.45 \text{ m}^3/\text{day}$$

5.7.1.3 Drainage Quantity of Single Wellpoint

The drainage quantity of single wellpoint is

$$q = 65\pi dl \cdot \sqrt[3]{K} = 65 \times 3.14 \times 0.038 \times 1.5 \times \sqrt[3]{0.39} = 8.5 \text{ m}^3/\text{day}$$

5.7.1.4 Number of Wellpoint Pipe

The number of wellpoint pipe is

$$n = 1.1\frac{Q}{q} = 1.1 \times \frac{213.45}{8.50} = 27.6$$

In this project, 28 wellpoints are arranged around the open caisson pit, as shown in Fig. 5.24.

The distance between wellpoints is

$$D = \frac{(22.60 + 16.40) \times 2}{28} = 2.80 \text{ m}$$

Note: During dewatering, it should be paid close attention to ground settlement to prevent ground subsidence issues. This ejector wellpoint design worked well in this project.

5.7.2 Tube Wellpoint Case

The foundation pit is for a three-hole box-type ferroconcrete culvert. This culvert has 19 parts and totally 316.00 m long. According to the landform, elevation, and hydrological and geological conditions, the tube wellpoint system is designed by parts. For one part, the ground elevation is +3.20 m and the designed elevation of pit bottom is −5.90 m. The required bottom and top excavation width are 18.00 m and 41.00 m, respectively. Design a row of wellpoint at both sides of the pit, which is 1.00 m away from the pit, as shown in Figs. 5.26 and 5.27. Thus, the distance between the two rows is 43.00 m.

The hydrogeological conditions are: at the top range of 7.00–10.00 m is silty clay and clayed silt, and the layer below is thick silty sand layer, and the pit bottom

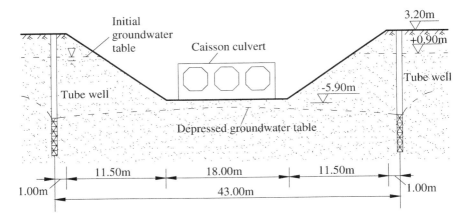

Fig. 5.26 Schematic of tube wellpoint dewatering

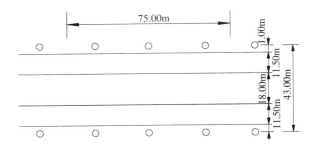

Fig. 5.27 Plane arrangement of tube wellpoints

is about at the boundary of these two layers. The hydraulic conductivity of the silty sand layer is about 2.60 m/day.

Take 75.00 m length of the pit as a calculation unit. The diameter of the filter pipe is 0.34 m. The hydraulic gradient is 0.3.

(1) The reference radius is

$$r_0 = \sqrt{\frac{F}{\pi}} = \sqrt{\frac{75.00 \times 43.00}{3.14}} = 32.04 \text{ m}$$

(2) The drawdown s in the well pipe is calculated as follows:

$$h_1 = 3.20 - 2.30 = 0.90 \text{ m}$$

$$h_2 = 0.90 + 5.90 = 6.80 \text{ m}$$

$$\Delta h = 0.5 \text{ m}$$

5.7 Design Cases of Dewatering Projects

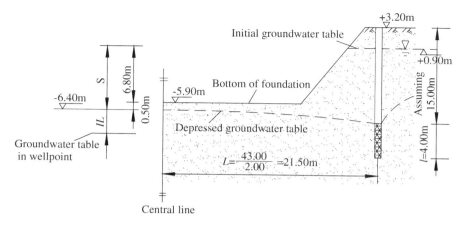

Fig. 5.28 Calculation schematic of tube wellpoint dewatering

$$L_1 = 43/2 = 21.50 \text{ m}$$

$$s = h_1 + h_2 + \Delta h + IL_1 = 0.90 + 6.80 + 0.50 + 0.3 \times 21.50 = 14.65 \text{ m}$$

So take the drawdown as 15.00 m.

(3) The calculation of effective depth.

For partially penetrated tube well in thick aquifer, the effective depth H_0 can be calculated as in Table 5.12.

$$\frac{s}{s+l} = \frac{15.00}{15.00 + 4.00} = 0.79$$

$$H_0 = 1.85(s+l) = 1.85 \times (15.00 + 4.00) = 35.15 \text{ m}$$

where l is the submerged depth of filter pipe when the water table is stable, which in this case is 4.00 m, as shown in Fig. 5.28.

(4) The drawdown in the pit center is

$$s' = 6.80 + 0.50 = 7.30 \text{ m}$$

The radius of influence is

$$R' = r_0 + R = \sqrt{\frac{F}{\pi}} + 1.95 s' \sqrt{H_0 K} = 32.04 + 1.95 \times 7.30 \times \sqrt{35.15 \times 2.60}$$
$$= 168.12 \text{ m}$$

(5) The total drainage quantity of the tube well system is

$$Q = \pi K \frac{(2H_0 - s')s'}{\ln R' - \ln r_0} = 3.14 \times 2.60 \times \frac{(2 \times 35.15 - 7.30) \times 7.30}{\ln 168.12 - \ln 32.04}$$
$$= 2266.12 \, \text{m}^3/\text{day}$$

(6) The maximum drainage quantity of single wellpoint is

$$q = 65\pi dl\sqrt[3]{K} = 65 \times 3.14 \times 0.34 \times 4 \times \sqrt[3]{2.60} = 381.88 \, \text{m}^3/\text{day}$$

where d is the diameter of filter pipe, which in this case is 0.34 m.

(7) The number of wellpoint pipe is

$$n = 1.1\frac{Q}{q} = 1.1 \times \frac{2297.44}{381.00} = 6.5$$

In this project, six wellpoints are arranged, three for each side. The distance between well pipes is 25.00 m.

(8) Check the drawdown of pit center. First of all, calculate the distances from all wellpoints to the pit center (Figs. 5.29 and 5.30).

$$r_1 = r_2 = r_3 = r_4 = 33.00 \, \text{m}$$

$$r_2 = r_5 = 21.50 \, \text{m}$$

$$\frac{1}{n}\ln r_1 r_2 \cdots r_n = \frac{1}{6}(2\ln 21.50 + 4\ln 33.00) = 3.35369$$

Calculate the water level from the effective region by Eq. (5.25), as shown in Fig. 5.29.

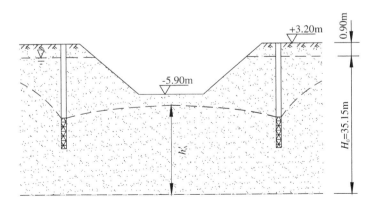

Fig. 5.29 The drawdown in the pit center

5.7 Design Cases of Dewatering Projects

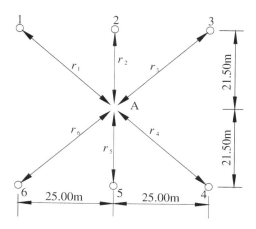

Fig. 5.30 The arrangement of wellpoint system

$$h'_A = \sqrt{H_0^2 - \frac{nq}{\pi K}\left[\ln R' - \frac{1}{n}\ln(r_1 r_2 \cdots r_n)\right]}$$

$$= \sqrt{35.15^2 - \frac{6 \times 381.88}{3.14 \times 2.60} \times \left[\ln 168.12 - \frac{1}{6}(33^4 \times 21.5^2)\right]} = 27.18 \text{ m}$$

The initial water level is $3.20 - 2.30 = 0.90$ m.
The effective elevation is $-(35.15 - 0.90) = -34.25$ m.
The top point elevation of h'_A is $-(34.25 - 27.18) = -7.07$ m
The distance of the top point of h'_A below the pit bottom is $7.07 - 5.90 = 1.17 > 0.50$ m.
Thus, this wellpoint design can meet the dewatering requirements.
If h'_A is calculated as

$$h'_A = \sqrt{35.15^2 - \frac{6 \times 381.88}{3.14 \times 2.60}(\ln(168.12 + 32.04) - 3.35369)} = 26.27 \text{ m}$$

then the top point elevation of h'_A is $-(34.25 - 26.27) = -7.98$ m.
The distance of the top point of h'_A below the pit bottom is $7.98 - 5.90 = 2.08 > 0.50$ m.
Thus, this wellpoint design can meet the dewatering requirements.
(9) The length of the wellpoint pipe is

$$H = 2.30 + 15.00 + 4.00 = 21.30 \text{ m}$$

In construction, the length of wellpoint pipe is taken as 27.00 m.
(10) The suction head of the sump pump is 18.00 m, which is larger than $15.00 + 2.30 = 17.30$ m.

The flow of this pump is 50 m³/h, which is larger than $\frac{381.88}{24} = 15.91$ m³/h. Thus, during dewatering, control the tap to adjust the drainage quantity. The pump with a flow of 25.00 m³/h is more suitable.

5.8 Common Issues of Wellpoint Dewatering Methods and Their Solutions

5.8.1 Light Wellpoint

The common issues of light wellpoint and their causes, prevention, and solutions are shown in Table 5.20.

Table 5.20 Common issues of light wellpoint and their causes, prevention, and solutions

Common issues	Causes	Prevention and solutions
Vacuum degree disorders: (a) Vacuum degree is small, the pointer of vacuum gauge shakes violently, and groundwater extraction amount is little (b) Vacuum degree is abnormally large, but the unusually large groundwater extraction amount is little (c) The water table does not drop down, instability of pit slope instability, and quicksand	(a) Bad sealing of wellpoint leads to large amount of air leakage (b) Abrasion or failure of dewatering unit components (c) Wellpoint filter screens, pipes, and water colleting tube are blocked by sand and clay, resulting in large reading of vacuum degree and little pumped water (d) The soil permeability is too small, improper choice of wellpoint type, or put the filter in the impermeable layer	(a) The pipes of wellpoint system should be strictly sealed in order to keep the vacuum degree larger than 93 kPa when idling. Check the sections one by one to find out the leaking point. Tightening bolts, covering with white painting or other material (b) Before stalling, clean the well pipes and remove rust and mud inside the pipe (c) The punched borehole should be 0.5 m deeper than the lower end of the filter pipe and the diameter should be larger than 30 cm. After completing installation, test pumping should be taken out in time to check the pump capacity and whether the pipe leaks. The issues like leaking or "dead well" should be solved in time (d) The methods for finding out whether the pipe is blocked include: feel the pipe temperature and it is not cool in summer and not warm in winter; there is no

(continued)

5.8 Common Issues of Wellpoint Dewatering Methods and Their Solutions

Table 5.20 (continued)

Common issues	Causes	Prevention and solutions
		moisture on the belt pipe; cannot listen the flow sound through a steel; cannot see water flow in transparent pipe; pull water into the pipe and no infiltration (e) Before excavation, punching the filter pipe with high-pressure water. Pull out the well pipe and clean it if needed
Water turbidity: (a) The pumped water is not clear, with lot of sand (b) The settlement of nearby ground is large	(a) Filter screen damage (b) The holes of filter screen are too large or the grain size of sand filling layer is too large, so that the fine particles of aquifer can flow away (c) The thickness of sand filter layer is not enough	(a) Before put in the well pipe, check the filter screen carefully and repair the damaged filter screen in time (b) The size of the filter screen and filled sand should be suitable for the soil conditions (c) The wellpoint should not be used if the pumped water is always turbid
Partial abnormities of dewatering: (a) Quicksand happens in some place in the pit (b) Cracks occur on the wall of the pit	(a) Many wellpoints are blocked or the vacuum degree is too small on the unstable side of the pit (b) The water in the nearby river or drainage ditch infiltrates into the pit, making the water head increased (c) The overloading on nearby ground or mechanical vibration may cause ground cracks or collapses, which will further lead to quicksand or slope failure	(a) The wellpoint system must be sealed strict, and the vacuum degree should be greater than 93 kPa during idle running (b) Arrange more wellpoints on the side with more water recharge. Prohibit digging drainage ditch near the pit slope (c) Avoid overloading on nearby ground or heavy mechanical vibration (d) Seal the ground cracks and drainage away the surface water from the pit (e) Take necessary measures to prevent further damage

5.8.2 Ejector Wellpoint

The common issues of ejector wellpoint and their causes, prevention, and solutions are shown in Table 5.21.

Table 5.21 Common issues of ejector wellpoint and their causes, prevention, and solutions

Common issues	Causes	Prevention and solutions
The failure of pump: (a) Abnormal pressure difference (b) Water gushing and surface sand boiling occurs near the wellpoint (c) Some soil is wet unmorally, and the slope of the pit is unstable	(a) The nozzle is clogged by debris, thus there is no or just little reading on the pressure gauge when closing the wellpoint (b) The nozzle is worn even perforated, and the nozzle splint is cracked, thus the reading on the pressure gauge is very large when closing the wellpoint	(a) Check the pump quality strictly, especially for concentricity and weld quality, and measure the performance. The vacuum degree should be larger than 93 kPa when idling (b) Before stalling, clean the well pipes and remove rust and mud inside the pipe (c) The working water should be kept clean. After 2 days test pumping, it should be changed. During dewatering, the water should be regularly replaced depending on the degree of turbidity. The working pressure should be adjusted to meet the requirements of dewatering, in order to reduce the wear of the nozzle (d) If the nozzle is blocking, it should be purged in time. First, close the wellpoint, loosen the fastener, and pull the pipe lightly. Then click the inner pipe to let the stemming fall into the sediment pipe. If the stemming is too tight to fall, full out the inner pipe and then remove it (e) If the nozzle splint is cracked, or worn even perforated, pull out the inner pipe and change the nozzle
Wellpoint blocking: (a) The water pressure is normal, but the vacuum degree is higher than the standard value (b) Pull water into the pipe and no infiltration	(a) After filling sand layer around wellpoint pipe, the test pumping is not carried out in time, making the mud and sand precipitated in well pipe and blocking the suction port of the inner pipe	(a) After installation completing, proceed test pumping in time (b) The filter pipe is better buried in the layer with large hydraulic conductivity. If necessary, enlarge the diameter of sand filter layer,

(continued)

5.8 Common Issues of Wellpoint Dewatering Methods and Their Solutions

Table 5.21 (continued)

Common issues	Causes	Prevention and solutions
(c) Some soil is wet unmorally, the pit slope is unstable, and quicksand occurs	(b) The hole necks or collapses during putting the well pipe, or the clay layer is not handled properly, thus the filled sand layer quantity is not good and the filter screen easily been blocked by mud	deepen the borehole or add sand wells (c) The borehole should be vertical. Its diameter should be no less than 400 mm, and it should be 1.0 m deeper than the end of filter pipe. Before pulling out the punching pipe, turn off the high-pressure water gun to avoid borehole collapse (d) If the filter pipe is blocked by sand and mud, pull out the inner pipe a little, and push water into the filter pipe through the annular space between the inner and outer tube to wash it and drain the muddy water from the inner pipe. The reverse way all can work (e) If the filter screen or sand filter layer is blocked by mud, push water into the inner pipe to wash them. The time for washing is about 1 h. After stopping, the suspended sand gradually deposited around the wellpoint filter pipe to recompose the filter layer (f) If the filter pipe is buried not deep enough, add some sand wells according to specific conditions, in order to improve the permeability of the soil layer. Also can bury another filter pipe in better permeable layer, or pull out the wellpoint and rebury it

(continued)

Table 5.21 (continued)

Common issues	Causes	Prevention and solutions
General malfunctions: (a) Water pulls out around the wellpoint (b) The pressure of working water is not high enough, so that the vacuum degree is little (c) The short connecting pipe for returning water cracks (d) The water level of circulating pool continuously drops	(a) Failure of pumping machine, inner pipe base leaking due to bad sealing or fastener looseness, the leaking on the joint position of inner and out pipes, and very low working water pressure all can lead to water pull out (b) The well pump is overburdened, or the sediment in circulating poor blocks the pump suction port, thus the water pressure cannot raise up and the vacuum degree is very small (c) The short pipe burst is caused by incautious operation (d) The location of circulating poor is too close to the pit, which makes it easily cracks due to ground settlement. The water in circulating poor will leak into the ground	(a) Check the pump quality strictly, especially for concentricity and weld quality, and measure the performance. The vacuum degree should be larger than 93 kPa when idling. Before installation, check the inner pipe base carefully and fasten the screws (b) If the water flows back, turn off the wellpoint immediately and then find out the reason. Check all parts successively according to the reading of the pressure gauge and in the order of easy issue first (c) If the pumped water quantity is not enough, add more pumps and clear the sediments in circulating poor. Find out the reason of large amount of sediment (d) When the short pipe burst, turn off the wellpoint immediately and change a spared one. To turn on the wellpoint, turn on the backwater valve first then the inflow valve (e) If the circulating poor crack, reinforce it and plug the leaking position, or replace it by circulating tank if necessary

5.8.3 Tube Wellpoint

The common issues of tube wellpoint and their causes, prevention, and solutions are shown in Table 5.22.

5.8 Common Issues of Wellpoint Dewatering Methods and Their Solutions

Table 5.22 Common issues of tube wellpoint and their causes, prevention, and solutions

Common issues	Causes	Prevention and solutions
The drainage amount is small, however the pump capacity is enough	(a) The well is not washed carefully, thus the mud content of sand filter layer is relatively high. The mud cover on the borehole wall is not cleared away, therefore the groundwater cannot flow into the well smoothly, which badly influences the drainage capacity of single well (b) Hydrogeological data does not match with the actual situation; the actual location of the filter pipe is not at the aquifer with good permeability (c) The depth, hole diameter, and verticality do not meet the requirements; there are excessive sediments in wells, making the borehole blocked	a) Wash the well in time after filling the sand layer. Then carry out the single well pump test to wash out the mud that in around layer. b) The filter pipes should be installed in all aquifers that need to be drained. The specification of filter screen and sand filter layer should be chosen based on the aquifer's grain size. c) For the complex layers or the layers without enough hydrogeological data, specialized drilling should be carried out according to dewatering requirements. For critical engineering project, field dewatering test should be done. In the drilling process, take soil samples from every borehole and recheck the original hydrogeological data. Before putting in the well pipe, measure the borehole depth again, and collocate the well pipe and filter pipe according to design requirements and actual hydrogeological data. d) Before installing or exchange pump, measure the actual borehole depth and the thickness of sediment on the bottom of the well. If the well is not deep enough or the sediment is too thick, wash the borehole before installation.
The drawdown is not enough	(a) The number of wellpoint is not enough (b) The pump is unsuitable that the pumping capacity is scarce (c) The wellpoint is not brought into full play (d) The hydrogeological data is incorrect, and the actual drainage amount is larger than the calculated value	(a) According to hydrogeological data, calculate the total flow quantity, the flow capacity of single well, the length of filter pipe, the number of wellpoints, the distance between wellpoints, etc. Recheck the length of filter pipe, the flow capacity of single well, and the required drawdown (b) The pump should meet all requirements for different dewatering stage

(continued)

Table 5.22 (continued)

Common issues	Causes	Prevention and solutions
		(c) Improve the flow capacity of single well. Set a suitable filter pipe and sand filter layer (d) Add more wells in the region that the drawdown does not meet the dewatering requirements (e) Change a bigger pump within the single well drainage capacity (f) Rewash the well if the former time is unqualified to improve the single well drainage capacity

5.9 Impact of Wellpoint Dewatering on the Environment and the Prevention

When dewatering starts, the well water level gradually drops, and the around water continuously flows to filter pipe. In unconfined aquifer, after a period of time dewatering, a conic water table will occur, and the stabilization of which normally takes a few days. The water table decrease will definitely result in ground settlement, which is differential. This differential settlement development requires a certain period of time. In engineering projects, due to poor structure of filter screen and filtering sand layer, the clay, silt, and even fine sand particles may flow away. This will lead to differential settlement and damage the ground buildings and underground pipelines with varying degrees. This damage can be minimized by raising the dewatering efficiency.

5.9.1 Ground Deformation Near a Dewatering Wellpoint

In California Santa Clara River Valley, 190 times of dewatering and recharge test had been carried out in 16 wells with different depth. Vertical and horizontal ground deformations had been measured. The depth of these wells vary from 5.00–10.00 m to 500.00–700.00 m. The pump flow of former wells was small, about 25 L/min. For the later ones, it could be 700 L/min. The aquifer consists of silt and clay with gravels. In short time dewatering (2 h or less), the ground deformation around every well could be measured immediately. The vertical deformation varied from 1–100 μm and the horizontal deformation could be 15 μm per meter. Since the deformation of the ground was quickly and almost elastic, the observed deformation may be caused by the compression of coarse particles in aquifer.

According to detailed observation data, the deformation near the dewatering well is generally the compressive displacement and the ground surface is concave

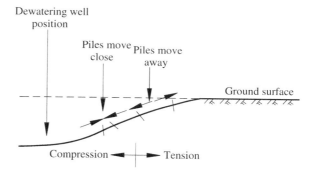

Fig. 5.31 Ground deformation around dewatering well

upward. The deformation a little far away from the dewatering well is generally the stretched displacement and the ground surface is convex upward. Between them, there is an antibending zone, in which the displacement decrease a little and the stretched displacement will turn into compressive displacement in a few minutes. The antibending zone will move outward when the drained water amount is relatively large. Figure 5.31 shows the ground deformation around dewatering well in short time dewatering. The distances from well to the outer edge of the concave band are different, which vary from 100.00 m to hundreds of meters. These distances mainly depend on the burial depth of the aquifer, groundwater seepage velocity, flow and the elastic characteristics of the aquifer and the overlying layer. The settlement area around dewatering well is usually a circular or elliptic caulbron. The difference between aquifer's thickness and lithology makes it not necessarily of radial symmetry.

In engineering construction, wellpoint dewatering is usually used to eliminate the threats caused by groundwater. However, the dewatering will lead to ground settlement, ground cracking, underground pipeline rupture, nearby buildings cracking, interior floor collapse, which will affect the normal production. Therefore, in wellpoint dewatering design, this phenomenon should be fully considered, and appropriate measures should be taken.

In one engineering project, ejector wellpoint is used to dewater. The length of the wellpoint is 21.00 m, average distance between wellpoints in one row is 2.00 m, and the distance between two rows of wellpoint is 13.30 m. The filter pipe is laid 2.00 m under the ground surface in confined aquifer. The layout is shown in Fig. 5.32a. The dewatering demand is to lower the water level 18.00 m below the surface. After 1–20# wellpoint dewatering 24 h, the water level in observation well 1# and 2# is 19.00 m below the surface. 84 h later, large amount settlement occurs. The settlement of 5# settlement observation point is about 90 mm (Fig. 5.32b), and the largest settlement occurs in 3# settlement observation point, where it's 133 mm, making the nearby interior floor and brick walls cracked. To reduce the amount of ground settlement caused by dewatering, recharging while dewatering method has been taken. Wellpoint 21–29# are used as recharging well, and the total recharging amount of them is 6.00 m^3/h. The total dewatering amount of 1–20# wellpoint is

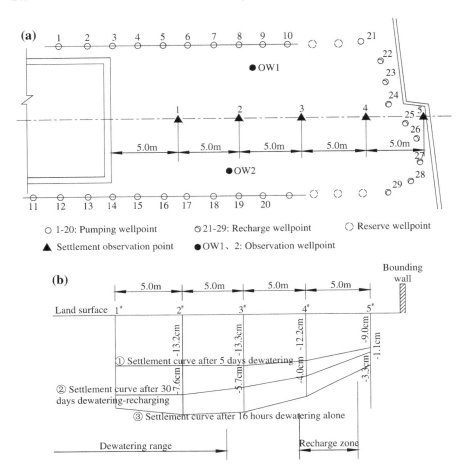

Fig. 5.32 Dewatering program **a** Layout of wellpoints. **b** Settlement observation

9.00 m³/h. Thus, the actual amount of pumped water is 3.00 m³/h, 6.00 m³/h less than that before recharging. Through observation, groundwater level can be maintained at a position 18 m below the surface, and the settlement has been significantly controlled. During 30 days recharging, the settlement of 5# settlement observation point is about 11 mm, and the maximum settlement of 2# settlement observation point (inside the dewatering area) is about 76 mm.

To further verify the effect of recharging, another 16 h dewatering with original wellpoints had been carried out. The ground settlement near the enclosure is 33 mm, which is three times of the settlement in last 30 days. So recharging during dewatering can significantly reduce the settlement. The effect of recharging is mainly controlled by ejector rate. Too much injected water may low the performance of dewatering, while too little injected water may not control the settlement well.

5.9 Impact of Wellpoint Dewatering on the Environment and the Prevention

Therefore, the settlement areas of groundwater extraction have the following characteristics:

(1) The settlement occurs after a certain time of dewatering.
(2) The water body in the formation could have been in a relatively closed condition, with considerable pressure. After draining of part of it, the pressure drops.
(3) The age of affected stratums are generally not earlier than Paleogene, which means they are unconsolidated.
(4) The settlement time, range, and degree are corresponding to those of water pressure decrease, respectively.
(5) In some major ground subsidence area, the layer compaction caused by extraction of groundwater mainly occurs in unconsolidated and semiconsolidated loose sediments of late Cenozoic, most of which are alluvial and lacustrine layers.
(6) From the ground settlement observations and tests in dewatering area, it can be concluded that the settlement is resulted from the layer compaction caused by water pressure decrease. However, in which layer does the compaction occur, the permeable aquifer or the relatively impermeable saturated clay layer? In 1959, Polland in America did consolidation tests of medium sand, silty sand, and clay, and found that the settlement mainly occurs in silty sand layer and clay layer. Other researchers also have drawn the same conclusion that the consolidation of saturated clay layer is the major reason for ground settlement. Sand compression also has some impact.

After years of research on Shanghai subsidence, ten factors that may cause subsidence can be summarized: rising sea level, new tectonic movement, static load, dynamic load, natural gas extraction, mining of groundwater, underground digging, deep sand extraction, artificial filling, and dredging of Huangpu river. After a comprehensive discussion, excessive exploitation of groundwater is considered as the dominant external factor, and the compressible saturated clay layer is the dominant internal factor. Static and dynamic loads, underground digging and rainfall are secondary factors.

5.9.2 Mechanism of Settlement Caused by Dewatering

5.9.2.1 The Basic Mechanical Effects on Soil Layer During Dewatering

Settlement caused by dewatering can also be calculated by Terzaghi effective stress principle and consolidation equation.

$$\sigma = \sigma' + u$$

where σ is the total stress, kPa; σ' is the effective stress, kPa; u is the pore-water pressure, kPa.

When draining the aquifers above or below the saturated clay aquitard, the water pressure will drop, while the total stress of soil layer will remain constant, the little change that caused by groundwater diffusion and moisture transfer can be ignored. The decrease of pore-water pressure will lead to the increase of effective stress, and further lead to the compaction of soil layer. Because the sand aquifer has good permeability, in this process, its effective stress will increase with the decrease of water pressure. Therefore, the aquifer can be regarded as elastomer and the compaction takes place instantaneously. If the water pressure regained, most of its compaction will recover. This compaction is generally small.

In saturated clay layer that can be referred as aquitard or aquifuge, the vertical penetration and the change of pore water pressure is extremely slow. This makes the estimation or prediction of the aquifer's compaction more complex. The increased effectiveness generates two kinds of mechanical effects, which are the changing of uplift force between soil particles caused by groundwater level fluctuation and the osmotic pressure generated by the change of confined water head (Fig. 5.33).

1. Uplift force

Dewatering in the aquifer above the aquitard could easily lead to uplift force decrease. In accordance with the different above boundary conditions, two cases may occur:

(1) Disappearance of uplift force

The disappearance of uplift force is due to the decrease of water level, which makes the buoyant unit weight replaced by saturated unit weight or wet unit weight. This part of the weight difference is the effective stress of the soil, its value is

$$\Delta\sigma' = \gamma_\mathrm{w} \Delta h \tag{5.43}$$

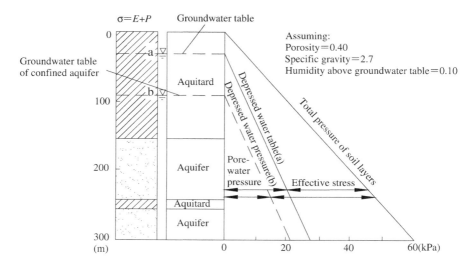

Fig. 5.33 Stress change during water head decrease in confined aquifer

5.9 Impact of Wellpoint Dewatering on the Environment and the Prevention

where $\Delta\sigma'$ is the increase of effective stress after dewatering, kPa; Δh is the thickness of drained aquifer, m.

Or

$$\Delta\sigma' = \frac{(1+eS_r)}{1+e}\gamma_w\Delta h \tag{5.44}$$

where S_r is the saturation degree of soil; e is the void ratio; γ_w is the unit weight of water, kN/m^3.

The disappearance of uplift force usually occurs in the compression layer (aquitard) below saturated sand layer. During shallow wellpoint dewatering, the decrease of phreatic water table will lead to ground settlement because of the disappearance of uplift force.

(2) Reduction of uplift force

During dewatering, the pore-water pressure of up boundary decreases, transferring the load from pore water to the oil skeleton as effective compression load. This is equivalent to add an extra load on the top of the compression layer. This situation occurs when the compression layer is under a thin sand layer that can be regarded as aquitard. Dewatering cannot directly cause the change of unit weight of soil, but it can lead to the reduction of uplift force. Equation (5.44) is still available for the calculation of the decrease. However in this case, $S_r = 1$, so

$$\Delta\sigma' = \frac{(1+e\times 1)}{1+e}\gamma_w\Delta h = \gamma_w\Delta h \tag{5.45}$$

The meanings of the symbols are the same with the former.

Equation (5.45) shows that the maximum effective stress of compression layer caused by uplift force is the product of water level change and the unit weight of water.

2. Osmotic pressure

The decrease of water pressure in aquifer results in the water pressure difference between the boundaries of aquifer and saturated clay layer (aquitard), and breaks the balance of the original pore-water pressure. In this case, the pore-water in clay layer will penetrate into sand aquifer until the water pressure reaches a new balance. This pressure that is applied to the clay particles skeleton is called osmotic pressure or hydrodynamic pressure. Osmotic pressure is a body force with directionality. When the water head decreases by Δh, the average osmotic pressure within the range of compression is

$$\Delta P = \frac{\gamma_w\Delta h}{2} \tag{5.46}$$

The meanings of the symbols are the same with the former.

Osmotic pressure also is the average increment of effective stress on the entire clay layer.

5.9.2.2 Effect of Osmotic Pressure on the Process of Soil Compaction

There is a clay layer with thickness of H. The pore-water pressure distribution with depth is shown in Fig. 5.34 as ABCD. When the dewatering of the underlie layer begins, the water pressure CD suddenly decreases to CE, and the water pressure AB remains unchanged. Therefore, the water pressure of planes that is represented by b, c, d, and e cannot decrease in time, and a large hydraulic gradient is formed.

When the water of plane a discharges, the pore-water pressure distribution between plane a and plane b is uneven, leading to pore water draining from plane b to plane a and soil layer compression. Then the uneven pore water pressure distribution between plane b and plane c leads to pore water draining from plane c. In a similar way, the pore water pressures of plane d and plane e decrease with the increase of effective stress. The pore water drains gradually to the CD direction with time and the soil layer is gradually compacted.

Theoretically, when the time reaches t_∞, pore water pressure distribution will approach to AE, reaching a new balance state. The hydraulic gradient of each point is constant, so that no water is discharged, and the compaction effect of osmotic pressure on soil is also temporarily ended.

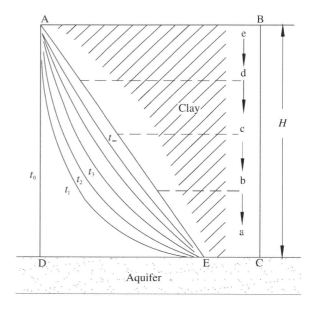

Fig. 5.34 Pore-water pressure change of clay layer during one-side dewatering

5.9.3 Impact of Changes in Groundwater Level on Soil Deformation

In loose or semiconsolidated marine or continental sediment alluvium or diluvium, aquifers are usually formed by coarse sand, medium sand, and fine sand, while aquitard or aquifuge are formed by clay layers between sand layers. If dewatering in these layers is on a larger scale and in a very long time, the confined water head will drop and the regional cone of depression will form. The decrease of confined water head will result in the decrease of pore water pressure of sand aquifer and of the clay layers above and below it, which will further lead to the increase of effective stress so that these layers will be compressed. This will ultimately lead to ground settlement. Apparently, the stratum structure, lithofacies characteristics, dewatering last time, and the value of water level decrease determine the settlement range, degree, and rate of that area. In most cases, the settlement is nonelastic and permanent.

According to Terzaghi one-dimensional consolidation theory, the completion of the consolidation process and the adjustment of confined water head in the fine-particle soil layer will take some time. The consolidation time t is determined by the following formula:

$$t = \frac{T_v H^2}{C_v} \text{ (year)} \qquad (5.47)$$

where T_v is the time factor, year; H is the thickness of aquifer (in m), when the aquifer is drained for two sides, it is $H/2$; C_v is the coefficient of consolidation or hydraulic conductivity:

$$C_v = \frac{K'}{\mu_s} \left(\text{cm}^2/\text{year}\right) \qquad (5.48)$$

where K' is the hydraulic conductivity of saturated clay soil layer, m/day; μ_s is the specific storage. According to Ukraine Waldron,

$$\mu_s = m_v \gamma_w \left(\text{m}^{-1}\right) \qquad (5.49)$$

where m_v is the coefficient of volume compressibility, kPa^{-1}; γ_w is the unit weight of water, kN/m^3.

And

$$\mu_s = m_v \gamma_w = \frac{a_{0.1-0.2}\gamma_w}{1+e} = \frac{\gamma_w}{E_{0.1-0.2}} \qquad (5.50)$$

where $E_{0.1-0.2}$ is the volume compression modulus, kPa; $a_{0.1-0.2}$ is the compression coefficient, kPa^{-1}; e is the void ratio.

μ_s is the ratio of water density to the volume compression modulus. According to Eq. (5.48), C_v is proportional to soil hydraulic conductivity. The lower the hydraulic conductivity is, the smaller the value of C_v is. Therefore, the compression rate in silty clay and clay layers depends on the pore water discharge rate. The time required for complete consolidation in thick clay layer often takes hundreds of years. A thin layer of clay soil also needs decades or longer. The main compaction process of sand and silty sand aquifers can be completed in a relatively short time after the water head dropping.

5.9.4 Differences Between Load Consolidation and Osmotic Consolidation

In 1925, Terzaghi raised the consolidation theory of soil. It was first applied to estimate the consolidation caused by ground loading (load consolidation), and later applied to estimate the consolidation caused by dewatering (osmotic consolidation). The later can be estimated in a relatively deep range according to the former method. However, there are some differences between them.

5.9.4.1 Features of Load Consolidation are as Follows

(1) Area of loading is small, and the stress decreases with depth.
(2) The load gradually increases since the beginning of construction, and then remains unchanged.
(3) For compressible fine-particle layer with low permeability, the increased stress is initially borne by the pore-water pressure, which is similar with the process of standard loading.
(4) During loading, the ultrahydrostatic pressure is generally allowed to fully dissipate until balance, so that the effective stress can be basically up to the final value if ignoring the secondary consolidation.
(5) The main consolidation layers are relatively less, so that enough undisturbed soil samples can be obtained for laboratory tests, and the change process of pore-water pressure can be easily observed in natural soil layers.

5.9.4.2 Features of Osmotic Consolidation are as Follows

(1) The area with significant settlement generally extends to around tens of meters, even up to several hundred meters in large-scale dewatering area.
(2) The load generally increases during whole construction period, and its value also often changes, the possible maximum can be 1–2 kPa. The change of stress that is caused by seasonal water level change may reach several times to the average annual increase in value.

(3) The total soil stress is unchanged except in the condition that water lost due to compaction. However, if the water level changes, soil stress will also change.
(4) The seepage pressure is referred to the pressure change caused by confined water head decrease and phreatic water level drop.
(5) Because the water head fluctuates over time, in aquitard pore-water pressure is hard to achieve equilibrium with the adjacent aquifer.
(6) If the entire compressed stratum is thick and with a lot of aquitard, for example, in one of the Shanghai projects the drawdown is more than 30 m, its vertical and horizontal permeability vary considerably. Therefore, horizontal and vertical undisturbed soil samples should be taken for laboratory tests, and different piezometers should be placed to observe the changes in pore-water pressure, which generally cost a lot.

5.9.5 Relationship Between Settlement Rate and Groundwater Pressure

The research in Japan and America shows that:

(1) There is a linear relationship between the settlement rate in clay soil layer and the groundwater pressure of aquifer, as shown in Fig. 5.35:

$$U = k(H - H_0) \qquad (5.51)$$

where U is the settlement rate, mm/d; H_0 is the initial water level, m; H is the average water level during observation period, m; k is the slope of the correlation curve.

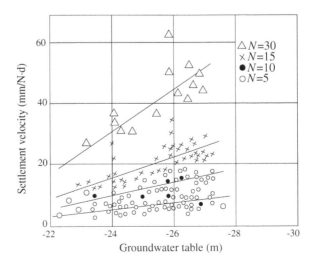

Fig. 5.35 Settlement rate versus water level curve, Kameido, Tokyo

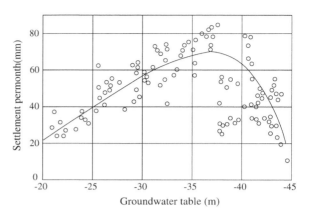

Fig. 5.36 Relationship between month settlement rate and water level

When water level is lower than a certain limit, the slope of the correlation curve will decrease, and the above relationship no longer meets the linear correlation, as shown in Fig. 5.36. This is due to the confined aquifer pressure decay. The soil stress will lead to the rearrangement of soil particles, with the change of stress–strain constant. So when the water table is reduced to a certain level, month settlement rate will decrease.

(2) The total ground settlement is calculated by adding layered settlement of different soil layer with different thickness. For each layer, its compression rate is related to the total water level decrease amount $(H - H_0)$ and the water level difference ΔH for the same period. The total compression rate is

$$U = \sum_{i=1}^{n} a_i(H - H_0) + \sum b_i \Delta H \qquad (5.52)$$

where a_i, b_i are characteristic coefficients assigned to the ith layer. Other symbols have the same meaning with the former.

(3) The relationship expressed by former two formulas is not reversible. That is every time the reduction of groundwater level will lead to the consolidation of soil layer, however the recovery of water level will not lead to the same resilience amount as consolidated. In most cases, the recovery of water level just stops or slows down the ground settlement. The explanation for this phenomenon is: the soil is a plastic or viscoelastic material that its deformation includes not only the relative displacement of soil particles but also the change of soil microstructure and the rearrangement of soil particles. Thus, under a certain stress, the settlement not only occurs in shallow saturated soft clay layer, but also occurs in the deep hard clay layers and sand layers.

5.9 Impact of Wellpoint Dewatering on the Environment and the Prevention

5.9.6 Calculation of Wellpoint Dewatering Influence Range and Ground Settlement

The calculation of wellpoint dewatering influence range and ground settlement can be learnt from the existing examples of similar projects, or adopt some simple estimate methods.

5.9.6.1 Calculation of Wellpoint Dewatering Influence Range

The radius of dewatering influence range R can be estimated by Coosa-King formula.

The radius of influence is greatly influenced by the stratified soil layers. In soft soil area, the radius of influence can achieve up to 84 m in sandy silt soil layer. The radius of dewatering influence in important project should be determined by dewatering tests.

5.9.6.2 Calculation of Ground Settlement

Calculation methods of ground settlement caused by dewatering are basically of three types:

(1) Using classic formula of consolidation theory.
(2) Using the stress–strain relationship and correlation.
(3) Using semiempirical formula.

Here are several formulas and methods that adopted domestic and overseas before.

1. Settlement calculation of clay soil

(1) In Tokyo Japan, one-dimensional consolidation theory is adopted to calculate the total settlement amount and predict settlement amount in a number of years, shown in Fig. 5.37. The basic form of this formula is

$$S_{1+2} = H_0 \frac{C_c}{1+e_0} \lg \frac{P_0 + \Delta P}{P_0} \quad (5.53)$$

where S_{1+2} is the total settlement amount (in m), including primary and secondary consolidations; e_0 is the void ratio before consolidation; C_c is the compression index of soil; P_0 is the vertical effective stress before consolidation, kPa; ΔP is the increment of vertical effective stress after consolidation, kPa; H_0 is the thickness of the soil before consolidation, m.

In practical engineering project, the calculated settlement is close to the observed settlement value.

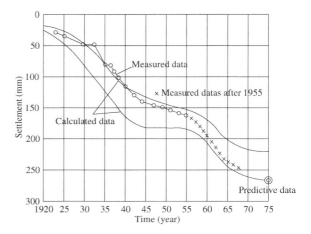

Fig. 5.37 Settlement prediction curve of Mimamisuna Machi, Tokyo

(2) In Shanghai area, one-dimensional consolidation formula is used to calculate the settlement, which is superposition of the settlements generated by different water pressures. For the calculation parameters, first take test data as reference and revise it by trial method, then backcalculate it by field-measured data. The main steps are as follows:

(1) Analyze the stratum structure of subsidence area, group them according to engineering geological and hydrogeological conditions, and then determine the settlement layer and stable layer.

(2) Draw measured and predicted curves of groundwater level changes over time (months or 10 days).

(3) Calculate the final settlements S_∞ (cm) of every water head difference.

$$S_\infty = \sum_{i=1}^{n} \frac{a_{0.1-0.2}}{1+e_0} \Delta PH \tag{5.54}$$

or

$$S_\infty = \frac{\Delta P}{E_{0.1-0.2}} H \tag{5.55}$$

where e_0 is the initial void ratio; H is the thickness of calculated soil layer, m; ΔP is the stress increment of soil layer caused by water level change, kPa; $a_{0.1-0.2}$ is the compression coefficient (in kPa^{-1}), for the condition of recharge, replace it as resilience coefficient a_s; $E_{0.1-0.2} = \frac{1+e_0}{a_{0.1-0.2}}$ is the compression modulus (in kPa), for the condition of recharge, replace it as resilience modulus $E'_s = \frac{1+e_0}{a_s}$.

5.9 Impact of Wellpoint Dewatering on the Environment and the Prevention

(4) Calculate the settlement S_t (mm) of every water level difference as selected time lag (months or 10 days):

$$S_t = u_t S_\infty \tag{5.56}$$

where $u_t = f(T_v)$ is the consolidation degree, for different stress situation, there are different approximate solutions:

For rectangular stress distribution (infinite uniform load)

$$u_{t,0} = 1 - \frac{8}{\pi^2} e^{-\frac{\pi^2}{4} T_v} \tag{5.57}$$

For triangular stress distribution

$$u_{t,1} = 1 - \frac{32}{\pi^4} e^{-\frac{\pi^2}{4} T_v} \tag{5.58}$$

where T_v is the time factor.

(5) Superimpose the settlement or resilience value of every water level different by a month or 10 days will get the total settlement or resilience value in that time, and the settlement versus time curve can be drawn.

(3) Back calculate the parameters according to stress–strain relation.

In America, Lyle calculated the settlement of a certain dewatering area in the central North American continent according to measured stress–strain relation on the basis of one-dimensional consolidation theory.

(1) Back calculate the unit storage μ'_s of aquifers according to measured data.

$$\mu'_s = \frac{S_{ke}}{m} \tag{5.59}$$

where m is the thickness of aquifer, m; $S_{ke} = \frac{\Delta m}{\Delta h}$ is the unit deformation of soil layer in elastic stage, cm/m; Δh is the water level change, m; Δm is the soil deformation (compression or resilience), cm.

(2) Calculate the coefficient of volume compressibility m_v or coefficient of resilience m_s of aquifers.

$$m_v = \frac{S_s}{\gamma_w} \tag{5.60}$$

Lyle backcalculated the m_v of whole aquifers of California Pixar area.

$$m_v = (3.6 \sim 5.9) \times 10^{-3} \, (\text{kPa}^{-1})$$

For aquitard

$$m_v = 7.5 \times 10^{-3} \, (\text{kPa}^{-1})$$

Predict the settlement according to m_v based on the previous principles.

2. Settlement calculation of sand soil layer

Sand aquifer generally has good permeability, in which deformation can be completed in a short time, so there is no need to consider the hysteresis. Settlement can be calculated by the elastic deformation formula.

One-dimensional consolidation formula is

$$\Delta S = \frac{\gamma_w \Delta h}{E_{0.1-0.2}} H \qquad (5.61)$$

where ΔS is the deformation of sand soil layer, cm; Δh is the change of water level, m; H is the initial thickness of aquifer, m; $E_{0.1-0.2}$ is the compression modulus of aquifer, kPa.

3. During dewatering, soil layer deformation usually may not occur in the soil that blow the dewatering table significantly, but will occur in the soil between the initial water table and the dewatering table, because here the added weight stress will soon lead to soil layer settlement due to good drainage condition. This part deformation is the main settlement caused by dewatering. Thus the settlement can be estimated by the following formula:

$$S = \Delta P \Delta H / E_{0.1-0.2} \qquad (5.62)$$

where ΔH is the drawdown, m; ΔP is the additional weight stress caused by dewatering (in kPa), which can be calculated as $\Delta P = \frac{\Delta \bar{H} \gamma_w}{2}$, and $\Delta \bar{H} = \frac{1}{2} \Delta H$; γ_w is the unit weight of water, g/cm^3; $E_{0.1-0.2}$ is the compression modulus of aquifer (in kPa), the value can be determined by soil test data or foundation code of Shanghai area.

For example, the wellpoint dewatering is used in the perpendicular shaft excavation in Pudong Tangqiao, Shanghai. This area is silty sand layer, where $E_{0.1-0.2} = 4$ MPa, drawdown is $\Delta H = 12$ m, $\Delta P = \frac{\frac{1}{2}\Delta H \gamma_w}{2} = 30$ kPa, so the calculated settlement is $S = \frac{\Delta P \Delta H}{E_{0.1-0.2}} = 9$ cm. The measured real settlement is 8.4 cm in 70 days.

4. Estimate impact of deep well dewatering on the environment.

The drawdown of deep well can be larger than 15.00 m, and the filter pipe is usually set in the sand soil layer, whose hydraulic conductivity is greater than 10^{-4} cm/s. The suction port of deep well pump is appropriately 1.00 m higher than well bottom and 3.00 m lower than the dynamic water level in well.

The purpose of the deep wellpoint dewatering in soft soil area mostly is to reduce the confined water head in deep sandy soil. Its impact on the environment depends largely on the distribution of soil, which can be estimated by following basic principles:

(1) When there is a hard clay layer on top of dewatering sand soil layer, it can be regarded as closed boundary state to calculate the settlement, which is only considered the settlement of this sand layer. Engineering practice shows that in this case, the impact of deep wellpoint dewatering on the environment is relatively small.

(2) The settlement of dewatering sand layer can be calculated by Eq. (5.62). Because this layer is generally deep, its compression modulus usually can be more than 100 MPa. If the aquifer's thickness is 2.00 m and the drawdown is 20.00 m, so that $\Delta P = 200$ kPa and the settlement is $S = \frac{200 \times 2.00}{100 \times 10^3} = 4 \times 10^{-3}$ m $= 4$ mm. It can be seen that the settlement of sand soil layer is relatively small.

(3) If there is no hard clay layer on top of dewatering sand soil layer and the dewatering time is relatively long, the settlement should be considered as the overlaying soil consolidation deformation that is caused by reduced water head ΔP. The specific calculation method should be carried out according to the actual distribution of the soil and with reference to the calculation method in soft soil area. In this case, the impact of deep wellpoint dewatering on the environment is relatively larger.

5.9.7 Precautions of Adversely Affects on Environment Caused by Wellpoint Dewatering

Wellpoint dewatering plays an important role in the construction of municipal engineering, but necessary precautions should be taken to prevent adverse affects on environment caused by wellpoint dewatering.

5.9.7.1 Doing Research Work on the Surrounding Environment

(1) Make a thorough investigation of engineering geology and hydrogeology conditions, and write detailed geological prospecting report, which should include stratigraphic distribution, conditions of permeable layers and lenses and their connection with other aquifers, variation of water level, hydraulic conductivity, void ratio, compression coefficient of different stratums, and so on.
(2) Identify underground water storage bodies, such as the distribution of surrounding ancient underground rivers, ancient pools, to prevent the penetration of wellpoint and underground water storage bodies.
(3) Identify the distributions, ages, and bearing capacity of differential settlement of water pipes, sewer pipes, gas pipes, telephone cables, electric transmission lines, to consider whether they need advance reinforcement.
(4) Make a thorough investigation of surrounding aboveground buildings and underground structures, including foundation type, upper structure type,

location in the dewatering area, and bearing capacity of differential settlement. Before dewatering, find out the settlement over past years and current damage degree of those buildings and structures to determine whether they need pre-reinforcement.

5.9.7.2 Use Wellpoint Dewatering Reasonably to Minimize Its Impact on the Surrounding Environment

Dewatering is bound to form a cone of depression, resulting around ground settlement. However, as long as wellpoint dewatering is used reasonably, such effects can be controlled within the surrounding environment-bearing capacity.

(1) Prevent the loss of fine particles. Always pay attention to whether the extracted groundwater is mudflow. The loss of fine soil particles will not only increase the ground settlement, but also block the well pipe and invalidate the wellpoint. To avoid this, an appropriate filter must be chosen based on the property of the surrounding soils, and meanwhile, more attention should be paid to the bore and the sand filter quality. Silt layers of soft soil area always extend horizontally, in which agitation should be minimized during drilling. In soft soil area, filter pipe should be laid in sandy layer. Casing method can be used, if necessary, and sand filter should be carefully prepared according to gradation.

(2) Decrease the slope degree of the cone of depression properly. On the premise of the same drawdown, the gentler the slope is, the greater the influence area is and the smaller differential settlement is, so there is smaller damage to the underground pipes and structures. The filter pipe should be horizontal continuously arranged in sandy soil layer based on the geological prospecting report to get a gentle depression curve, so that to minimize the influence on surrounding environment.

(3) Wellpoint should run continuously to avoid repeated and intermittent dewatering. In principle, light wellpoint and ejector wellpoint should be laid in sandy soil layer, of which the settlement caused by dewatering is small except for loose ones. However, field and laboratory tests show that repeated and intermittent dewatering can lead to a large amount settlement, though it will tend to be a steady value. Thus the repeated and intermittent dewatering should be avoided.

(4) Avoid quicksand and large settlement caused by confined water during pit excavation. Figure 5.38 shows that there is a thin layer of clay and a relatively thick layer of silty sand under the foundation bottom, and the wellpoint only reach the bottom of the pit. In this condition, this layer of clay will bear the water pressure difference as, $\Delta P = (h - h_1)\gamma_w$ so that quicksand may occur. In this case, it is better to make the wellpoint punctured into silty sand layer to lower the water head. By this way the confined water pressure will be released to make the pit bottom more stable.

(5) Avoid the penetration between wellpoint and nearby water storage body, which may lead to water head decrease then further lead to quicksand. In this

5.9 Impact of Wellpoint Dewatering on the Environment and the Prevention

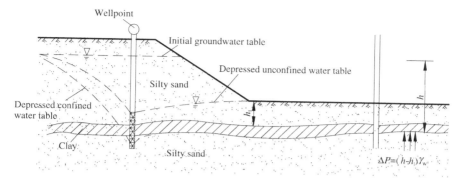

Fig. 5.38 Prevention of quicksand

condition, water-resisting wall should be built between the wellpoint and nearby water storage body.

(6) Inside wellpoint dewatering can reduce the adverse impact on the surrounding environment. The wellpoints that circlewise set inside the sheeting piles and underground diaphragm wall are called inside wellpoints. Sheeting piles that as lateral supporting should be close enough to each other and 2.00 m deeper than the wellpoints, so that the adverse impact on the surrounding environment can be greatly reduced and the dewatering can go well.

(7) Closed ejector wellpoint system can be used in step-slope excavation pit, where the sandy soil layer is at a certain depth below the pit bottom. Wellpoints can be placed on the surrounding platform that is 2.00–3.00 m lower than the ground surface, or wellpoints can be placed closer to the pit center to reduce the range of influence. The outside depression slope that is caused by ejector wellpoint system is steeper and the influence area is smaller so that the settlement region can be controlled in the range of 2.00–3.00 m, as shown in Fig. 5.39. This arrangement is more reasonable than arranging ejector wellpoint on the ground surface surrounding the pit or using multistage wellpoint, for not only reducing the impact on environment but also saving the dewatering cost.

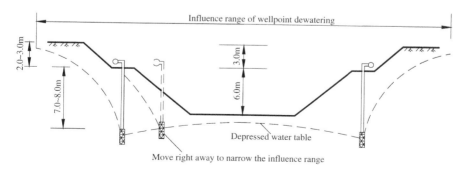

Fig. 5.39 Wellpoint arrangement for reducing settlement influence range

(8) Not apply wellpoint blindly to those soil that is not suitable for wellpoint dewatering, especially to those clay layers without sand interlayers. The hydraulic conductivity of these clay layers is always no more than 10^{-7} cm/s, which means they are impermeable and light wellpoint and ejector wellpoint are useless. Meanwhile, their own shear strength is enough to maintain the stability of pit wall during excavation. The needs to increase the depth of the excavation can be meet by decreasing the slope degree or deepening the lateral supporting sheeting pile.

5.9.7.3 Set Recharge Well System at the Edge of the Conservation Area

Wellpoint dewatering will lead to unavoidable water table decrease and differential settlement, so that lots of ground buildings and underground structures and pipelines will subject to different degrees of damage. To eliminate such effects as much as possible, set recharge well system at the edge of the conservation area.

(1) The arrangement of recharge wellpoint is similar to the arrangement of dewatering wellpoint. Recharge wellpoint system also includes one pump and one water storage tank. Water that pumps out from dewatering well is sent to storage tank, then to main water ejector tube. Redundant water can be drained away by drainage ditch.

The filter pipe of recharge well is usually 2.00–2.50 m, which is longer than that of dewatering well. Coarse sand can be filled between well casing and borepipe as filter layer.

(2) The recharge water often includes $F_e(OH)_2$ precipitate, active rusting and insoluble substances that can accumulate inside the recharge well pipe. Thus during recharging, the ejector pressure should be escalated to maintain stable recharge quantity. For a longer period recharging project, apply a chemical coating on the well pipe and set filter screen on the entrance and the exit of the water storage tank to slow down the blocking. During recharging, water should be kept clean.

(3) Observation wells should be set in recharging area to continuously record the change of water table. Adjust the ejector water pressure to keep the water table in the original state.

5.9.7.4 Set Water-Proof Curtain

Set water-proof curtain around the excavation area can reduce the impact on the surrounding environment to a small degree. Common water-proof curtain types are the following:

(1) Deep mixing piles water-proof curtain

Deep mixing piles water-proof curtain is constructed by overlap method. Mixing cement and soil will generate chemical reaction. The permeability of this mixing body is no more than 10^{-7} cm/s, to form a continuous water-proof wall. It can be arranged behind the steel sheet piles, or be used as a lateral supporting structure.

(2) Mortar cutoff sheet piling

Drive a row of H-shape steel piles in the position of water-proof wall. Inject cement mortar and at the same time pull out the steel piles to form a circle of mortar cutoff sheet piling. The H-shape steel piles can be 20–30# and the key to success construction is keeping the piles vertical and closely contact.

(3) Root pile water-proof curtain

The root piles are with a diameter of 100–200 mm. Pressing the pure cement paste to form piles as water-proof curtain and there are no rebar in them. General engineering geological drilling rig can be used in construction, and use hop beat process to prevent perforation. The key to success construction is keeping the piles vertical and no collapse and necking phenomenon. The casing can be used as assistive means if necessary.

5.10 Case Study

Shown as Fig. 5.40, a rectangle foundation pit with underlying soil layers as in Table 5.23, should consider the dewatering design during excavation. The hydraulic conductivity of sands K is around $K = 0.002$ m/s; the groundwater table is 1.5 m at depth from the ground surface; the requirement of drawdown is 4.5 m for safety. The curve of drawdown in the foundation pit center is 0.5 m lower than the foundation base. As hydraulic conductivity of sands is relativity large, deep well wellpoint dewatering is needed and partially penetrated well is applied in diameter of 15 cm. Construction site surrounding is farmland.

The plane arrangement of dewatering well is presented in Fig. 5.40 and Table 5.24. Please determine

Table 5.23 Properties of soil layers

Soil type	Depth (m)	Void ratio	Specific gravity	Liquid limit (%)	Plastic limit (%)	Plasticity index	Granulometry		
							Sand	Silty	Clay
Gray silty clay	0–1.07	0.88	2.73	30.1	21.3	8.8	13	76	11
	1.07–1.86		2.70	29.4	21.7	7.7	10	80	10
Gray silty sand	1.86–2.59		2.72				16	80	4
	2.59–3.14	0.94	2.69				16	78	6
	3.14–3.81	0.86	2.68				28	70	2
	3.81–4.66		2.68				38	60	2
	4.66–5.48		2.68				38	65	2
Gray silty clay	5.48–6.40		2.69	28.1	23.2	5.5	15	78	7
	6.40–7.62		2.70				96	4	0

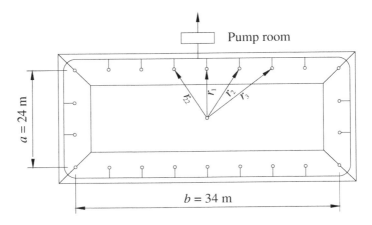

Fig. 5.40 Plane arrangement of dewatering well

Table 5.24 The distance of each well to the foundation pit center

$r_1 = 12.00$ m	$r_6 = 17.46$ m	$r_{11} = 12.73$ m	$r_{16} = 20.81$ m	$r_{21} = 14.71$ m
$r_2 = 12.73$ m	$r_7 = 17.46$ m	$r_{12} = 12.00$ m	$r_{17} = 17.46$ m	$r_{22} = 12.73$ m
$r_3 = 14.71$ m	$r_8 = 20.81$ m	$r_{13} = 12.73$ m	$r_{18} = 17.46$ m	
$r_4 = 17.51$ m	$r_9 = 17.51$ m	$r_{14} = 14.71$ m	$r_{19} = 20.81$ m	
$r_5 = 20.81$ m	$r_{10} = 14.71$ m	$r_{15} = 17.51$ m	$r_{20} = 17.51$ m	

1. The buried depth of dewatering well;
2. Check the drawdown of the pit center by calculation;
3. Estimate the water discharge when the drawdown equals to 4.5 m in the center.

5.11 Exercises

1. What is the principle of light wellpoint dewatering? How about the application conditions? How to design the light wellpoint dewatering program? What are the common problems and how about the reasons? How to prevent them?
2. What is the principle of ejector wellpoint dewatering? How about the application conditions? How to design the light wellpoint dewatering program? What are the common problems and how about the reasons? How to prevent them?
3. In what conditions are the tube wellpoint dewatering usually used? What are the common problems and how about the reasons? How to prevent them?

5.11 Exercises

4. What is the principle of electroosmosis wellpoint dewatering? How about the application conditions?
5. What is the principle of recharge wellpoint dewatering?
6. What should be monitored during the dewatering engineering?
7. What is the mechanism of ground settlement caused by dewatering? How to prevent it?

Chapter 6
Dewatering Well and Requirements of Drilling Completion

6.1 Structural Design of Dewatering Well

Compared with pumping-test well, dewatering well has the same basic structure and function of each part. In loose sediments, both are composed of well pipe and surrounding back-fill material. Well pipe consists of wall, filter tube (filter) and sedimentation tube, and other parts.

Generally, requirements of well pipe are as follows: well pipe should not bend; its inner wall should be smooth, round, and even. The thickness of the wall should be suitable with a certain tensile strength, compressive strength, shear strength, and bending strength; the porosity of the filter should be large enough.

6.1.1 Determination of Well Pipe, Depth and Diameter of Drilling

1. Drilling depth
   According to specific requirement of dewatering or embedded depth and length of the filter, the depth of drilling is determined.
2. Well pipe and drilling diameter
   Casing structure of dewatering well can use a type of diameter (as shown Fig. 6.1). Based on the premise of dewatering requirement, the designed casing structure should be as simple as possible in order to save pipes and make construction convenient.

Drilling diameter is mainly dependent on designed discharge of dewatering well, ability of well-digging equipment, types of well pipe, filter diameter, and thickness of artificial gravel pack.

In loose sediments as shown in Fig. 6.1, drilling diameter should be the sum of well pipe diameter and double diameter of gravel pack. Well pipe diameter depends

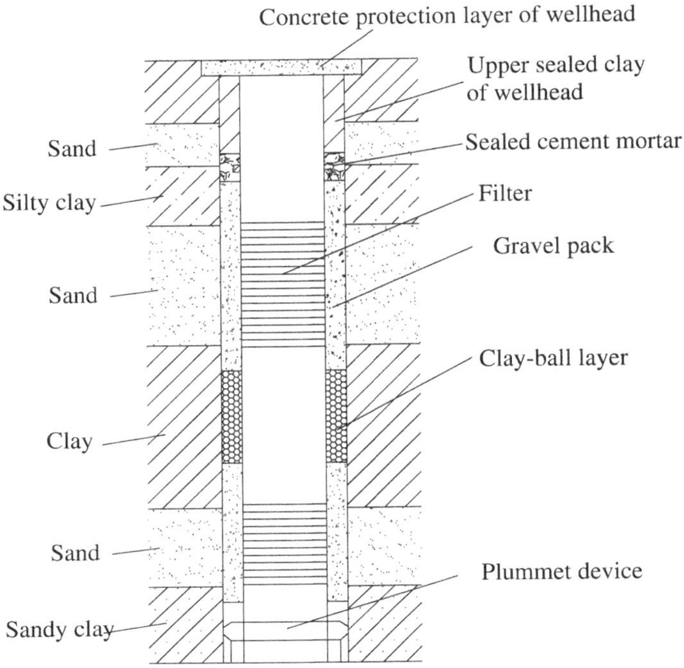

Fig. 6.1 Structural design of drilling well

on pump type determined by designed water discharge, and thickness of gravel pack depends on aquifer lithology.

Example

Designed discharge of dewatering well is 80 m³/h, and installing deep-well pump inside well pipe is required. Choose 200JC80-16 × 3—type deep-water pump (or 200JC$_K$80—16 × 3—type deep-well pump, or 8JD80 × 10—type deep-well pump). This type of deep-well pump requires the well pipe diameter to be $\geq$ 200 mm (or 8 in).

Specific descriptions of deep-well pump model:

200	minimum diameter of the pipe applicable to the pump, mm;
JC	long axis centrifuge deep-well pump (the impeller is closed);
JC$_K$	long axis centrifuge deep-well pump (the impeller is semi-open);
80	flow discharge, m³/h;
16	single-stage lift head, m;
3	stage number (the number of impeller);
8	minimum diameter of the pipe applicable to the pump, in;
J	deep-well multistage pump;
80	flow discharge, m³/h;
10	impeller number.

6.1 Structural Design of Dewatering Well

Table 6.1 Allowable flow velocity into the pipe

Permeability coefficient of aquifer (m/d)	Allowable flow velocity into the pipe (m/s)
>122	0.030
82–122	0.025
41–82	0.020
20–41	0.015
<20	0.010

In addition, when determining the pipe diameter in loose aquifer, we should also use allowable flow velocity into the pipe to recheck pipe diameter. In other words, filter diameter should meet the following requirements:

$$D \geq \frac{Q}{\pi \ln v'_{allowable}} \tag{6.1}$$

where D is the filter diameter, m; Q is the designed discharge, m³/s; l is the working length of filter, m; n is the effective porosity of water-inlet surface in the filter surface, which is generally considered as 50 %; $v'_{allowable}$ is the allowable flow velocity into the pipe, whose value can be determined according to Table 6.1.

6.1.2 Design of Filter in Well Pipe

In addition to preventing the collapse of the well wall and prolonging the life of pipe well, its main role is to prevent fine particle of outer-well aquifer into the well, reduce the flow resistance, increase the catchment area, and increase the water discharge.

In order to increase the amount of water discharge, flowing resistance of groundwater into the pipe must be minimized. Turbulence friction loss near wellhead and friction loss of groundwater flowing through the filter section are the maximum among the resistance. To reduce friction loss, the permeability of well walls must be increased. The most effective way is to make the correct choice of filter types, increase the thickness of gravel pack, and select filled gravel type which adapts to the nature of the aquifer. Thus, on one hand, proper selection of the filter type is the key to get the maximum water discharge and minimum sand content in the water; on the other hand, it is directly related to the efficiency and lifetime of well pipes.

From Figs. 6.2 and 6.3, reasonable length of filter is obviously related to drawdown and water yield; it also relates to the thickness and permeability of the aquifer, diameter of the filter, and other factors. Reasonable length of filter can be

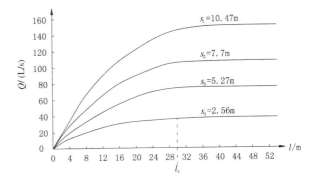

Fig. 6.2 Relationship curve between water yield and filter length ($Q - l$)

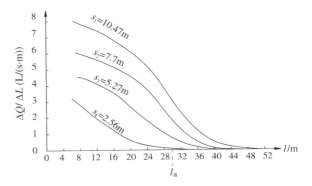

Fig. 6.3 Relationship curve between increasing intensity of water yield and filter length

directly determined by stage pumping tests, and l_a in Fig. 6.2 can be calculated according to the empirical formula (found in the relevant manuals) established by pumping tests. Based on local stage pumping experience, when the thickness of the aquifer is large, the reasonable length of filter is generally between 20 and 30 m.

6.1.2.1 Several Commonly Used Filter Types and Selection

There are many types of filters, and the following lists the commonly used types:

1. According to the type of material, there are steel or iron-cast filters, gravel cement filters, rigid plastic (polyethylene) filters, etc.
2. According to the form of filter pores, there are circle-hole filters, band-hole filters (as shown Fig. 6.4), bridge-hole filters, brim-hole filters (as shown Fig. 6.5), etc.

6.1 Structural Design of Dewatering Well

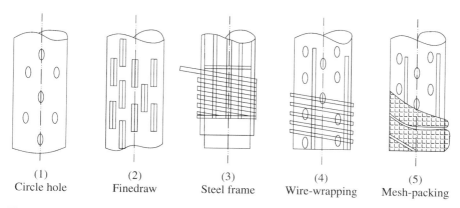

(1) Circle hole   (2) Finedraw   (3) Steel frame   (4) Wire-wrapping   (5) Mesh-packing

Fig. 6.4 Types of filters

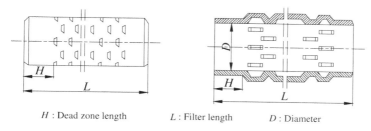

H : Dead zone length L : Filter length D : Diameter

Fig. 6.5 Structure of a molded filter

3. According to the filter structure, there are steel frame filters, wire-wrapping filters, mesh-packaging filters, cage or basket-shaped filters, gravel-attaching filters, molded filters etc., as shown in Fig. 6.4.

The commonly used iron-cast and wire-wrapping filter structure is shown in Fig. 6.6, and molded filter structure is shown in Fig. 6.3. For the technical specifications refer to the manual.

According to Table 6.2, different aquifers can select applicable or available filter types.

6.1.2.2 Structure of Gravel-Pack Filter

Gravel-pack filter consists of drainage pipe and gravel pack of outside well. Wire-wrapping filter, mesh-packaging filter, and molded filter are commonly used as drainage pipe, and they are made of steel pipe, iron-cast, reinforced concrete.

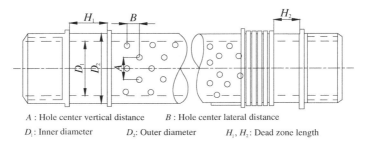

A : Hole center vertical distance   B : Hole center lateral distance
D_1: Inner diameter D_2: Outer diameter H_1, H_2: Dead zone length

Fig. 6.6 Iron-cast and wire-wrapping filter structure

Table 6.2 Filter types applicable for different aquifer

Aquifer lithology	Applicable filter types		Available filter types
Fine sand, silty sand	Double-layer gravel pack filter	Molded-hole filter	Single-layer gravel pack filter
Medium sand, coarse sand, gravel sand, debris soil with $d_{20} < 2$ mm	Single-layer gravel pack filter		Wire-wrapping filter
Debris soil with $d_{20} > 2$ mm	Skeleton filter or single gravel pack filter		
Bedrock fissure cave (with sand filling)	Single-layer gravel pack filter		
Bedrock fissure cave (without sand filling)	Skeleton filter		
Not broken bedrock	Without filter		

Gravel pack specification can be determined by the following formula:
When $\eta < 10$, and aquifer is sand, then

$$D_{50} = (6-8)d_{50} \tag{6.2}$$

where η is the uniform coefficient of aquifer; D_{50} and d_{50} is, respectively, particle diameter of 50 % cumulative percentage weight in the grain size distribution curve of gravel pack and aquifer.

When $\eta \geq 10$, coarse particles of screening samples should be first removed until $\eta < 10$. According to the grain size distribution curve, d_{50} is determined; and gravel diameter is determined based on Eq. (6.2).

6.1 Structural Design of Dewatering Well

For gravel aquifer, in which d_{50} is more than 2 mm, then:

$$D_{50} = (6-8)d_{20} \tag{6.3}$$

where d_{20} is corresponding particle diameter of 20 % cumulative percentage weight in the grain size distribution curve of aquifer.

When d_{20} is more than 2 mm, gravel of 10–20 mm diameter or no gravel is necessary. Outer gravel specification of double-layer gravel pack filter can be determined by the above method. Diameter of inner gravel is usually four or six times that of outside gravel diameter.

Thickness of single-layer gravel pack:

In gravel aquifer, it is not less than 75 mm; in coarse sand aquifer, it is 100 mm; in silty, fine or, medium sand aquifer, it is not less than 100 mm.

Gravel pack specification of well and screen slot size requirement of filter are determined as in Table 6.3.

Table 6.3 Specification of gravel pack in the well pipe and screen slot size of filter

Type of aquifer	Gradation result	Specification of uniform gravel pack (mm)	Thickness of gravel pack (mm)	Screen slot size of filter (mm)	Specification of semi-uniform gravel pack (mm)
Cobble	It accounts for 80–90 % as particle diameter is more than 3 mm	15–20	75–100	2–3	12–25
Gravel	It accounts for 80–90 % as particle diameter is more than 2.5 mm	10–12	75–100	2–3	8–20
Gravel	It accounts for 80–90 % as particle diameter is more than 1.25 mm	6–8	75–100	2–3	5–12
Gravel	It accounts for 80–90 % as particle diameter is more than 1 mm	5–6	75–100	2–3	4–10
Coarse sand	It accounts for 60 % as particle diameter is more than 0.75 mm	4–5	100	1.5–2	3–8
Coarse sand	It accounts for 60 % as particle diameter is more than 0.6 mm	3–4	100	1.5–2	2.5–6
Coarse sand	It accounts for 60 % as particle diameter is more than 0.5 mm	2.5–3	100	1.5	2.0–5

(continued)

Table 6.3 (continued)

Type of aquifer	Gradation result	Specification of uniform gravel pack (mm)	Thickness of gravel pack (mm)	Screen slot size of filter (mm)	Specification of semi-uniform gravel pack (mm)
Medium sand	It accounts for 50 % as particle diameter is more than 0.4 mm	2.0–2.5	100–200	1	1.5–4
Medium sand	It accounts for 50 % as particle diameter is more than 0.3 mm	1.5–2.0	100–200	1	1–3
Medium sand	It accounts for 50 % as particle diameter is more than 0.25 mm	1.5–2.0	100–200	1	1–3
Fine sand	It accounts for 50 % as particle diameter is more than 0.2 mm	1.0–1.5	100–200	1	0.75–2
Fine sand	It accounts for 50 % as particle diameter is more than 0.15 mm	0.75–1.0	100–200	0.75	0. 5–1.5
Silty sand	It accounts for 50 % as particle diameter is more than 0.1 mm	0. 5–0.75	100–200	0.75	0. 5–1.0

Thickness of double-layer gravel pack:

Thickness of inner gravel is 30–50 mm; thickness of outer gravel is 100 mm. Four spring-steel plates and other protective devices should be set up at the top and bottom of inner cage in the double-layer gravel pack filter.

Circular or nearly elliptic quartz gravel is used as optional gravel at best, and uniform coefficient of optional gravel is usually <2. Height of gravel pack is determined by filter location; at the same time the end elevation should be 2 m lower than filter bottom and its top should be 8 m higher than filter top.

At the top of first gravel-pack layer, high-quality clay ball is used to seal it to 3–5 m, then clay block is used to infill it to wellhead. When highly mineralized water layer exists at the top of filter, thickness of clay ball should increase.

When smelly sludge or other adverse aquifer exists at the non-working part of filter, high-quality semi-dry clay ball must be used to seal this layer with 3–5 m extended up and down, and the fill volume is calculated by well volume and compression of the clay balls.

6.2 Technical Requirements of Dewatering Well Completion

6.2.1 Water Sealing Requirement for Drilling

Sealing parts should have good watertight performance, large thickness, and relatively complete drilling segment. Water sealing method and material selection should be in accordance with the requirements of water sealing and drilling geological conditions. Clay, kelp, rubber, leather, etc., can be used for temporary water sealing, and clay, cement, asphalt, and other materials can be used for permanent water sealing.

Commonly used methods in water sealing can be divided into two types, i.e., water sealing out of well pipe and water sealing in the well pipe. Although water sealing out of well pipe with the same diameter or combination method of water sealing out of well pipe with the same diameter and water sealing in well pipe with the same diameter has bad effect and is hard to check, they have the following advantages: their drilling structure is simple; they have high drilling efficiency and; material usage is small.

6.2.2 Demands of Drilling Flushing Fluid

Theoretically, drilling of dewatering well is best made by water washing. However, in order to save hole-retaining pipe and improve efficiency, mud drilling is often used in practice. If mud consistency is too large or wash drilling holes after the work is not enough, it may have serious adverse impact on the permeability of aquifer. Therefore, we must strictly control the consistency of mud, which is generally less than 18 s at best.

6.2.3 Requirements of Drilling Inclination

To ensure pumping smoothly and normally, when drilling depth is <100 m, drilling deviation shall not exceed 1° in inclination; when drilling depth is more than 100 m, drilling deviation shall not exceed 3°.

6.3 Well Washing

The completion technology of pipe well is quite complex, which contains many processes and can be summarized into six main steps:

(1) Drilling into the ground until the borehole meets the designed requirement;
(2) Conducting geophysical prospecting for the borehole completion;

(3) Measuring the slope of borehole;
(4) Installing well pipes (it contains: the inspection of well roundness and depth; dilution of the slurry in borehole; inspection of the quality of well pipes; measurement and arrangement of well pipes; installation of well pipes, and so on);
(5) Packing gravel and sealing the gaps outside the pipes;
(6) Washing well. These processes are all important, which means improper operations in any process can lead to the poor quality of pipe wells. These flaws can affect the water yield and even disable the pipe well.

Well washing is the last and also the most important process in the construction of a pipe well. The objective of well washing is increasing the water yield and reducing the sand content of water. For realizing this, well washing can bring out the fine particles in the aquifer and clear the mud cake attached on the well wall as well as the slurry, which remains in the well or infiltrates into the aquifer. The dredged aquifer forms a natural filter around the well and largely increases the permeability of the material around pipe well. Besides, well washing is also a crucial measurement to recover the water yield of a pipe well after completion in construction.

There are two main methods for well washing: mechanical washing and chemical washing. The former method is more widely applied currently, while the latter is more promising.

6.3.1 Mechanical Methods for Well Washing

Of all the mechanical methods of well washing, piston washing and air-compressor washing are most widely used, followed by water pump washing, puncher washing, and their combined application. The principles of different mechanical well washing are similar: through changing the pressure and producing washing effect in the pipe well, the washing equipment can increase the pressure difference and accelerate the flow velocity of underground water until the well has been cleaned up.

6.3.1.1 Piston Washing

If the drilling mud is applied in the process of pipe well construction, the piston washing should be conducted immediately after completion of installing well pipes and packing gravels. Common pistons are made of wood and iron; their structure, as shown in Fig. 6.7, can be different between those with valves and those without.

The mechanism of piston washing is utilizing the up-down movement of the drill pipe in the well to clean. Piston washing is usually conducted for each aquifer in steps. It should be noted that the movement speeds of pipe drills should be specifically controlled for pipe wells constructed by different materials. The duration of piston washing is determined by the well depth and aquifer state.

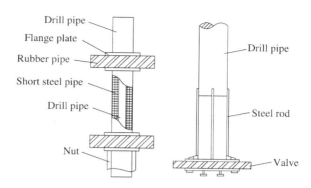

Fig. 6.7 Structure diagram of iron pistons with (*right*) and without (*left*) the valve

The efficiency of piston washing is good. Meanwhile, the required equipment for it is less and the whole processes are simple, which makes this method more popular. However, the well pipes can be damaged in the clean process if their strength is relatively low, which could cause sand leaks in the well when the pipe well is buried in the aquifer of fine soil.

6.3.1.2 Air-Compressor Washing

Air-compressor washing is usually applied after piston washing. This method can be conducted in two common ways: Concentric washing and jet back-washing. Based on the pipe well structure, water yield, water level, and so on, we can choose the appropriate way.

1. Concentric washing

   Concentric air-compressor washing is the most widely used well washing method. "Concentric" means the water pipe and air pipe are arranged concentrically, the installation instruction is shown in Fig. 6.8. Aquifers are flushed every 2–5 m each time. The bottom of the water outlet is commonly fixed in the treating aquifer until the slurry and fine sand in this aquifer are cleared, then it is moved to another position. Repeat this process from the well top until all the expected aquifers are clean.

2. Jet back-washing

   As shown in Fig. 6.9, the air pipe and jet are required to be installed inside the well. One end of the air pipe is connected to the air-compressor by high-pressure rubber pipe; the length of air pipe is determined through the pressure produced by air-compressor. Generally, for the air-compressor that can produce 0.686 MPa pressure, the submerged part of the air pipe should be <70 m. If the well depth is beyond this value, more air-compressors or air-compressors of higher pressure are required to be applied. Commonly, each 10 m water column can be calculated as 0.098 MPa in design.

Fig. 6.8 Installation instruction of air-compressor washing

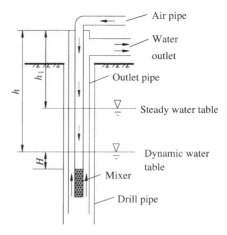

Fig. 6.9 Installation instruction of jet back-washing

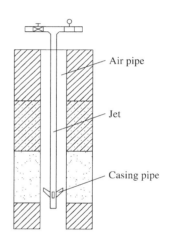

In this method, the washing process should be conducted by segments, which is similar to concentric washing. The flow velocity of air in the well is quite high; it tends to break the wall of borehole. Therefore, the structure of aquifer should be less disturbed as possible in the whole washing process. Otherwise, the packing gravel of filter could be damaged. This method is not suitable for (1) pipe wells whose well pipes are weak; (2) pipe well whose packing gravel layer is thin and; (3) pipe wells drilled in the silty sand aquifer.

In addition, there are closed back-washing and excitation back-washing; we can choose the appropriate way to install air-compressor based on the engineering requirement.

The air-compressor washing is safe and effective, however, the cost is high and its application is limited by the groundwater depth. Therefore, it is not suitable for the well whose dynamic water level is deep or the well depth is shallow.

6.3 Well Washing

6.3.1.3 Water Pump and Slurry Pump for Well Washing

There are some other methods available as air-compressor washing is not appropriate for a pipe well, as follows:

1. Horizontal centrifugal pump method
   For the pipe well with large diameter, large water yield, and shallow water level, horizontal centrifugal pump with wear-proof wing wheels can be applied to clean well after thorough piston well washing. The water yield in this process is required over the designed water yield in the well's usage period. During well washing, the centrifugal pump should be used intermittently to clean out the soil in the well until the well water is clear and the water level is stable.
2. Deep-well pump method
   For the pipe well which has deep groundwater, deep-well pump of different sizes can be applied. The specific processes are similar to the horizontal centrifugal pump.
3. Slurry pump combined with piston method
   For the pipe well which is drilled by rotary drilling rig, slurry pump combined with piston can be applied to well washing after the filter is installed (Fig. 6.10). After the gravel is filled, washing equipment can be connected to the bottom of drill pipe. The water or slurry sprayer is used to wash while the piston is pulled along the well and the waste washing fluid soil is drawn out by slurry pump. When the filter of a section is cleaned, the drill pipe can be lifted to the next section for filter cleaning until the whole well is cleaned. Generally, this method is widely applicable and effective.

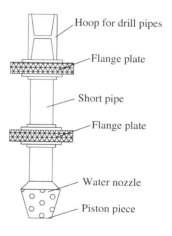

Fig. 6.10 Washing jet with piston for well washing

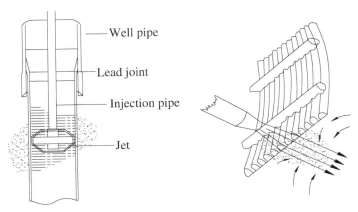

Fig. 6.11 High-velocity hydraulic water jetting for well washing. **a** Installation instructions **b** Schematic diagram of well washing process

4. High-velocity hydraulic jetting

The well washing equipment for high-velocity sprayed water method is shown in Fig. 6.11, the ejector commonly consists of two (angle 180°) or four jets (angle 90°). In this method, the high-speed flow in the injection pipe will enter into the filter or aquifer through these jets. Because the energy of the high-velocity flow is concentrated into a quite small area, it can wash each part of the filter and the adjacent aquifer thoroughly. Specifically, the high-velocity flow can break the slurry wall, remove the mud in the aquifer, and bring out the fine sand. Therefore, the high-velocity sprayed water method is effective and commendable for wells, which have thick slurry wall or excess leaking mud in the gravel stratum.

6.3.2 Chemical Methods for Well Washing

Chemical methods for well washing are recently developed worldwide. These methods are cost-effective and simple to operate. For wells jammed by chemical or biochemical deposits, chemical methods are much more effective than mechanical methods. Moreover, for some carbonate aquifers, the chemical methods can even expand the fractures and pores in the aquifers.

6.3.2.1 Liquid Carbon Dioxide (with Acid) Method for Well Washing

The mechanism of liquid carbon dioxide method can be summarized as: through the physical form changes of carbon dioxide, the pressure in the well can be changed

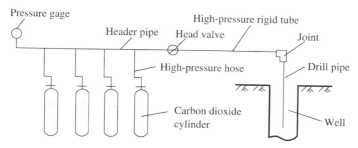

Fig. 6.12 Installation instruction of liquid carbon dioxide method for well washing

dramatically and cause water sprays out of the well. This method is commonly applied when the efficiency of mechanical well washing is poor, like in the old wells whose filter is jammed and water yield is low. For wells drilled in the carbonate aquifer with fractures, acid can be added. After chemical well washing, the water yield can increase to two or even more times of the original yield.

The sketch map of liquid carbon dioxide method is demonstrated in Fig. 6.12. Liquid carbon dioxide is transported into the well through the tubes under pressure; due to the high temperature and low pressure in the well, the liquid carbon dioxide will rapidly evaporate and generate high-pressure air and water flow which are powerful enough to break the slurry wall attached to the well pipe. Then, this flow can enter the aquifer around, dredge the pores and fractures in the rock, and bring out the fillings (like: rock debris, mud) to the ground. This is how this method can wash the well and enhance the water yield.

For tube wells constructed in the soluble rock stratum, a certain amount of hydrochloric acid can be added into the well, then after 2–5 h, liquid carbon dioxide can be injected into the well. The evaporated carbon dioxide can press the hydrochloric acid into fractures of the rock stratum and the acid can dissolve the soluble rock and expand the fractures; the dissolved materials are bought out of the well through the blowout. Sometimes, to wells drilled in carbonate rocks, only hydrochloric acid is sufficient to generate amounts of carbon dioxide to cause blowout.

A certain percentage of anticorrosive, such as formaldehyde, butynediol [$C_4H_4(OH)_2$], sodium iodide [NaI], potassium iodide [KI], and so on, is required to be added into the acid to prevent the acid corrosion of metal pipes in the process of well washing. In addition, sodium polyphosphates can be applied to improve the washing effect through delaying the solidification of mud cake in the well, especially if the well has thick mud cakes.

Liquid carbon dioxide (with acid) method is an advanced method for well washing. It is simple and low cost in terms of time and money. Also, it is widely applicable for wells of different depths, materials, structures, and service time, and is especially effective for wells constructed in rock aquifers with pores and fractures.

6.3.2.2 Sodium Polyphosphates Method for Well Washing

At present, the most used sodium polyphosphates in well washing include sodium hexametaphosphate [$(NaPO_3)_6$], sodium tripolyphosphate [$Na_5P_3O_{10}$], sodium pyrophosphate [Na_3PO_4], tri-sodium phosphate [Na_3PO_4], and so on. Here, sodium pyrophosphate is applied as an example to illustrate the mechanism and procedures of this method.

Due to the complexity between the sodium pyrophosphate and clay particles, water-soluble complex ions are produced in this process; the equations are demonstrated as follows:

$$Na_4P_2O_7 + Ca^{2+} \rightarrow CaNa_4(P_2O_7)^{2-}$$
$$Na_4P_2O_7 + Mg^{2+} \rightarrow MgNa_4(P_2O_7)^{2-}$$

Complex ions [$CaNa_4(P_2O_7)$]$^{2-}$ and $MgNa_4(P_2O_7)^{2-}$ are inert ions, they can hardly generate the reverse chemical reaction and precipitate. This feature makes them apt to be ejected out in well washing or water drawing. Meanwhile, the complex ions, which have negative charges that can be absorbed by clay particles, can strengthen the electronegativity of the clay particle surface. The strengthened electronegativity will increase the repulsion between clay particles and decrease the viscosity and shear strength of slurry. This is why the sodium pyrophosphate can resolve and further break the mud cakes and sediments in the aquifer.

The processes of sodium pyrophosphate well washing: First, install the well pipe and fill the gravel to the designed position, slurry pump is used to eject the mud in the well. Second, inject sodium pyrophosphate whose concentration is 0.6–0.8 % into the inside and outside of the well pipe (inside first) through slurry pump. Third, after 5–6 h, the sodium pyrophosphate is combined with clay particles thoroughly, other well washing methods can be applied. Generally, the well can achieve its normal water yield in a short time.

The chemical well washing method can reduce the work time and largely decrease the workload of piston and air-compressor well washing. Moreover, it can increase the final water yield compared to using the mechanical method only.

All the well washing methods introduced above should be chosen according to the specific work circumstance, and the combination of different methods can enhance the well washing effect.

After well washing, the sand content of pump water is required to be measured; indeed, the sand content is an important index for the pipe well. The excess sand content can impair the dewatering effect and even cause engineering accidents. Therefore, the strict control of sand content is crucial.

The standard for the sand content:

(1) From the beginning of pumping to 30 s after the well is dry, the measured sand contents of water come from wells constructed in coarse sand or gravel aquifer should be less than the 1/50,000 of the water volume; while the sand

6.3 Well Washing

content for wells constructed in coarse sand or gravel aquifer should be less than the 1/20,000 of the water volume.

(2) Keep pumping for several hours till the well water is gradually clear and the sand content becomes stable. In the followed stable period, the sand content of living and industrial wells should be <1/2,000,000. For some special wells, this content may need to be <1/10,000,000.

6.4 Case Study

Groundwater assessment and management

Structure design of hydrogeological experimental borehole should consider the geological information as in Table 6.4.

200JC(8JD) × 80 × 10 type of pump is required to design the hydrogeological drilling; and the boring log should be drawn as in Fig. 6.13. The minimum diameter of grain size is around 0.15 mm of aquifer, above which the percentage is 55 %.

Table 6.4 Geological information

Depth (m)	Thickness (m)	Soil type	Legend
0–2.80	2.80	Silty clay	
2.80–14.06	11.26	Silty sand	
14.06–19.76	5.70	Mucky silty clay	
19.76–26.00	6.24	Silty sand	
26.00–36.60	10.60	Silty clay	
36.60–39.20	2.60	Peat	
39.20–67.70	28.50	Silty clay inter-bedded with silty sand	
67.70–90.80	23.10	Fine and medium sand	
90.80–97.80	7.00	Clay	

Scale: 1:500

Stratigraphic age	Depth (m)	Thickness (m)	Soil type	Soil layer profile	Construction profile

Fig. 6.13 Hydrogeological boring log scale: 1:500

6.5 Exercises

1. How many parts does dewatering well structure include? What are they?
2. How many methods does well washing have? What are they?

Chapter 7
Dewatering Types in Foundation Pit

7.1 Types and Effect of Dewatering in Foundation Pit

In general, the deep foundation pit is defined as follows: excavation pit is over 5 m (including 5 m), or even though the depth is no greater than 5 m, the geological and surrounding environment conditions are very complex, or the excavation affect the safety of adjacent buildings.

7.1.1 Effects of Dewatering in Foundation Pit Construction

Before the excavation of a foundation pit, dewatering can ensure the safety of excavation, increase the stability of pit support structures. The main effects on be described as below:

(1) Prevent the seepage in lateral or basal surface in the foundation pit. Keep the excavation be conducted in dry conditions. Be in favor of other constructions.
(2) Decrease the water content of soils in the pit and surroundings to strengthen the physical and mechanical properties.
(3) Decline the hydraulic gradient to prevent the quicksand phenomenon induced by lateral and bottom soil particle movement with groundwater seepage.
(4) Increase the safe stability of lateral overturning and basal up-heaving of foundation pit.

Above is all about the advantages of dewatering in foundation pit excavation. Apparently, it also has some disadvantages for projects, such as the dewatering device may induce difficulties for foundation pit construction; or ground settlement in surrounding buildings.

7.1.2 Different Types of Dewatering in Foundation Pit Construction

Dewatering method in foundation pit can be divided into two types, i.e., open pumping or wellpoint dewatering.

Open pumping is to set ditches and sumps in foundation pit to collect groundwater; and then make the water pumped away or out. For the stage-excavation foundation, as the excavation surface moves deeper, new ditches, and sumps should be re-dug. This method is suitable for the shallow foundation pit in the aquitard. When surrounding conditions is simple; the aquifer is thin and the required drawdown is relatively small, open pumping is most inexpensive.

Wellpoint dewatering is a kind of predrainage method, which is depressed the groundwater table by artificial discharging in pumping well before the excavation of foundation pit. According to the force-applied or pumping device, wellpoint dewatering can be described as light wellpoint, ejector wellpoint, electroosmosis wellpoint, and tube (deep well) wellpoint.

This chapter mainly introduces the tube (deep well) wellpoint dewatering in foundation pit.

7.2 The Seepage Properties of Dewatering in Foundation Pit

In the dewatering for foundation pit, the seepage properties of groundwater are greatly related with the in situ hydrogeological conditions, support structure (water-proof curtain), the location of dewatering well, the length of filter, etc. The corresponding analysis is conducted by calculation as followings.

7.2.1 Water-Proof Curtain

Case 1:
This is no water-proof curtain built during dewatering, shown as Fig. 7.1. The calculation results can be seen in Fig. 7.2.

Case 2:
There is a water-proof curtain. Dewatering is conducted outside the foundation pit. The plane layout and the calculation results are shown in Figs. 7.3 and 7.4.

Case 3:
There is a water-proof curtain. Dewatering is conducted inside the foundation pit. The plane layout and the calculation results are shown in Figs. 7.5 and 7.6.

Comparatively analyzed, the effect of water-proof curtain for the seepage of foundation pit is characterized as below:

7.2 The Seepage Properties of Dewatering in Foundation Pit

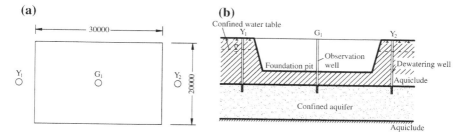

Fig. 7.1 The dewatering schematic in case 1. **a** Plane view of dewatering (unit: mm). **b** Cross sectional profile of dewatering

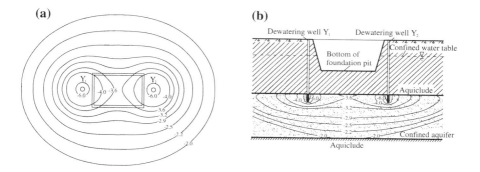

Fig. 7.2 The equipotential lines surrounding foundation pit in case 1. **a** Plane view of dewatering equipotential line (unit: m). **b** Cross sectional profile of dewatering equipotential line (unit: m)

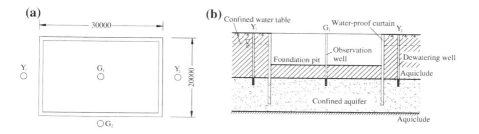

Fig. 7.3 The dewatering schematic in case 2. **a** Plane view of dewatering (unit: mm). **b** Cross sectional profile of dewatering

(1) Water-proof curtain changes the seepage flow state. Figure 7.2 shows the equipotential lines paralleling pass through the foundation pit bottom. It is plane seepage flow. Because the dewatering wells use partially penetrating wells, the groundwater around the filter presents a state of three-dimensional space seepage flow.

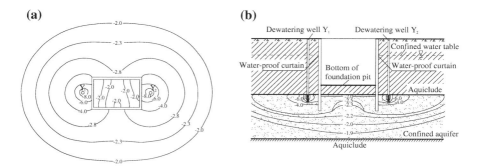

Fig. 7.4 The equipotential lines surrounding foundation pit in case 2. **a** Plane view of dewatering equipotential line (unit: m). **b** Cross sectional profile of dewatering equipotential line (unit: m)

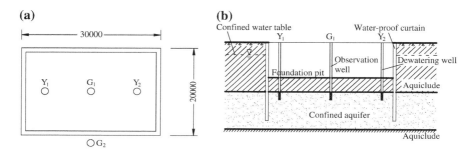

Fig. 7.5 The dewatering schematic in case 3. **a** Plane view of dewatering (unit: mm). **b** Cross sectional profile of dewatering

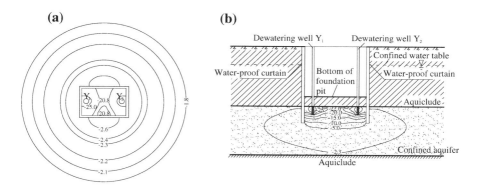

Fig. 7.6 The equipotential lines surrounding foundation pit in case 3. **a** Plane view of dewatering equipotential line (unit: m). **b** Cross sectional profile of dewatering equipotential line (unit: m)

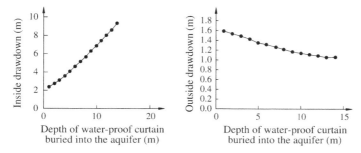

Fig. 7.7 The buried depth of water-proof curtain versus drawdown

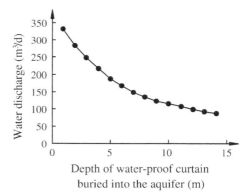

Fig. 7.8 The buried depth of water-proof curtain versus water discharge

(2) Water-proof curtain changes the hydraulic gradient. Shown in Figs. 7.4 and 7.6, the equipotential lines are densest around the bottom of water-proof curtains, i.e., the hydraulic gradient at these places are largest, where the cross-section declines sharply and the seepage velocity increases a lot.

(3) The buried depth of water-proof curtain influences the drawdown in dewatering. Figure 7.7 indicates the drawdown inside the foundation pit increases as the depth of water-proof curtain buried into the aquifer gets larger. While in the circumstance of drawdown outside the pit, it is decreased by larger buried depth of water-proof curtain.

(4) The buried depth of water-proof curtain influences the water discharge of dewatering wells. Figure 7.8 indicates the dewatering wells in the foundation pit have smaller water discharge when enlarging the buried depth of the water-proof curtain.

7.2.2 Length of Filter

The influence of filter is also discussed in limited thickness of aquifer. Some results by numerical simulation can be obtained in Fig. 7.9. It is seen that water discharge and drawdown per unit length of filter both sharply decrease with increasing length of the filter. In case of 14 m filter, the unit water discharge is just 23 % of that in 2 m filter condition. It means longer the filter, smaller the discharge efficiency.

7.2.3 Vertical Hydraulic Conductivity of Aquifer

With water-proof curtain, the relationship of drawdown with vertical hydraulic conductivity is presented in Fig. 7.10, including drawdown inside the pit (a) and drawdown outside the pit (b). As vertical hydraulic conductivity increases, the drawdown inside the pit decreased soon, while outside the pit, the drawdown just slightly changes, even it is a little bit increased.

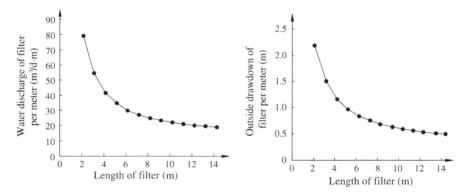

Fig. 7.9 The influence of the length of filter on water discharge and drawdown

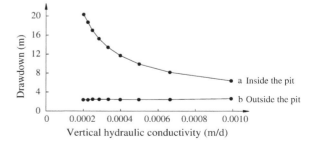

Fig. 7.10 Vertical hydraulic conductivity versus drawdown

7.3 The Classification and Characteristics of Dewatering in Foundation Pit

According to the cut off effect of water-proof curtain, the foundation pit dewatering can be classified into following four types.

7.3.1 The First Class

Shown in Fig. 7.11, the occurrence of the aquifer is very shallow, the water-proof curtain is totally buried into the aquiclude. Apparently in this circumstance, providing the water-proof curtain is conducted well without problem of leakage, there is no hydraulic connection of groundwater in and outside the pit. Dewatering well or light wellpoint can only installed inside the pit. The aim is to drain away the groundwater above the foundation pit bottom (Fig. 7.11a); or initially decrease groundwater table and then conduct drainage (Fig. 7.11b). The water discharge of dewatering well (light wellpoint) is totally from the groundwater inside the foundation pit. The amount is very limited. The target is easy to implement. During the dewatering, there is no influence on the groundwater outside the pit, so the impact on surrounding environment is very small. The groundwater flow inside the pit is generally plane two-dimensional flow with impermeable boundary.

This kind of dewatering is usually used in following two circumstances. First is the occurrence of aquifer is very shallow, generally unconfined aquifer. The excavation depth of foundation pit is small. Second is the aquifer is not shallow with a certain buried depth, but the thickness is not very large; and the excavation depth is relatively deep. Usually diaphragm wall is constructed as foundation pit water-proof curtain and support structure. For the stability of overturning in the lateral sides of the pit, the diaphragm wall should be buried below the base of the aquifer. In addition, some other conditions, from the safety consideration of foundation pit, the dewatering just need to be designed as the third class, unnecessary to cut off the imperious base, but the surrounding environmental requirement is very strict, the additional larger water-proof curtain should be constructed into the

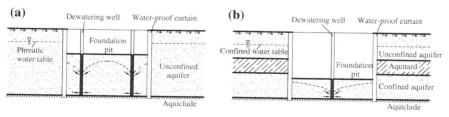

Fig. 7.11 The schematic of the first class dewatering. **a** Water-proof curtain into the bottom of unconfined aquifer. **b** Water-proof curtain into the bottom of confined aquifer

base of the aquiclude. Then, this kind of third class dewatering case is turned into the first class.

According to the old experience, the leakage problem of water-proof curtain always appeared. Particularly, when the circumstance is very serious, it makes great difficulties on the drainage efficiency. This kind of problem is not easy to be fixed. Though the existence can be easily confirmed by practical dewatering, the specific location of leakage could not be simply found out.

In total, the first class dewatering is relatively easy to implement and the dewatering duration is short. Respective working quality is not necessarily high. It depends on the practice. Light wellpoint, tube wellpoint, open pumping can all be employed in this class of dewatering.

7.3.2 The Second Class

Providing the occurrence of the aquifer is very deep; the water-proof curtain can only reach the overlying confining bed and have not penetrated into the dewatering aquifer (Fig. 7.12), the hydraulic connection is rarely influenced by the cut-off of water-proof curtain. The dewatering of wellpoint is aimed to depress the water head of the confined aquifer, prevent the up-heave of the pit bottom, or quicksand. The seepage properties in this condition are:

Due to no influence by water-proof curtain, the confined water inside and outside the foundation pit has good hydraulic connection, which presents two-dimensional flow state. It is called plane seepage without boundary. When partially penetrated well is employed in large thickness aquifer, the groundwater flow state is three-dimensional space flow. This influence range of this class dewatering is very large with relatively slow cone of drawdown. The resulted ground settlement by dewatering is mostly even.

Comparing first and second classes, it can be found that mostly the first class dewatering is a kind of drainage dewatering, which is drained out all the groundwater cut off by water-proof curtain and aquiclude to generate a dry environment for excavation and other underground construction. While the second class dewatering is mainly depressed the water pressure to lower the water head beneath the excavation surface. Since there is no lateral obstruction, the horizontal water supply is

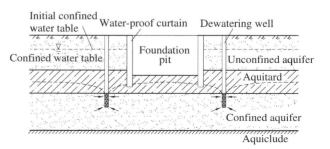

Fig. 7.12 The schematic of the second class dewatering

continuous, and then the water discharge is very huge in this dewatering type. Generally, tube wellpoint is necessary. The construction difficulties are more than the first class dewatering. The entire duration of dewatering is relatively longer until the pit bottom slab is concreted. If the project encounters the duration delay of some other construction part, such as pit excavation, the dewatering should be also prolonged for the working duration to conform with the entire project duration.

In addition, it should be noted that the second class dewatering once starts, it could not be stopped since the continuous lateral water supply. Emergency power supply should be reserved in case of accident power cut.

7.3.3 The Third Class

Shown in Fig. 7.13, the excavation depth is large, the water-proof curtain is already penetrated into the dewatering aquifer, in which if the excavation depth is relatively small, the pit bottom has not already excavated into the dewatering aquifer, and the dewatering of wellpoint is just aimed to depress the water pressure of the confined aquifer (Fig. 7.13a); on the other hand, if the excavation depth is relatively deep, the pit bottom is already in the aquifer, the dewatering of wellpoint in this circumstance initially is also to depress the water head but in later stage the aim is drainage (Fig. 7.13b). Due to the existence of water-proof curtain, which is penetrated into the aquifer, the hydraulic connection is inevitably influenced, especially in the horizontal direction, although the water-proof curtain does not totally cut off the aquifer. The upper part of the confined aquifer is not continuous inside and outside the foundation pit. The groundwater must detour to the lower part of the aquifer.

In this dewatering type, the water-proof curtain is penetrated into the middle or lower part of the dewatering aquifer, so whether the dewatering well inside or outside the foundation pit, the groundwater flow is definitely obstructed. The seepage boundaries get much more complicated. The flow presents typical three-dimensional state. In case of calculation in dewatering design, this aspect should be fully considered. Moreover, this kind of dewatering always is accompanied with large-scale pit excavation. The dewatering scale is also very large. The requirement of dewatering duration is relatively long. Thus, the entire construction difficulties and uncertainties are significantly increase compared with the second class.

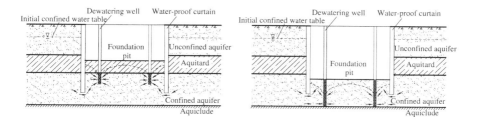

Fig. 7.13 The schematic of the third class dewatering

7.3.4 The Fourth Class

Shown in Fig. 7.14, in condition that the aquifer is buried very shallow; there is no water-proof curtain surrounding; slope excavation is constructed; the excavation depth is also relatively small, the hydraulic connection in this circumstance is similar to the case of second class dewatering. The groundwater presents natural plane flow state. Near the dewatering well, it is three-dimensional flow state.

This type of dewatering is usually constructed far away from the urban city or dense building areas, where no requirement for the environmental problem, such as land subsidence.

Different kind dewatering types can be combined to employ. Such as, for very deep excavation, when diaphragm wall is used as water-proof curtain, due to the large wall structure with the impermeable property, the unconfined aquifer, and the upper confined aquifer in the foundation pit belong to first class dewatering, while the lower part of the unconfined aquifer belongs to the second or third class dewatering. Or for the large-area foundation pit, the depth may be different in various locations for excavation. Corresponding requirement for drawdown, buried depth of water-proof curtain may all be different. In this type of foundation pit, there is always several different dewatering types combination.

Seepage properties vary with different foundation pit dewatering. Correspondingly, the design, construction, and operation management of dewatering have its own characteristics. In practice, the difference should be precisely distinguished. And it is better to coordinate with the design of support structure or other underground construction.

7.4 Dewatering Design of Foundation Pit Engineering

7.4.1 Design for the First Class Dewatering

As explained above, in first class dewatering, the lateral sides of dewatering aquifer is totally cut off by water-proof curtain, with no hydraulic connection to outside groundwater of foundation pit, the wellpoints should be arranged inside the

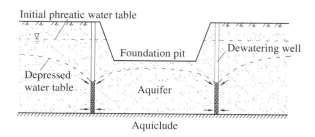

Fig. 7.14 The schematic of the fourth class dewatering

foundation pit. After the onset of dewatering, the water table in the pit is always on a unsteady state. The design calculation can follow the unsteady well group formula with impermeable boundary. The plane arrangement of wellpoint is usually determined based on local experience. When the hydraulic conductivity of aquifer is very small, vacuum pump is need for the drainage.

7.4.2 Design for the Second Class Dewatering

In the design of this class dewatering, the main principle is to ensure the stability of uplift of water pressure of confined aquifer from the starting of excavation to the foundation pit bottom (Fig. 7.15). That is:

$$\sum \gamma_{si} M_i \geq \gamma_w H \cdot K_s \qquad (7.1)$$

where γ_{si} is unit weight of each soil layer from foundation pit bottom to overlying confining bed, kN/m³; M_i is thickness of each soil layer from foundation pit bottom to upper confining bed, m; γ_w is the unit weight of water, kN/m³; H is the distance from depressed water table to the upper confining bed, m; K_s is the safety factor, according to specific standards. In practice, the determination of the safety factor value should be also consider some other influence factors, such as the sealing of geotechnical investigation drilling, the backfill of hydrogeological investigation well.

Once the aim is determined, the first step of consideration is the plan arrangement of wellpoints. As for the foundation pit with small area and loose environmental requirement, wellpoints can be arranged around the excavation periphery. Though the dewatering efficiency may be lower compared the inner dewatering in foundation pit, the arrangement is very favorable to the excavation construction, dewatering equipment, and other operation and maintaining. For the large-area foundation pit, the above arrangement generally could not meet the requirement of drawdown in the center of the pit, so dewatering inside the foundation pit is necessary. Due to the influence of support structure and engineering foundation piles, the plane layout in this condition is usually greatly limited.

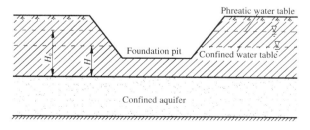

Fig. 7.15 The schematic of the second class dewatering calculation

The well structure and plane arrangement have great correlation. Increasing well depth can result in large dewatering influence range. The plane arrangement has more selections but the total efficiency is lowered. Meantime the water discharge has greatly increased, which is not favorable to the environment. Therefore, the plane arrangement and depth of wells should be comprehensively considered. Generally, according to previous hydrogeological investigation report or local design experience to calculate a single well-structured design, and then, considering the group well effect, water discharge amount of single well can be determined; lastly the entire foundation pit is simplified as a large well to calculate the total discharge amount, so the required number of wells can be obtained. For safety factor, the final number is increased by 10 % (the number of emergent well is no less than 1 to reserve). After all the wells are arranged, some confirmation calculation should be conducted in the foundation center and four corners or other weak place of wells (no need for the reserve well). In case of unsatisfaction, it is necessary to adjust the well arrangement or increase the well number to make the requirement of drawdown in any place can meet. Because many hydrogeological parameters are needed in the calculation, if there is no previous professional hydrogeological investigation and just depends on the local design experience, some dewatering construction is necessary for the conformation, to prevent large error in design.

The well structure strength is the key problem in most cases. For the arrangement of dewatering well, the deformation induced by excavation may result influence on the well structure. In addition, real relatively deep foundation pit, the excavation is conducted in stages. Replacing H in Eq. (7.1) into H_0, i.e., the natural water table of the aquifer, the minimum excavation depth can be calculated as Eq. (7.2) (Fig. 7.15):

$$\sum \gamma_{si}M_i + \sum \gamma_{sj}M_j \geq \gamma_w H_0 \cdot K_s \tag{7.2}$$

where H_0 is the natural water table of aquifer, m; γ_{si} is unit weight of each soil layer from foundation pit bottom to upper confining bed, kN/m^3; M_i is thickness of each soil layer from foundation pit bottom to upper confining bed, m. Others have the same meaning with Eq. (7.1).

As the excavation surface moves down, H_0 can be replaced by H to calculate the safe depth for different excavation depth.

7.4.3 Design for the Third Class Dewatering

Since the hydraulic connection of aquifers is apparently influenced by the water-proof curtain, in principle the dewatering design should arrange the well inside the foundation pit. It has several advantages as follows:

The dewatering efficiency of single well can be very high, especially for the well close to the water-proof curtain; the dewatering effect is very effective. The reason is that the water-proof curtain greatly influences the movement of groundwater into foundation pit; the groundwater in the upper part of aquifer should detour to the lower part and then can be move into the dewatering well inside the pit (shown as Fig. 7.13). Thus, under the identical well structure, the water discharge is mostly attributed from the groundwater inside the pit with water-proof curtain. Likewise, the water pumped out in the dewatering well outside the pit is mainly from the aquifer outside the pit. That is to say, the dewatering well set inside the pit is much in favor of pumping out the groundwater inside the pit, while the dewatering well set outside the pit can hardly have much effect to the purpose.

It is much advantage to diminish the environment effect by dewatering. The existence of water-proof curtain extends the seepage path. Under the same other conditions, the drawdown outside the pit is much smaller than that in the circumstance without the cut off of water-proof curtain. This decrease is directly related to the buried depth of water-proof curtain, well structure, the properties of aquifer, etc. Some basic rules are deeper the buried depth of the water-proof curtain, smaller the depth of wellpoint, poorer the vertical hydraulic conductivity of the aquifer, then less the influence of dewatering on the surrounding environment outside the pit.

In addition, even there is no water-proof curtain, the dewatering wells are set inside the pit have better dewatering efficiency than those outside the pit. Apparently, when dewatering wells are arranged outside the pit, the total plane area of wellpoints is larger than the foundation pit size; otherwise it is close or smaller. Under the same water discharge amount, definitely latter has much larger drawdown. But the inner arrangement of dewatering well brings in another inconvenience in the excavation and other underground construction. This need comprehensive coordination and operation management.

The seepage flow calculation in this class dewatering is a difficulty. Due to the influence of water-proof curtain, groundwater presents complicated three-dimensional flow state. Currently, there is no respective analytical solution or approximate analytical solution. Only through the numerical simulation based on specific three-dimensional seepage model the specific reference can be provided.

7.4.4 Design for the Fourth Class Dewatering

This type of dewatering is mostly employed in the unconfined aquifer foundation pit excavation. During the excavation the water table generally should be controlled beneath the foundation bottom by 1 m. When the excavation depth is very small and the foundation pit size is also relatively small, light wellpoint, open pumping both can be used for dewatering.

7.5 Case Study

7.5.1 Case 1—The Second Class Foundation Dewatering Engineering of Small Area and Large Drawdown

7.5.1.1 Project Summary

Diaphragm wall structure with a diameter of 29.60 m and thickness of 1.00 m is employed as cylinder supporting in a foundation pit construction, which is 27.6 in diameter and 0.80 m in thickness. The buried depth of this supporting structure is 54.20 m, while the excavation depth is 33.50 m (all the values are calculated from the designed ground elevation +4.85 m). The support structure design requires the water table should be depressed 26.30 m below the ground surface after the excavation down to 22 m.

The construction site is located in the alluvial plan of Yangtze River. The ground elevation is 3.90–4.10 m. The soil layers distribution along depth is shown as Table 7.1.

In construction site, there are two types of groundwater: unconfined–slightly confined water and confined water. Unconfined-slightly confined water exists in the layers of ① artificial fill and ③$_2$ sandy silt. The stable groundwater table is approximate 0.50 m (correspondingly 3.5 m in Wusong elevation).

Confined water is buried in two soil layers. One is ⑤$_1$ sandy silt in depth of 22.20–33.60 m. The average thickness is 8.85 m. The buried depth of water

Table 7.1 Soil layer distribution in site

Layer number	Soil type	Base elevation (m)	Thickness (m)
①$_{1-3}$	Plain fill	4.31 to 2.15	0.20–6.50
①$_2$	Creek fill	3.21 to −0.05	0.40–1.65
②	Silty clay	2.70 to −0.46	0.50–3.40
③$_1$	Mucky silty clay	1.41 to −2.66	0.30–3.50
③$_2$	Sandy silt	−1.09 to −5.05	0.50–8.30
③$_3$	Mucky silty clay	−2.91 to −7.92	0.6–4.70
④	Mucky clay	−13.32 to −18.87	6.4–13.95
⑤$_1$	Silty clay	−15.92 to 38.81	1.10–21.30
⑤$_{1-p}$	Sandy silt	−18.70 to −40.30	1.00–21.80
⑤$_3$	Silty clay	−46.02 to −60.28	8.10–25.40
⑧$_2$	Sandy silt interlayed with silty clay	−48.83 to −64.67	0.50–14.80
⑨$_1$	Silty and fine sands	−61.54 to −78.19 (un-penetrated)	5.90–24.30
⑨$_{1-p}$	Sitly clay	−65.18 to −72.59	0.90–8.30
⑨$_2$	Coarse and medium sands	Un-penetrated	Un-penetrated

7.5 Case Study

pressure is stable at 5.70 m (corresponding −1.68 m in Wusong elevation), in lentoid distribution. The lateral hydraulic connection is cut off by diaphragm wall. The second is ⑧₂ sandy silt interlayered with silty clay, ⑨₁ silty and fine sands, ⑨₂ coarse, and medium sands, in depth of 50 m. The stable groundwater is buried approximate at 7.50 m (corresponding −3.45 m in Wusong elevation).

7.5.1.2 Dewatering Design

1. The influence of groundwater on excavation of foundation pit

According to the practical soil drilling holes and geotechnical tests data, the checking of piping is conducted as Table 7.2.

The results show that the safety factor is far from 1 when excavation of foundation bottom. The groundwater table should be largely depressed.

The admissible groundwater tables under different circumstances can be seen in Table 7.3.

According to the investigation report of hydrogeology, there was an observation well (D3, shown in Fig. 7.16) in foundation pit. It was completely sealed before

Table 7.2 The calculation results of anti-piping

Hole No.	⑤₃ Base elevation (m)	Thickness of each layer beneath the excavation bottom (m)		The unit weight of each layer beneath the excavation bottom (kN/m³)		Anti-buoyancy (kN/m²)	Static water pressure (kN/m²)	Safety factor
		⑤₁	⑤₃	⑤₁	⑤₃			
K427	−46.22	7.17	10.40	17.9	18.0	316	428	<1
K422	−51.70	7.22	9.90	18.1	18.1	310	423	<1
Average	−50.37	13.77	14.15	17.8	17.8	313	426	<1

Note The design ground elevation is 4.85 m; the elevation of foundation bottom is 4.85 − 33.50 = −28.65 m; Unit weight of water is selected as 10 kN/m³; The groundwater table of soil layer ⑧₂, ⑨₁ is −3.45 m (buried depth is 8.30 m)

Table 7.3 Requirement of drawdown under different circumstances

Anti-buoyancy (kN/m²)	Safety factor	Admissible static water pressure (kN/m²)	Natural groundwater table (m)	Admissible groundwater table (m)	Drawdown requirement (m)
313	1.10	284	−3.45	−17.6	14.2
313	1.15	272	−3.45	−18.8	15.4
313	1.20	260	−3.45	−20.0	16.6

Note The elevation of layer base of ⑤₃ is determined as −46.00 m from the information of drilling hole of K427 and K422

excavation starting. The depth was 70 m, penetrating the lower part of soil layer ⑨$_1$. In case of the sudden piping, quicksand, etc., the requirement of drawdown in this project is to depress the groundwater table in confined layers of ⑧$_2$–⑨$_1$ below the excavation bottom by 1 m (the elevation is −29.65 m), i.e., 26.2 m.

2. Dewatering design

Based on the similar project experience, the upper unconfined water has no influence on excavation, so this aspect is not considered in dewatering designation.

Because the total area of the foundation pit is small, and the environmental requirement for surrounding condition is not very strict, the dewatering wells are all arranged outside the pit. Considering the influence to the other surrounding constructions, the well arrangement is shown in Fig. 7.16. The well structure can be found in Fig. 7.17.

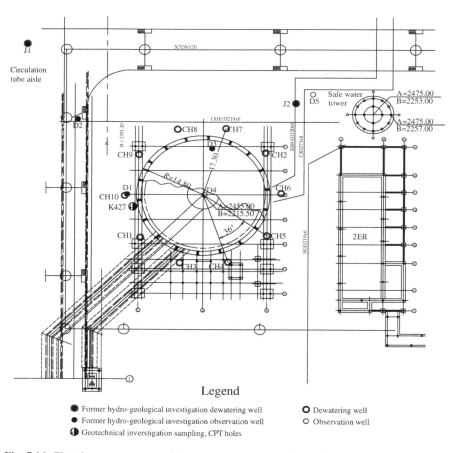

Fig. 7.16 The plane arrangement of dewatering and observation wells

7.5 Case Study

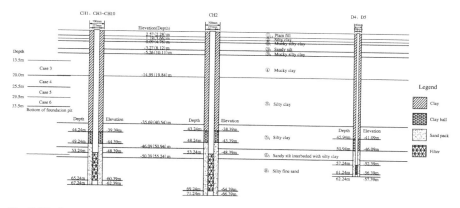

Fig. 7.17 Structure of dewatering and observation wells

Even though the site investigation has been provided in this project, in situ dewatering tests are conducted since there is large difference to the experience of designers. It should be noted that the trail-testing wells are also used as dewatering well after testing. There are two testing wells (CH1, CH2) and two observation wells (D4, D5). The structure of testing well has two types.

CH1—67.24 m in depth (calculated from the design ground elevation of 4.85 m), 700 mm in diameter, 325 mm in well pipe diameter, 8 mm in well thickness, 12 m in length of filter (buried from 53.24 to 65.24 m);

CH2—71.24 m in depth (the same with CH1), 700 mm in diameter, 325 mm in well pipe diameter, 8 mm in well thickness, 16 m in length of filter (buried from 53.24 to 69.24 m);

The observation wells are unified as: 62.24 m in depth, 350 mm in diameter, 108 mm in well pipe diameter, 4 mm in well thickness, 4 m in length of filter (buried from 57.24 to 61.24 m).

Finally, CH1 is selected as the well-structured type, in which the length of the filter is determined as 60 % of the value in the investigation report of hydrogeology. In practice, theory on unsteady flow of partially penetrated well is employed as Eq. (7.3).

$$s = \frac{Q}{4\pi KM} \left\{ W\left(u, \frac{r}{B}\right) + \frac{2M}{\pi L} \sum_{n=1}^{\infty} \frac{1}{n} \left[\cos\left(\frac{n\pi z}{M}\right) \right] \right. \\ \left. \times \left[\sin\left(\frac{n\pi(L+d)}{M}\right) - \sin\left(\frac{n\pi d}{M}\right) \right] W\left[u, \sqrt{\left(\frac{r}{B}\right)^2 + \left(\frac{n\pi r}{M}\right)^2} \right] \right\}$$
(7.3)

where K is hydraulic conductivity of the aquifer, m/d; B is the leakage factor, m; Q is the water discharge, m^3/d; s is the drawdown, m; M is the thickness of the aquifer, m; T is the coefficient of transmissibility (m²/d), $T = KM$; r is the distance from the dewatering well to the observation well or preset point, m; L is the length of filter in dewatering well, m; d is the distance from the top of the filter to the upper base of the aquifer, m; $W\left(u, \frac{r}{B}\right)$ is Well Function, in which u is the independent variable of Well Function, $u = r^2 \mu^* / 4Tt$; μ^* is the coefficient of elastic storage; t is duration of dewatering, d.

The parameters for calculation are: $T = 129.3$ m²/d, $\mu^* = 8.754 \times 10^{-4}$, $B = 235$ m, $M = 58$ m, designing $L = 12$ m, $d = 2.3$ m, $Q = 840$ m³/d (35 m³/h), so in results, the drawdown in the center of the foundation pit ($r = 17.3$ m) is 3.69 m after 12 h dewatering.

Thus, 10 dewatering wells are applied in practical construction (eight in operation and two for reserve) (Fig. 7.16).

According to this program, under regular operation, eight dewatering wells can depress the water table to the elevation of −32 m in duration of 24 h, which meets the requirement of drawdown to make the water table 1 m lower than the foundation pit, i.e., −29.65 m.

After the completion of well construction, trail testing is conducted. Table 7.4 shows the results.

Totally, from the above results, it is shown that after 30 h dewatering, the water table was depressed to the elevation −34.5 m related to the design ground elevation, which was really 1 m lower than the foundation pit bottom. The requirement was met. The testing results also match well with the previous design calculation.

Table 7.4 The trail testing results of group dewatering wells

Duration (h)	Well No.	Water table in CH1 (m)	Water table in CH2 (m)	Water table in D5 (m)
0		7.293	7.325	8.084
24	CH4, CH8	16.59	16.51	16.18
32	CH4, CH8, CH6, CH10	24.01	23.15	20.84
46	CH4, CH8, CH6, CH10, CH5, CH9	30.45	29.00	26.09
0		7.79	7.92	8.63
4	CH5, CH9, CH3, CH7	21.23	19.85	18.14
20	CH5, CH9, CH3, CH7, CH4, CH8	29.71	28.29	25.91
30	CH5, CH9, CH3, CH7, CH4, CH8, CH6, CH10	34.68	32.53	29.22

Note The observation well of D3 was previously destroyed by foundation pit construction. It was refound after the excavation. After 28 h dewatering, the water table in CH1 and CH2 are respectively −34.10 m. The relative elevation in D5 is −0.328, −0.311 and 0.428 m

7.5.2 Case 2—The Third Class Foundation Dewatering Engineering of Large Drawdown and Double Aquifers

7.5.2.1 Project Summary

This foundation pit engineering has excavation depth of 35.50 m, 31.60 m diameter of diaphragm wall with thickness of 1 m and buried depth of 53.00 m. After multistage excavation, 0.80 m lining is constructed.

The construction site is flat and smooth. The ground elevation is 4.18–4.64 m. The underground soil layers are presented in Table 7.5.

There are two kinds of groundwater type in this construction site, unconfined water and confined water.

The occurrence of unconfined water is shallow artificial fill (①$_1$, ①$_2$) and silty clay (②$_3$, ③$_2$). The stable water table is buried at depth of 0.30–3.30 m. The main supply is from precipitation.

The deep confined water is located in the ⑦ silty-fine sands and ⑨ silty-fine-medium sands. The upper part of ⑦ soil layer is the sandy silt with 15 m thickness, while the lower part is the silty sand with 12 m thickness. The elevation of the static water table is around −3.50 m. The upper confining bed is buried in 72.00 m. The investigation report of hydrogeology and dewatering tests both confirm that ⑧ is very impermeable.

Table 7.5 The soil layers in the construction site

Layer No.	Soil type	Base elevation (m)	Thickness (m)
①$_1$	Artificial fill	3.23 to −1.06	1.50–5.40
①$_2$	Creek fill	0.38 to −0.51	2.00–2.50
②$_1$	Silty clay	1.78 to −0.53	0.80–2.80
②$_3$	Silty clay	1.46 to −0.18	0.50–3.10
③$_2$	Silty clay	−1.98 to −6.15	3.00–6.70
③$_3$	Mucky silty clay	−4.53 to −7.28	0.50–4.00
④	Mucky clay	−14.50 to −19.46	8.90–13.70
⑤$_1$	Silty clay	−24.25 to −29.35	7.10–12.20
⑤$_3$	Silty clay	−34.60 to −40.14	6.90–13.30
⑦$_1$	Sandy silt	−48.04 to −68.25	7.90–31.90
⑦$_2$	Silty sand	−61.39 to −69.70	2.30–12.30
⑧$_1$	Silty clay	−64.69 to −75.71	1.90–13.10
⑨$_1$	Sitly sand	−73.01 to −77.11	1.00–11.60
⑨$_2$	Medium sands		

7.5.2.2 Dewatering Design

1. Data collecting and analyzing

Compared the information of geotechnical and hydrogeology investigation, it is found that:

There is an old observation well with 159 mm diameter and 70 m depth, left in the foundation pit, penetrating to the lower part of ⑦ soil layer. Because the construction site of foundation pit is very complicated, after the completion of diaphragm wall, this well could not be found any more, so that effective sealing measurement could not be conducted before foundation excavation. Thus some potential safety problem may threaten the success of this project, which should be seriously considered in dewatering design.

The thickness of ⑧$_1$ soil layer is revised from 9.9 m in the previous investigation report to 3 m in average (1.8 m for smallest) after professional expert confirmation. By recalculation, the safety factor of ⑨ soil layer (silty-fine sands, medium sands) is also smaller than 1, which could not meet the standard requirement, shown as Table 7.6.

2. Dewatering objective

For ⑦ soli layer (sandy silts, silty sands), since there is an observation well penetrating ⑦$_2$ soil layer. From the aspect of safety, the water table of confined aquifer should be controlled 1 m beneath the foundation pit excavation surface (36 m in the depth or more). The natural water table is about -3.5 m (depth is 8 m). Thus, the dewatering design is to ensure the water table depressed more than 28 m.

For ⑨ soil layer (silty-fine sands, medium sands), the water table should be lowered down for 1.5 m (shown as Table 7.7).

3. Difficulties in construction

The requirement of drawdown is very large, for the first confined aquifer the drawdown is needed for 28 m.

The diaphragm wall is buried into ⑦ soli layer (the upper base is 42 m in depth) as deep as 53 m. The horizontal hydraulic connection is highly influenced and three-dimensional flow states should be considered in calculation.

Two different dewatering types should be conducted in this foundation pit excavation for two separated confined water. For the first confined aquifer, over 28 m drawdown is required, while in the second confined aquifer, that is, only 1.5 m is needed.

Preliminary information shows that there are big differences between these two confined aquifers on hydraulic conductivity.

These two confined aquifers are just separated by a small thickness aquiclude (1.8 m smallest). Dewatering well drilling has difficulties.

7.5 Case Study

Table 7.6 The calculation of anti-piping

Hole No.	⑧₁ Base elevation (m)	Thickness of each layer beneath the excavation bottom (m)				The unit weight of each layer beneath the excavation bottom (kN/m³)				Anti-buoyancy (kN/m²)	Static water pressure (kN/m²)	Safety factor
		⑤₃	⑦₁	⑦₂	⑧₁	⑤₃	⑦₁	⑦₂	⑧₁			
A21	−66.30	8.00	14.00	11.40	1.80	18.2	18.3	18.7	17.9	647	613	1.05
A22	−67.20	8.90	12.50	11.70	3.00	18.3	18.4	18.6	18.2	665	621	1.06
A44	−68.55	6.15	15.60	11.70	4.00	18.2	18.4	18.4	18.4	688	635	1.07
A45	−67.92	5.22	17.20	11.40	3.00	18.4	18.6	18.5	18.4	682	628	1.08

Note The ground elevation is 4.20 m; the pit bottom elevation is −31.10 m; the unit weight of water is 9.81 kN/m³; the water table of ⑨₁ is −3.28 m (depth is 8.38 m)

Table 7.7 The drawdown requirement for each point

Hole No.	Drawdown under different safety factor (m)		
	$K_s = 1.10$	$K_s = 1.15$	$K_s = 1.20$
A21	2.53	5.14	7.53
A22	1.68	4.36	6.81
A44	0.97	3.75	6.29
A45	0.82	3.56	6.08
Average	1.50	4.20	6.68

4. In situ pumping test

Because the hydrogeological parameters of aquifers provided by previous investigation hydrogeological report are calculated without the influence of diaphragm wall, and the excavation depth is also assumed as 30 m, which is 6 m shallower than practice, thus the in situ pumping test is necessary before the submission of dewatering design. It has several purposes:

(1) To determine the hydrogeological parameters and water table of the first confined aquifer after the completion of diaphragm wall;
(2) To determine the dewatering well structure and single well water discharge;
(3) To know about the water table variation after the completion of diaphragm wall for specific dewatering design;
(4) To provide very specific controlling parameter of the construction of dewatering wells.

Three pumping wells and two observation wells was designed previously. After all testing wells were completed, due to the construction of foundation pit; real design program has been changed that three dewatering wells are also constructed inside the pit, i.e., CH1–CH6, G1, G2. The plane arrangement and testing wells are both adjusted correspondingly (Fig. 7.18).

The structures of pumping wells and observation wells are presented in Fig. 7.19.

Pumping well—CH4, 61 m in depth, 700 mm in diameter, 273 mm in well pipe diameter, 8 mm in well thickness, 15 m in length of filter (buried from 44 to 59 m); CH2, CH3, 68 m in depth, 700 mm in diameter, 273 mm in well pipe diameter, 8 mm in well thickness, 20 m in length of filter (buried from 46 to 66 m).

Observation well—G1, 59 m in depth, 350 mm in diameter, 108 mm in well pipe diameter, 4 mm in well thickness, 5 m in length of filter (buried from 54 to 59 m); G2, 61 m in depth, 350 mm in diameter, 108 mm in well pipe diameter, 4 mm in well thickness, 5 m in length of filter (buried from 54 to 59 m).

According to *Standard for Hydrogeological Investigation of Water Supply*, the filter should be back-filled surrounded by 0.5–2.0 mm gravel pack, which needs extra 6 m up. Then 5 m clay ball for sealing. Close to the well head, additional clay is also needed. Finally, concrete or cement is poured.

7.5 Case Study

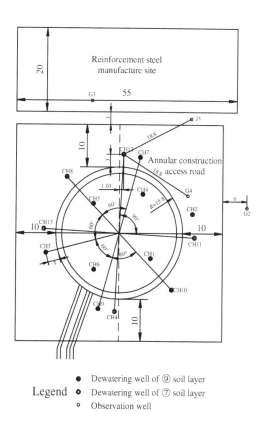

Fig. 7.18 The arrangement of dewatering and observation wells

Table 7.8 presents the results of pumping tests. The group pumping test has conducted for 3 days. First was CH1, CH5. After 24 h CH4, CH6 were added. Finally, CH2 outside the pit also started pumping.

According to the above results, the existed five group wells simultaneously pumped; there was still 5.162 m gap to the drawdown requirement of 28 m (the water table should be depressed to −36.00 m). The pumping of CH2 outside the pit has great influence on the drawdown inside the pit (2.191 m), so three more pumping wells are needed to pump outside the pit. Eight wells pump at the same time can meet the requirement. In addition, two more wells are for reserve. In total there are 10 pumping wells together (eight in operation and two for reserve).

5. Dewatering design
(1) The first confined aquifer of ⑦

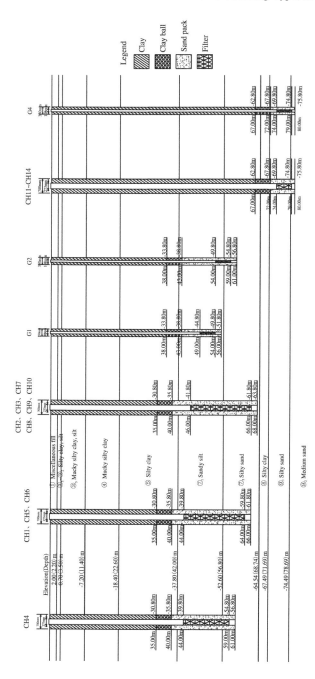

Fig. 7.19 The structure of dewatering and observation wells

7.5 Case Study

Table 7.8 The drawdown requirement for each point

Pumping well No.	Initial water table (m)	Water table after 24 h pumping (depth from design ground surface) (m)	Drawdown (m)	Accumulative drawdown (m)
CH1, CH5	7.665	20.676 (21.856)	13.011	13.011
CH1, CH5, CH4, CH6		27.467 (28.647)	6.791	19.802
CH1, CH5, CH4, CH6, CH2		29.658 (30.838)	2.191	21.993

Firstly, back analysis of hydrogeological parameters by three-dimensional unsteady flow is conducted as follows:

$$\begin{cases} \frac{\partial}{\partial x}\left(K_{xx}\frac{\partial h}{\partial x}\right) + \frac{\partial}{\partial y}\left(K_{yy}\frac{\partial h}{\partial y}\right) + \frac{\partial}{\partial z}\left(K_{zz}\frac{\partial h}{\partial z}\right) - W = \mu_s \frac{\partial h}{\partial t} & (x,y,z) \in \Omega \\ K_{xx}\frac{\partial h}{\partial n_x} + K_{yy}\frac{\partial h}{\partial n_y} + K_{zz}\frac{\partial h}{\partial n_z}\Big|_{\Gamma_2} = q(x,y,z,t) & (x,y,z) \in \Gamma_2 \\ h(x,y,z,t)|_{t=t_0} = h_0(x,y,z) & (x,y,z) \in \Omega \end{cases} \quad (7.4)$$

where K_{xx}, K_{yy}, K_{zz} are hydraulic conductivity along x, y, z direction, respectively, cm/s; h is the water table at location (x, y, z) at time t, m; W is the water recharge, d^{-1}; μ_s is storage coefficient at location (x, y, z), m^{-1}; t is time, h; Ω is the seepage area; Γ_2 is the third boundary condition; n_x is the unit vector along x direction of the exterior normal of boundary Γ_2; n_y is the unit vector along y direction of the exterior normal of boundary Γ_2; n_z is the unit vector along z direction of the exterior normal of boundary Γ_2; q is the lateral supply amount per unit area of boundary Γ_2, m^3/d.

Based on the parameter values from above analysis, the drawdown variations dewatering by wells CH1, CH4, CH5, CH6 inside the pit and wells CH2, CH3, CH7, CH9 outside the pit are presented in Figs. 7.20 and 7.21.

According to this calculation results, combined the three-dimensional numerical simulation, the final dewatering design is to drill 10 wells (eight in operation and two for reserve). Considering the complicated construction site, CH1, CH4, CH5, CH6 are set inside the pit, while CH2, CH3, CH7, CH9 are set outside the pit. They are working simultaneously.

After all pumping well completion, to check the dewatering design program and the well-completion quality, pretesting of engineering dewatering results are shown in Table 7.9.

From above table, after 24 h dewatering of eight pumping wells, the water table is buried at the depth of −36.12 m, which is very close to the requirement of drawdown, 1 m lower than the foundation pit excavation surface (buried depth is 36.30 m). The results prove the dewatering program is effective and reasonable.

(2) The second confined aquifer of ⑨

Due to big difference on the previous hydrogeological investigation report the ⑨ soil layer structure, in dewatering design of the second confined aquifer, the well

308   7 Dewatering Types in Foundation Pit

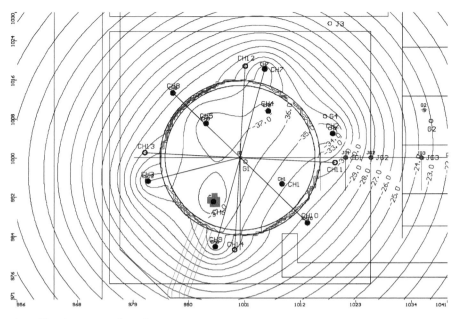

Fig. 7.20 The contour line of water table when design drawdown is met

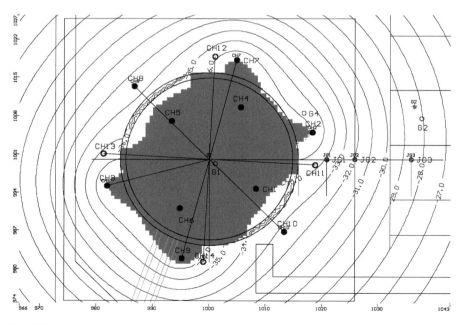

Fig. 7.21 The contour line of water table after 1 month dewatering

7.5 Case Study

Table 7.9 The results of trail-testing

Duration (h)	Pumping well No.	Water table of observation well (from the top of diaphragm wall) (m)	
		G1	G3
0	CH1, CH4, CH5, CH6	19.30	16.35
12	CH1, CH4, CH5, CH6, CH7, CH9	28.15	18.75
16	CH1, CH4, CH5, CH6, CH7, CH9	29.36	19.65
20	CH1, CH4, CH5, CH6, CH7, CH9	30.46	20.25
24	CH1, CH4, CH5, CH6, CH7, CH9	31.42	20.75
36	CH1, CH4, CH5, CH6, CH7, CH9, CH8, CH10	33.30	22.12
40	CH1, CH4, CH5, CH6, CH7, CH9, CH8, CH10	33.69	22.95
44	CH1, CH4, CH5, CH6, CH7, CH9, CH8, CH10	34.36	24.07
48	CH1, CH4, CH5, CH6, CH7, CH9, CH8, CH10	34.88	24.85
60	CH1. CH4, CH5, CH6, CH7, CH9, CH8, CH10	36.12	26.00

structure has been adjusted for improvement. Thus pumping test is conducted in situ.

According to the in situ construction duration, based on the relative experience and the calculation results in Table 7.8, four pumping well and one observation well are arranged in the dewatering design (the plane arrangement and well structure are shown in Figs. 7.18 and 7.19), in which four dewatering well CH11, CH12, CH13, CH14 are 8 mm in thickness, the observation well G4 is 4 mm in thickness.

Before construction, CH12 and G4 are first drilled for trail pumping test to check the well quality and dewatering effectivity. In case of unusual circumstance, timely adjustment can be made.

Practical pumping test results indicate CH12 has 140 m³/h water discharge amount. The drawdown observed by G4 in 24 h duration is 0.559 m. It meets the project requirement.

7.5.2.3 Dewatering Operation

1. The first confined aquifer of ⑦

Due to the foundation pit excavation, the observation well in the center of pit has been broken, so the operation of dewatering started earlier than predesign. Once the excavation of the third soil layer was onset, two pumping wells were conducted dewatering. When the excavation extended to the fourth layer, the four wells inside the pit were all turned on. Until the fifth layer, the additional three wells outside the pit were also involved. During the construction, the water table inside the pit is always kept under the excavation surface. After the completion of foundation pit

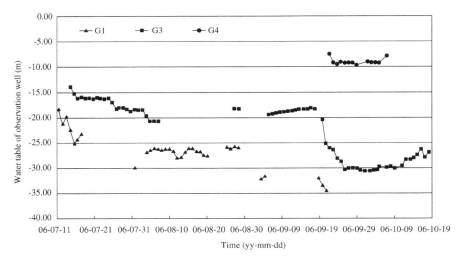

Fig. 7.22 Drawdown variation of observation wells

concrete pouring, gradually decrease the number of dewatering wells. After the first support is completely constructed, the dewatering can be terminated.

In the later period, just seven dewatering wells are in operation, which can also meet the drawdown requirement.

2. The second confined aquifer of ⑨

After calculation, it indicates when excavation moves to fifth soil layer, the dewatering of the second confined aquifer should be started gradually. Two dewatering wells were open on September 20, 2006; next day one well was added into dewatering. Hereafter, these three wells take turns in operation (totally the safety factor is 1.10). 5 days after the completion of pit bottom slab concreting, the dewatering well number was reduced to 1; and next day dewatering operation was terminated (Fig. 7.22).

Through calculation, the water tables of observation well G4 and the center of the pit under three or four group wells operation are almost the same. The variation of drawdown and theoretical calculation match very well.

7.6 Exercises

1. What are the effects of dewatering in foundation pit engineering?
2. What influences of diaphragm wall on seepage properties in foundation pit?
3. How the length of filter influences the groundwater seepage in foundation pit?
4. How many types of foundation pit dewatering according to the influence of diaphragm wall on the aquifer? What are they and how about the properties?

Chapter 8
Engineering Groundwater of Bedrock Area

8.1 Concepts and Classifications of Groundwater in Bedrock Area

8.1.1 Concept of the Bedrock Groundwater

The concept about fissure water in hydrogeological literature was first introduced by former Soviet Union scholar, Г.Н. Каменский and А.Н. Семихатов. In their book named *Soviet Hydrogeology* written in 1932, the groundwater in fissure and limestone was unified to the term of "groundwater flow," which was defined as follows: it can flow along the hard rock fissures, cracks, holes, or other channels according to a certain rule and has larger flow velocity and flux and higher temperature. Based on this, the "groundwater flow" was further divided into two subclasses, "groundwater flow in fissure rocks," and "cave river in limestone areas." The groundwater in fractured rocks and in porous rocks has been distinguished, and the following scholars put forward the basic concept of fissure water and karst water according to this concept mentioned above.

8.1.2 Classification of the Bedrock Groundwater

The clear concept of fissure water in hydrogeology was formally put forward by Ф.П. Саворенский, the well-known former Soviet Union scholar, and fissure water was defined as an independent type of groundwater. In the book of *Hydrogeology* written in 1935, for the parts about classification of groundwater, Ф.П. Саворенский divided the confined water into two subcategories of fissure water and karst water. The fissure water is the groundwater that buried in the fracture fissure of geologic structure. In the hydrogeological textbook written by Ф.П. Саворенский in 1939, the classification of groundwater and basic concepts of various types of groundwater were made further correction. The groundwater was divided into five categories in this book, namely: soil water (including swamp water and perched water), phreatic water, karst water, artesian water, and vein water (fissure water). Besides the soil water, the author gave definitions for the various groundwater types as the following:

Phreatic water—the groundwater in the surface sediments and upper layer of weathering crust;

Artesian water—the groundwater in the sedimentary structure (basin);

Karst water (cavern water)—the groundwater in limestone, dolomite, and other soluble rocks;

Vein water (fissure water)—the groundwater that mainly in structural fissure.

From present point of view in evaluating the groundwater classification, the main problem is that both the hydraulic properties of aquifer (or buried characteristics of aquifer) and the medium type constitute the basis for classification. As such, the characters of groundwater types were mutual tolerance and uniqueness. For example, under certain buried conditions, karst water and fissure water can also be the phreatic water or artesian water. In addition, the definitions of some groundwater types are not scientific and rigorous. Artesian water not only exists in sedimentary basin, and fissure water not always distributes as the shape of vein and only in structural fractures. Although there are some defects in this classification discussed above, it is the first and the most comprehensive classification scheme of groundwater in the discipline history of hydrogeology. Based on Саворенский classification, the Russia and China scholars developed various classification schemes for groundwater.

Following the Саворенский classification, А.Н. Овчиников, the former Soviet Union scholar, proposed a more comprehensive and precise classification of groundwater in 1949 to overcome the defects of Саворенский classification scheme. In his book of *Common Hydrogeology* (Soviet higher school teaching material) written in 1949, according to the burial conditions, groundwater was clearly divided into three types, namely upper perched water, phreatic water, and artesian water. According to the aquifer lithology, the above three types of groundwater were further divided into two subclasses of pore water and fissure water. In the description of the subclass, the karst water belonged to fissure water. But the fissure water and karst water were as the separate chapters in this book. In other words, the groundwater was actually divided into three types of pore water, fissure water, and karst water according to the aquifer lithology. In addition, the author gave a widely accepted definition for fissure water which was the groundwater in fracture rock, such as igneous rock and sedimentary rock. This definition was obviously more comprehensive than the Саворенский classification of "fissure water is the groundwater that mainly in structural fracture."

The classification of А.Н. Овчиников has many advantages. It not only summarized the basic types of groundwater, and also reflects the two main characteristics of hydraulic properties and medium types for various types of groundwater. Thus, this classification scheme was widely accepted and used in the field of hydrogeology. Various groundwater classification schemes are basically following the А.Н. Овчинников classification, in which the classification of groundwater was divided according to the hydraulic properties and medium type.

Hydrogeologists around the world have a similar view about the groundwater classification based on hydraulic properties. The groundwater of which has a higher pressure above the aquifer was referred to as artesian water. However, this definition has some limitations and according to the establish basis of the two differential equation types of groundwater dynamics. From the view of hydraulics, almost all hydrogeologists have the opinion that the groundwater should be divided into two types of phreatic water and confined water.

In addition, due to the hydraulic properties of groundwater that mainly depend on the burial condition, some hydrogeological literature classified the groundwater according to the hydraulic properties referred to as buried groundwater types. The groundwater was divided into the three types of upper perched water, phreatic water, and confined water. Some literatures also added a groundwater type between two stratums. In fact the groundwater in aquifers between two aquicludes which do not have pressure head should also be the phreatic water from the view of hydraulic properties.

About groundwater classifications according to the aquifer medium, there are two kinds of classification schemes at present.

The first classification scheme origin from Soviet Union is used in Russia, China, and some other countries. The divided groundwater types are based on the gap types of aquifer medium. The basic idea of this classification is that there are completely corresponding relation between basic types of rock and gap types. A certain gap type (including intergranular pores, fracture, and corrosion pores) corresponds to a certain type of groundwater. According to this view, the groundwater can be divided into three classifications, namely pore water in loose uncemented rock, fissure water in non-soluble solid rock, and karst water in soluble rocks (limestone, dolomite, etc.). This classification can directly reflect the interdependent relationship among rock types, gap types of water storage, and the groundwater types (see Table 8.1). Therefore it becomes the theoretical basis for finding, exploration, evaluation of groundwater resources and has been widely used in hydrogeology textbooks and various rules of groundwater exploration, and hydrological geological scientific research and production.

The second classification scheme of groundwater classified by aqueous medium can be represented by the European and American countries, and can be seen in the book of *Hydrogeology* written by Davis and Dewiest (1966, USA), *Groundwater* written by R.A. Freeze and J.A. Cherry (1979, Canada), *Dynamics of Fluids in Porous Media* written by J. Bear (1979, Israel), *Groundwater Hydrology* written by H. Yamamoto (1992, Japan), and some other monographs. The types of groundwater are directly controlled by the rock types. Although there are no specific chapters about groundwater classification in these books, the characteristics of groundwater were all described based on the rock types of magmatic rock, metamorphic rock, volcanic, sedimentary rock (or further divided into sandy rock and carbonate rock), alluvium and permafrost rock types. The named aquifers also accord to the rock type (such as igneous rock and metamorphic rock aquifer,

Table 8.1 Aquifer classifications in bedrock

Rock types	Basic types of groundwater		Subclasses in rock types	Space types for water storage	Subclasses of groundwater	Names of the groundwater subclasses
Unconsolidated rock	Groundwater in unconsolidated rock (I)		Unconsolidated porous rock	Intergranular pore	Groundwater in unconsolidated porous rock (I_1)	Pore water
			Some loess and loess rocks	Intergranular pore, diagenetic pore and diagenetic fracture (vertical fissure)	Loess pore and fissure water (I_2)	Pore and fissure water
			Some clay	Diagenetic fracture (Consolidation)	Clay fissure water (I_3)	Clay fissure water
Half-hard rock	Bedrock groundwater	Half-hard rock groundwater (II)	Clastic rock of weak cementation	Pore between layers, bedding and structural fissures	Pore-fissure water in half cemented rocks (II)	Fissure-pore water
Insoluble solid rock		Groundwater in insoluble solid rock (III)	Basite volcanic ash layers of Cenozoic	The hole formed by diagenetic and weathering	Pore groundwater in volcanic ash (III_1)	Pore groundwater water in volcanic ash
			Basic lava (basalt) of Cenozoic	Diagenetic large holes and horizontal conduit	Groundwater in lava holes (III_2)	Groundwater in lava holes
			Insoluble solid rock	Tectonic, diagenetic and weathering fissures	Fissure water in bedrock (III_3)	Fissure water in bedrock (fissure water)
Soluble rock		Groundwater in soluble rock (IV)	Soluble rocks (various kinds of carbonate rocks and clastic rock with solute elements)	Tectonic rock fracture and fissure, cavity, caves of karst rock	Fissure-karst water (IV_1)	Fissure-karst water
			Soluble rock	Corrosion conduit and cave	Groundwater in corrosion conduit (IV_2)	Karst water

8.1 Concepts and Classifications of Groundwater in Bedrock Area

carbonate aquifer, clastic rock aquifer, etc.). The advantage of this classification scheme is intuitive, and easy to understand. But the rock types are various, thus the classifications of groundwater are multifarious and lack of systematicness. At the same time, this classification could not reflect the important hydrogeology properties, such as storage and transport of groundwater.

Compared to the above two kinds of groundwater classification schemes, it is obvious that the classification according to the rock gap types is more scientific. But in recent years, with the deepening development and exploration of the groundwater, the single groundwater classification scheme based on pore types of aqueous medium is still imperfect. There are several aspects as the following:

1. There is no absolute correspondence relation among the rock types, gap types, and groundwater types. For example, fracture gap is not only in the non-soluble solid rock. A large amount of fracture space can also exist in some loose rocks such as the loess and some kind of clay soils. Large-size space of holes is also not only in soluble carbonate rocks, but also in some of clastic rock with soluble components, such as cement or soluble breccia. Even there are various holes and conduits in volcanic rocks.

2. Some transitional rock types are among the three basic types of rocks (loose rock, the soluble solid rock, and soluble rock). They often have two types of space systems for water storage (i.e., double porosity media). There are many half cementation (hard) clastic rocks in Mesozoic and Cenozoic tertiary strata of China, which have both intergranular pores and diagenesis and the structural fissures. Thus both pore water and fissure groundwater exist in these rocks. Some clastic rocks with soluble compounds may have various space of diagenesis and the structural fissures, solution fissures and holes and even pipeline space, which contains both fissure water and karst water. The loess plateaus in northwestern China that both have pore water and fissure (vertical fissure) characterize as the dual-pore medium. For the present groundwater classification that based on aquifer medium types, the position of groundwater types in interim rocks and dual-pore medium is not clear.

3. During the groundwater exploration and development in recent years, some new types of water storage space were found. Water storage in large tunnel, shaft, and hole of the basic lava, and in some big holes layer of basalt (possibly the buried volcanic ash) also has great significance. However, these gaps and groundwater types in the current classification based on the general medium are in no position. In conclusion, the groundwater classifications simply according to rock types and its characteristics are not fit for the actual situation of groundwater existing forms and cannot summarize all groundwater types existed in nature. Therefore, the current widespread classification of groundwater is necessary to be further improved. The concepts of the three basic types of groundwater, especially the concept of fissure water also needs to be redefined.

Based on the above problems existing in the groundwater classification, Zisheng Liao added the interim type to the current three types, and named the groundwater classifications reflected characteristics the rock and space types. The improved groundwater classifications (according to the classification of aquifer medium types) are shown in Table 8.1.

The basic features of this classification are as the following. First, the groundwater can be divided into two categories of "unconsolidated rock groundwater" and "bedrock groundwater" according to the structure characteristics of the rock. The bedrock groundwater can be further divided into three basic types of "half-hard rock groundwater," "groundwater in insoluble solid rock," and "groundwater in soluble rock" according to the rock structure and soluble features. Then according to the space characteristics of aquifer medium, unconsolidated rock groundwater can be further divided into three subclasses and the bedrock groundwater can be divided into six subclasses. Besides the traditional bedrock fissure water (III_3), this classification of bedrock fissure water actually includes the types of fissure-karst water (IV_1) in soluble rock and fissure-pore water in half-hard rock (II). Because fissure is the main space for the storage and transport of the three groundwater types, the regularity of groundwater movement and enrichment is mainly controlled by tectonic conditions. General bedrock fissure water, therefore, is the groundwater in hard and half-hard rocks and fissure is the main space for water storage.

8.2 Forming Conditions, Characteristics, and Storage Regularities of the Bedrock Fissure Water

8.2.1 Forming Conditions of the Bedrock Fissure Water

The formation, storage, and transport of the bedrock fissure water are influenced by various inside and outside factors. A combination of three basic conditions for the formation of bedrock fissure water is hard or half-hard rock, abundant water, and rock fissure resulted from the long-time tectonic movement.

Geological tectonism results in a large number of structural fissures in hard and half-hard rocks, which is the advantageous condition for groundwater storage and transport. The storage space for groundwater was formed at the fault fracture zone and its related fracture belt during the process that stratum or rock mass moved along the fracture surface. The intrusive body caused the deformation of surrounding rock at different rock contact zone, and fracture was formed or the original space of the fracture was increased. During the condensation of the dyke invasion and the influence of the later tectonic movement, a large number of protogenesis and secondary cracks was formed on dyke and both sides of the rock mass, providing favorable conditions for groundwater storage; Soluble rocks are mainly distributed in carbonatite area. The strong interaction would occur between carbonatite and hot water solution, resulting in the gradual increase of protogenesis and structural fissure and finally the various formations of karst structure.

8.2.2 Characteristics of the Bedrock Fissure Water

Distribution and movement of the bedrock fissure water (including buried karst water) has its uniqueness. Its main features are as follows:

1. The burial and distribution of bedrock fissure water are extremely uneven and are completely controlled by various kinds of fracture belts. Thus the aquifer is irregular.
2. The shape of the bedrock fissure aquifer is variety, and size and shape of the bedrock fissure are controlled by the geological structure and landform conditions. Thus burial and distribution condition of the bedrock fissure water are complex.
3. The bedrock fault vein aquifer buried deeply, but the quantity of groundwater is not abundant in normally.
4. The bedrock fissure water is obviously controlled by geological structure. The formation and distribution of various gaps in rocks are normally associated with the geological structure. Geological structure factor plays a leading role in the formation of the abundant aquifer in bedrock.
5. Dynamic properties of bedrock fissure water have their particularity. The groundwater buried in the same bedrock may not have the unified water table, and may alternately characterize as pressurized and confined water. Water movement is also complex, including laminar and turbulent flow. There may be pipe flow and open channel flow in the underground karst cave, which are not the forms of seepage flow. This is determined by the rock fissure and the special form of the karst cave.

8.2.3 Occurrence Regularity of the Bedrock Fissure Water

Occurrence regularity of the bedrock fissure water is the result of comprehensive shaping of many factors. Generally, lithology, geological structure, recharge, runoff and drainage, terrain, landform, and climate play a certain role in occurrence and distribution of the bedrock fissure water, and the lithology, structure, and supply factors play a main role.

The influence of lithology on bedrock fissure water is through the effective control on the fracture. The different mechanical properties of the rock develop the different size of the structural fissures under a certain tectonic stress. Take the plastic stratum for illustration, tectonic stress results in plastic deformation of the rock with the poor water occurrence. However, tectonic stress results in rupture of the brittle stratum with the rich water occurrence.

Under the effect of tectonic stress, various forms of deformation, such as fold and fault, are formed on the strata, increasing the space for water storage and the favorable conditions for water catchment. Due to the high fissures destiny of the thin layered strata, the occurrence conditions for rich water are significant beneficial. For the thick-block strata, oblique stratifications are development with few and same scale fissures which are basically in closed state, and water occurrence conditions are poor. For the extraordinary development of the fracture structure, the squeeze fault resulted from squeeze and mylonitization is water-blocking, and cannot have the abundant groundwater. The faults with water occurrence and transport performed as water storage structures are usually extension or shear faults in the low class or order. These faults tend to have strong conductivity properties, but can't form a water-rich hose. As such, spatial variability of water occurrence condition was strong in the different parts of a fold. A certain scale of water-rich segment was usually formed at the wings of the anticline, steep turn slow parts of the stratum and syncline area under the negative terrain condition.

8.2.4 Flow Regularity of the Bedrock Fissure Water

The seepage of fissure water is much more complex than the pore water. The rock matrix of pore water is composed of particles with different sizes, and pores connect with each other between particles. Thus a unified free water surface can exist in the aquifer, and obey the Darcy's law for permeability. The fissure water is in the matrix of fractured rock. According to the distribution of structural planes, the fractured rock can be divided into five types: the whole structure, block structure, layered structure, fracture structure, and loose structure. In fact, the distribution of fractures in rock mass is inhomogenous, and some fissures are not connected. Thus the fissure water characterized as the shapes of slice, fasciculation, or vein. The permeability of the rock mass is controlled by the opening size of the fracture, geostress conditions and the rock properties.

Fractured rock mass is a multiphase discontinuous medium resulted from the incision of different scales, direction, and properties of the fissures due to the various geological processes. Geological structure, terrain, and hydrological factors resulted in the uneven and strong directional rock permeability for the fissure water with the extremely complicated seepage law.

1. Seepage law of fissure water in blocky rock mass

Blocky rock mass mainly refers to the rock mass with the relatively uniform properties, and has the blocky structure. The deep intrusive rocks, such as granite, and some volcanic and subvolcanic rocks are blocky rock masses. Except the strong tectonism areas, blocky rock mass general develops normal joint system and small fault with no regional fracture. However, the shear fractures are developed. The three kinds of the shear, squeezing, and tensile fractures are developed in the strong tectonism areas. Squeezing fractures have the weak permeability with no water occurrence. The permeability of shear fracture is medium. Tensile fracture has the high permeability with a large space for water storage.

In general, the permeability of rock mass is poorer with small space for water storage in blocky rock distribution area. Groundwater seepage in blocky rock mass depends on the structural plane (especially the fissure). Fissure development of the blocky rock mass is generally in weathered zone, fault or fracture zone, and intrusive contact zone.

2. Seepage law of the layered fractured rock mass

Layered rock mass mainly refers to the layered sedimentary rocks and sedimentary metamorphic rocks. The bedding, schistosity, and joint are developed in layered rock mass. The development characteristics and occurrence changes of the bedded fissures have a close relationship with the development degree of fold strata and stress condition.

(1) The thin layered rock mass is beneficial for water storage due to the high density of bedded fissures. Oblique beddings are developed in thick blocky rock mass with little quantity and small size of bedded fissures which are basically in closed state. Therefore, the water occurrence conditions are poor.
(2) For the same rock thickness, the plastic rock mass (clay rock, shale, marl, etc.) has the higher fissure density than the brittle rock mass (sandstone, limestone, igneous rock, etc.).
(3) For the stratums with bedding developed, anticlinal structure is beneficial for groundwater transport with the developed penetrability fissures. Groundwater often flows off to the two wings, and do not exist in axis of the anticline.
(4) The syncline structure is generally beneficial for water collection. If a tunnel locates at the syncline axis, it often encounters the relatively large amounts of water gushing. Fracture part in the shaft section of the syncline structure, which is called interlaminar fracture, is mainly deep embedded. When geostress worked on the stratum and the syncline was formed, the upper part of syncline structure was under the by maximum extrusion pressure, and the lower part under the tensile stress. Thus for the hard brittle stratum, the interlayer rock was breaked with the developed fissure joints. But for the argillaceous stratum, such as clay shale, the situation was different. Under the ideal circumstances of lithology, topography, geomorphology and climate, the interlaminar fractures in hard brittle stratum of the syncline of are favorable for groundwater accumulation.
(5) The seepage of fissure water in bedrock is along fissure networks of the strata with the characters of heterogeneity, anisotropy, orientation, and interaction with tectonic stress.

3. Seepage law of the cataclastic structured rock mass

Cataclastic structure mass was resulted from tectonic broken, fold broken of the rock, and interspersed extrusion of the magmatic rocks, joint, fault, fault infected zone, cleavage, bedding, schistosity, interlayer sliding surface, etc., are the main structural planes, developing the weak structural plane filled with some mud.

Groundwater is characterized as vein and fissure water, and often with local vein pressure.

The faults that have effects of block water are usually resulting from squeezing. Due to the extrusion and mylonitization, these faults are impossible occurrence of groundwater. However, the developed tensile or shear fractures at one or both sides of the fault effect zones are beneficial for groundwater occurrence. Tensile or shear faults with the properties of water transport and storage may form the local part for water rich, and sometimes with high-confined water head tend to have strong conductivity properties but no water-rich slices. If these faults communicate with the leaking limestone, fissure water of the sand and mudstone layer would be unwatering, and pore water in unconsolidated layer would be under the strong discharge with water table declining sharply.

4. Seepage law of unconsolidated structure

Unconsolidated structure is mainly fault fracture zone and strong weathering fracture zone resulted from tectonic action. Vein and pore water exist in the unconsolidated structure. Fracture rate at fracture surface or fracture zone of a single lithology, is usually higher in tensile fracture than in shear fracture, and squeezing fracture has the lowest fracture rate. The pure tensile fracture has the highest fracture rate. Thus, the squeezing fracture has the character of anti-water, but some parts of the fault may also be water rich. Tensile fracture is water rich, and tensile fracture is more beneficial for water storage. Water-rich capacity of the shear fracture is between the squeezing and tensile fracture.

8.3 Groundwater Seepage Model of Fractured Rock Mass

People have gradually realized the importance and urgency of fractured rock mass seepage since the arch dam break of Malpasset, French in 1959. Fractured rock mass seepage model is the basis of the fractured rock mass seepage analysis, although various fracture seepage models have been put forward, each one has its shortcomings. Perfect fractured rock mass permeability model still needs to be further established. Models of present mainly developed along the two directions, one was the fracture-pore dual-medium model considering the water exchange between fractures and matrix in the system, and the other was the non-dual-medium model which ignored the water exchange process.

8.3.1 Dual Model of Fracture-Pore

The dual-medium model of fracture-pore was considered that fractured rock mass is the unity formed by a fracture system with poor porosity but good hydraulic conductivity, and a pore system with good porosity but poor hydraulic conductivity.

It takes into account the water exchange process between the two kinds of systems. Fist, based on Darcy's law, water flow equations of the two medium systems were set up respectively. Then the water exchange equation connected the two separate flow equations. According to the established method, it can be divided into quasi-stable and unsteady flow models.

8.3.1.1 Quasi-stable Flow Model

The water exchange quantity of the fracture and pore rock system, which is implicit expression of time "t", is proportional to water head difference of the two systems for quasi-stable flow model. The main representative scholars of the model are Barenblattt, Warren and Rott, etc.

The concept of hydraulic dual medium was first put forward by Barenblatt with the main views as follows:.

(1) Fracture system and rock pore system are all throughout the region, forming an overlapping continuum. Each point of seepage field has two water head values, i.e., the average head value the pore system and the water head of fissure system.
(2) Permeability is several orders of magnitude smaller than the porosity of the rock. However, permeability of the fissures is several orders of magnitude larger than the porosity. Water flow in the rock mass is characterized by intense water exchange between the two different systems;
(3) It assumed that fissure and pore rocks are homogeneous and isotropic.

Although Barenblatt model was an important basis for the development of the dual-medium theory, the penetration mechanism of it was parochialism. Creak and pore system were all assumed to be the isotropic. Thus if the water exchange between two systems was ignored, the fractured rock mass could be seen as an isotropic porous media, and the desultorily fissures only worked as the pore channels. The extreme penetration mechanism may only appear in argillaceous rock affected by the intense tectonic movements, the rock mass suffered surface weathering. Therefore, the main disadvantage of Barenblatt model is that it did not reflect the anisotropic characteristics of the fractured rock medium and the fissure water flow.

Warren and Rott added new geometry and penetration restrictions to the assumptions of Barenblatt model of fissure system.

(1) Fissure system in rock mass is homogeneous, orthogonal, and interconnected. Permeability axis parallels to the orientation of fissure groups. Equal interval fracture groups with constant width are perpendicular to the main permeability axis. On the contrary, fracture groups along the main permeability axis may have various intervals and fissure widths for simulation of the anisotropy medium.

(2) The pore system divided by fissures is homogeneous and isotropic.
(3) Water exchange, which is proportional to the water head difference occurs between the two different systems, and water exchange quantity. Compared with Barenblatt model, the new model obviously considered the widespread anisotropy of permeability in fractured rock mass. However, it can only be applied to the uniform orthogonal fracture networks.

8.3.1.2 The Unsteady Flow Model

Unsteady flow model assumes that the process of water exchange of the two systems is the water in the pore system flows into the fracture system. According to the character of water flow in pore system, and the water exchange equation can be established. Due to the amount of water exchange is explicit formulation of time "t", it was called unsteady flow model. According to the space configuration of fracture system, the unsteady flow model present mainly includes parallel fractures unsteady flow model and the group fractures unsteady flow model.

The main assumptions in parallel fractures unsteady flow model are as follows:

(1) The fissure system is composed of a set of the width and interval of parallel fractures with infinite extension, and the rock was cut into columnar structures by fractures.
(2) The water exchange of the two systems is expressed as a vertical linear flow from pore system into the fracture system, which can be described by one-dimensional control equation with appropriate boundary and initial conditions. Obviously, the model is only appropriate for penetration space structure formed by the bedding fissure system.

The main assumptions of group fractures unsteady flow model are as follows:

(1) Fracture system consists of three intersection groups of fractures with the same crack width. The rock was cut into the massive body, and can be replaced with a series of equivalent homogeneous sphere with the same radius.
(2) Water exchange of the two systems is expressed as a radial flow from the center of rock matrix to the fracture. Compared to the parallel fractures unsteady flow model, this model obviously has made some improvements, but it still makes a certain restrictions on crack configuration.

The outstanding advantage of fracture-pore dual-medium model is that water exchange between the two different systems is considered, which is especially suitable for fluid storage in the fracture aquifer. The theory can be used for study of oil recovery from high pressure fractures of kilometers depth, or taking rare earth elements from ancient metamorphic water. However, for the two kinds of models based on this theory, the quasi-stable state flow model assumes that water exchange quantity is proportional to the he water head difference between the two systems, and is not directly explicit formulation of time "t" which would actually lead to big error. Zimmerman pointed out that this error would eliminate only after a long time, and it cannot be ignored in the initial stage. For the unsteady flow model, water exchange equation is related to the space configuration of the fissure system. In order to establish water exchange equation, the configuration and shape of fissure system are made certain restrictions, which limit the application of these models. It should be noted that the conditions of fractures development should be considered in model selection for actual engineering. The fracture-pore dual-medium model remains to be further perfected.

8.3.2 Non-Dual-Medium Model

Another kind of model for fractured rock mass seepage analysis is the non-dual-medium model considering the fracture permeability. Due to ignoring water exchange between the pore and fissure system, it is not limited by rock mass fracture configuration when the model is applied, and it also can reflect the heterogeneous and anisotropic properties in fluid seepage. Thus, the non-dual-medium model is the most studied and widely used model at present. The non-dual-medium model mainly includes the equivalent continuum medium model, discrete fracture network model and the discrete coupling model which combines the advantages of the former two models.

8.3.2.1 Equivalent Continuum Medium Model

The fissure water is equivalent averaged to the entire rock mass, which can be characterized as an anisotropic continuum with a symmetric permeability tensor, and in the classical continuum theory is used in the equivalent continuum medium model.

The prominent advantage of equivalent continuous medium model is that anisotropic continuum theory which has solid foundation and experience both in theory and problem-solving methods can be used. What's more, the model can work without the exact location and hydraulic characteristics of each fracture, which is useful for those engineering problems with difficulty to obtain a single fissure data. However, equivalent continuous medium model has two difficulties in application. One is the determination of equivalent permeability tensor for fractured rock mass, and the other is that the effectiveness of the equivalent continuum model cannot be guaranteed.

1. Determination of the equivalent permeability tensor

A given equivalent permeability tensor must be unconditionally applied in the dynamic fields with similar water systems. Otherwise, the following questions will appear in determination of the equivalent permeability tensor. (1) The equivalent permeability tensor obtained in a certain boundary condition may not correctly predict the flux through another boundary condition. (2) The equivalent permeability tensor decided by the flux may not be able to forecast the water head distribution accordingly. Therefore, when the model is applied to the fractured rock mass, the determination methods of the equivalent permeability tensor are very important.

Field water pressure test method, inversion method, and geometry method are the normal methods for determination of the equivalent permeability tensor.

(1) Water pressure test method. Because the permeability tensors of fractured rock mass have six independent parameters, a single-hole water pressure test is not sufficient to determine the permeability of fractured rock mass. In General, the permeability tensors of fractured rock mass in engineering are measured by three sections of water pressure test, group wells test, and cross wells of water pressure test. However, due to the large discrete degree of permeability of fractured rock mass, test results inevitably have the size effects. Considering the high cost of water pressure test, it cannot be performed in large area, so the field water pressure test is difficult to be applied widely.

(2) Inversion method. Inversion method is a kind of optimization methods. It determines the best collocation of divisional seepage parameters of rock mass based on the principle of the least difference between the analyzed and the tested groundwater table. The inversion method can be divided into two types: direct and indirect method. Due to the instability of direct method during calculating and high demands on the measured data, the indirect method is used more often. Inversion method is currently the most widely used method in engineering. However, because the permeability tensors of fractured rock mass have six independent parameters, determination of the permeability tensors may encounter the problems of nonuniqueness and instability. The selection of initial permeability parameter and some optimized coefficients is normally depended on the experience to a large extent, and inappropriate choice may not only affect the calculation speed but even the convergence of the results.

(3) Geometry method. Permeability of fractured rock mass mainly depends on the geometry parameters of fracture system, such as fracture azimuth, width and density of the fissures, etc., and also closely related to the size of the crack and connectivity. Therefore, for a known fracture system, geometry method can be used to determine the permeability tensors. For an actual rock mass, due to the random fracture distribution, statistical analysis on the fractures is needed firstly to divide the fractures into several groups with typical fracture surface, and then the equivalent permeability tensors can be obtained. Because it is difficult to measure the geometry parameters of fissure system accurately in engineering, and also difficult to quantify the influence of crack size and connectivity on the permeability tensor of rock mass, geometry method can only determine the initial value of permeability tensor which should be finally corrected by hydraulic test or inversion method.

2. Effectiveness of the equivalent continuum model

Whether the continuum seepage theory can be used to analyze the fractured rock mass seepage is a controversial issue. Many scholars have put forward some criterions. Louis gave a view that for an engineering rock mass, when the number of fractures in rock mass is more than 1000, the equivalent continuous medium model can be used. Maini found that if the ratio of average crack spacing of rock mass and building size was less than 1/20, the equivalent continuum model can be used. In the study by Wilson and Witherspoon, the ratio of largest joint spacing of the rock mass and the construction size is less than 1/50 is the criterion of using equivalent continuum model. But all these criterions are obtained from the specific project or theory analysis, and they are difficult to be applied in actual engineering.

Long did the further research, and pointed out that two conditions were needed to characterize the fractured rock mass as a continuous media.

(1) Sample volume is suitable for simulation, i.e., equivalent permeability only has a small change with the slight increase or decrease of the size of test body.
(2) The symmetrical equivalent permeability tensors, which can be determined by the measured directional permeability, exist in the fractured rock mass. Supposing K_J as the permeability in hydraulic gradient direction and K_f as the permeability in flow direction, if $K_J^{1/2}$ and $K_f^{1/2}$ can be plotted as an ellipse in polar coordinates, the medium should have symmetrical permeability tensors. In addition, the effect of fracture geometry parameters on fissure permeability was also studied by Long. He gave the view that the random distribution of fracture directions and the constant crack width are beneficial to the effectiveness of the equivalent continuum model. The fracture rock mass with more intensive fracture networks, resulting a better connectivity, is more close to an equivalent anisotropic continuum medium.

8.3.2.2 The Discrete Fracture Network Model

Discrete fracture network model is on the premise of figuring out every spatial orientation, gap width of crack geometry parameters, and is based on the basic flow formula in a single fissure. It uses the principle of equal quantity between inflows and outflows of each fracture intersection to establish the equation and then by solving equations to obtain the crack head value of the intersection.

The network line element method was first put forward by Wittke and Louis, which is similar to loop the current method in circuit analysis. Mao Changxi described a crack hydraulic network model similar to pipe network of hydraulics. Wilson and Witherspoon simulated the fissures in rock mass using triangular elements or line units respectively. They showed a finite element technology to simulate the two-dimensional fracture network flow, and demonstrated that interference of fracture cross-flow can be neglected in an example to illustrate the advantages and feasibility of the line unit method. For the three-dimensional problems, a three-dimensional fracture network disk model was first put forward by Long, solved by the hybrid analytical–numerical method. Nordqvist showed a three-dimensional network model with the metabolic fracture width. Dershowitz described a three-dimensional polygonal fracture network model. In further research, Wanli combined it with the finite element method, and showed a polygon unit seepage model of three-dimensional fracture network.

In these methods of solving three-dimensional problems, finite element method is most effective and convenient to simulate the discrete fracture network. If the fracture cross interference flow can be neglected, for saving calculation, graphic element can be used to simulate fracture surface, and surface intersections will be the nodes. Thus the two-dimensional flow on the local surfaces of fracture network forms the overall three-dimensional flow. Because crack units are not on the same plane, first of all local coordinate system $o'x'y'$ for each fracture unit should be set up. In the local coordinate system $o'x'y'$, the cracks flow can be regarded as local two-dimensional isotropic flow. Then finite element control equation $[K]^e h^e = F^e$ of each fracture unit can be established respectively. According to flow balance in the node of fracture intersections, i.e., $\sum_e F^e \cdot b = 0$, the integral finite element equation is obtained.

It shows that the discrete fracture network model gives a specific simulation on each fracture in the network system and tries to get real seepage state of each point in the fissure system, which obviously has good simulation and high precision. However, when the number of fissures is large, the workload is tremendous or even impossible, especially for the three-dimensional problems. In addition, due to the randomness of fracture distribution, to establish a discrete real fissure system is also very difficult. Therefore, except the simple condition, it's difficult to use discrete fracture network model widely d in actual engineering.

8.3.2.3 Equivalent Discrete Coupling Model

Equivalent discrete coupling model is a new model combined with the advantages of equivalent continuous medium model and discrete fracture network model. As proposed above, discrete fracture network is characterized as the good simulation and high precision, but tremendous workload for simulation of larger number of fissures. Equivalent continuum model can overcome the difficulties, but its effectiveness is difficult to guarantee when fissure density is small. Thus, some scholars put forward a coupling model with combining of the above two models. A partition mixture model can be described as follow. Discrete fracture network model is used for the fracture areas with smaller fissure density and equivalent continuum model is used for large fissure density areas. The unified domain mixed model can be used for the an area that the discrete fracture network model is used for simulation of a few large and medium-sized cracks fissures which play an important part in permeability adopt, the equivalent continuum model is used for simulation of large destiny small cracks in the blocks divided by large or medium-sized cracks. Then coupling discrete equation can be established according to the equal heads (i.e., unified head) and node flow balance in the connected areas of the two medium types. This model with enough engineering precision can not only obviously avoid the huge workload in discrete fracture network model for simulating each fracture, but also guarantee the effectiveness of the equivalent continuum model.

8.4 Three-Dimensional Numerical Model for Bedrock Fissure Water

Adopting model technology into study on the migration regularity of groundwater in the rock mass is of great significance for analyzing the seepage law of fissure water, the calculation and evaluation of water resources, the seepage field analysis and forecast, and so on. Building a mathematical model to reflect the migration regularity of real fissure water in the space is still a difficult problem. A comprehensive model which has coupled a one-dimensional flow model specifically reflecting the flow in a water channel and a three-dimensional numerical model reflecting the seepage of fractured rock mass, which can also link the water corridor model in turbulent flow condition, has a practical meaning.

Fissure water model research began in the 1960s, and a certain amount of progress has been made. The study mainly developed along two directions. One is the double-medium model, but the establishment of the model needs to make some assumptions for the fissure and rock mass system, which limits its application. Another is the double-medium model, mainly includes the discrete medium model and equivalent continuum model. Building a discrete medium model requires the geometrical characteristics and permeability coefficients of all the fissures for transporting water. As its difficulty and big workload, it is hard to apply in practical. Widely used equivalent continuous medium model can simulate the fractured rock mass with symmetrical permeability tensors of anisotropic body, and the mature continuum theory can be used. At present, it is mainly applied in two-dimensional flow model, and few researches were about the three-dimensional flow model. Also, the equivalent continuum model is not always effective.

According to the research status of fissure water numerical models, the equivalent coupling continuum model based on the equivalent continuous theory, reflecting the one-dimensional linear flow in a transport channel and the three-dimensional of the seepage in fractured rock mass, and coupling water sink gallery model of the non-Darcy flow, should be established. It is valuable and important to seek effective solving method for coupling model, and make the research results into practical applications.

The fissure development characteristics are the main factors influencing the fissure water transport. Fissure water flows in the complex networks composed of banded faults, planar apertures and tubular and cavern fissures. Specially, the tubular fractures usually play a role of water corridor and transport channel. Thus, the vein seepage characteristics of fissure water are that the water mainly transports in the trunk fractures and stores in the microcracks and blowholes. Fissure water flow of the high heterogeneity and anisotropy, is generally the laminar flow in rock mass and conforms to the linear Darcy's law. However, there may be turbulent flows at the large space such as the karst channels.

According to characteristics of fissure water seepage, a coupling model was put forward for the three-dimensional flow. One-dimensional flow model based on the local coordinate system was established for specific simulation of water flow in the transport channels of fractured rock mass. Water exchange model was established for the turbulent state of flow in water transport channels. But for the whole fractured rock mass system with small scale and large density of fissures, the three-dimensional flow model should be established.

8.4.1 Equivalent Three-Dimensional Model

$$\begin{cases} \frac{\partial}{\partial x}\left(K_{xx}\frac{\partial h}{\partial x}\right) + \frac{\partial}{\partial y}\left(K_{yy}\frac{\partial h}{\partial y}\right) + \frac{\partial}{\partial z}\left(K_{zz}\frac{\partial h}{\partial z}\right) - W = \mu_s \frac{\partial h}{\partial t} & (x,y,z) \in G, t \geq 0 \\ h(x,y,z,t) = h_0(x,y,z) & (x,y,z) \in G, t = 0, \\ h(x,y,z,t)|_{\Gamma_1} = h_1(x,y,z,t) & (x,y,z) \in \Gamma_1, t \geq 0 \\ [K]\text{grad} h \cdot n|_{\Gamma_2} = q(x,y,z,t) & (x,y,z) \in \Gamma_2, t \geq 0 \end{cases} \quad (8.1)$$

where h is the groundwater head, L; μ_s is the specific storage of fracture media, L^{-1}; K is the equivalent anisotropic hydraulic conductivity, LT^{-1}; n is the unit outside normal vector of the boundary.

$$W = \sum_{j=1}^{N_w} \sum_{k=1}^{N_L} Q_{jk} \delta(x - x_j, y - y_j, z - z_{jk}) \quad (8.2)$$

where N_w is the number of pumping wells; N_L is the number of the layers; Q_{jk} is the water yield of the kth layer and jth well.

8.4.2 One-Dimensional Model

$$\begin{cases} \frac{\partial}{\partial x}\left(K_x \frac{\partial h}{\partial x}\right) + Q = \mu \frac{dh}{dt} & (0 \leq x \leq L, t > 0) \\ h|_{t=0} = h_0(x) & (0 \leq x \leq L) \\ h|_{x=0} = f_1(t), h|_{x=l} = f_2(t) & (t > 0) \end{cases} \quad (8.3)$$

where h is the water transport channel; Q is recharge-source; K_x is the permeability coefficient along the fracture, LT^{-1}, $K = \frac{b^2 \rho g}{\lambda \mu}$; μ is water release coefficient of the fractured medium (dimensionless); t is the time; T, x is the local coordinates along the fracture.

8.4.3 Water Catchment Corridor Model

The approximate calculation formula for water discharge considering the inside and outside water head difference of water transport channel and the conductivity performance between rock masses can be expressed as:

$$Q \approx T(h - d) \tag{8.4}$$

where Q is the inflow of water transport channel, L^3T^{-1}; h and d are the water heads of fractured rock mass and water transport channel, respectively, L; T is the transmissibility coefficient of the interface of fractured rock mass and water transport channel, L^2T^{-1}.

Based on principle of the of water head continuity at interface of water transport channel and the overall system of rock mass, and the flow equilibrium at unit nodes of the according parts, the coupling of two kinds of medium models is as the following. The respective permeation unit matrices of two medium models were formed after the discretization of the partial differential equation. According to the principle of unified node number at the interface of the two kinds of medium, permeation unit matrices of two medium were superposed to form the overall seepage matrix without having to make other transformation. Water exchange model of the turbulent water catchment corridor can be superposed to the overall matrix according to the unit node numbers of water catchment corridor. The algebraic equation of coupling seepage model of the fractured rock mass was obtained on the basis of above assembling. During the process of forming calculation program, the overall matrix of coupling model can be formed by cumsum with the same node number. The finally formed large algebraic equations can be solved by strong implicit iteration method with well stability.

8.5 Project Types and Instances of Bedrock Fissure Water

8.5.1 Groundwater of the Rock Slope Engineering

8.5.1.1 The Effects of Groundwater on the Rock Slope Stability

Groundwater has important effects on slope stability, especially on the reservoir bank slope. The influence of groundwater on the slope stability is mainly manifested in the following three aspects:
1. The influence of groundwater soil physical and mechanical properties of the slope rock

On the one hand, the saturated rock was soft after reservoir impoundment. Friction coefficient and cementing ability of the soil particles are lower as a result of the water lubrication. Shear parameters of the potential sliding surface are reduced, decreasing slope sliding resistance. On the other hand, during the reservoir operation, repeat rise and fall of the water table result in cycle seepage flow in the slope body. Groundwater seepage has the leached effects on slope, and the tiny particles also transport under the action of groundwater. Slope erosion phenomenon, the mesoscopic and macroscopic cavities on the potential sliding surface of slope body, reduce shear strength of the potential sliding surface.

2. Buoyancy

Rock body immersed in reservoir water is under the buoyancy effect of groundwater. Buoyancy is equal to the product of underwater rock volume and water weight of per unit volume, i.e., $A_w \gamma_w$. In general, the weight of underwater slope is equal to its weight under buoyancy. There are two aspects of buoyancy on the slope stability. On the one hand, buoyancy reduces the effective weight of the slope body. The sliding resistance of the sliding surface is reduced, which brings adverse effect to the slope stability. On the other hand, the weight decrease of the slope body reduces the slide force, contributing to the slope stability. Therefore, stability evaluation of the buoyancy on slope is not simply the pros and cons, and should consider the specific engineering geological conditions and comprehensive evaluation of mechanical parameters of the rock–soil body.

3. Seepage force

Instability of many slopes occurred while the reservoir water table dropped sharply. One reason for this was that the disappearance of the buoyancy increased effective weight of according part of the slope, leading to changes in slope stability. The other reason is that the dropped water table resulted seepage flow in rock–soil body, and the seepage force caused the slope unstability. The seepage force was controlled by shape of the saturation line, the permeability coefficient of the rock mass, the size of saturated area and dip angle the potential sliding surface, etc. Currently, there is no accurate calculation method for seepage force. The saturation line of slope body is complex to determine, is affected by permeability coefficient and specific yield of the rock and dropping speed of the reservoir water table. According to *The Engineering Handbook of Foundation Pit*, the average hydraulic grade is used for calculating the seepage force. And the average hydraulic grade is slope of the line connected the intersection points of saturation line and the landslide mass slope (as shown in Fig. 8.1, the slop of the AB line). The total seepage T of slide body can be calculated by the following equation:

$$T = \gamma_w A_w I \quad (8.5)$$

where γ_w is the water weight of per unit volume; A_w is the area beneath the saturated line; I is the average hydraulic grade.

Seepage force is along the direction of average hydraulic grade. Although the calculation of the hydraulic gradient is simplified in Eq. (8.5), the shape of the saturation line should be determined to calculate the saturated area. Thus the Eq. (8.5) is more suitable for the circular sliding surface. Transfer coefficient

8.5 Project Types and Instances of Bedrock Fissure Water

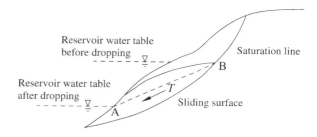

Fig. 8.1 Schematic of the hydraulic gradient for a slope

method can be used to analyze the slope stability. For the slip surface shaped as broken line, based on the simplified and safety consideration after the reservoir water table decline, the slope line is generally assumed to be the saturation line. Assuming the direction of seepage force is sliding surface tangential, the seepage force can be expressed as:

$$D_i = \gamma_w A_w \sin \omega_i \tag{8.6}$$

where D_i is the seepage force of the ith slice; ω_i is the dip angle of the ith slice.

8.5.1.2 Calculation Model of the Slope Stability, Considering the Role of Groundwater

Engineering stability of rock slope can be calculated by various methods, such as the finite element method, probability method and limit equilibrium method, and so on. However, the limit equilibrium method is the most commonly used method, and is described below for the broken line sliding surface.

Residual sliding force and stability coefficient of the slop are calculated by the transfer coefficient method. According to the broken line shape of the sliding surface, the sliding body is divided into vertical bars at line turning points of the broken line (Fig. 8.2). The transfer coefficient for the residual force of the vertical bars from top to down is expressed as:

$$\lambda_i = \cos(\omega_i - \omega_{i+1}) - \sin(\omega_i - \omega_{i+1}) \tan \varphi_i \tag{8.7}$$

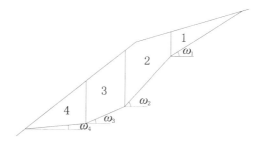

Fig. 8.2 Schematic of slice method for the slop

where λ_i is the transfer coefficient passed from residual sliding force of the ith slice to the $(i + 1)$th slice; φ_i is the friction angle of sliding surface of the ith slice, which is the saturated friction angle when soil is saturated.

Thus, the residual sliding force passed from the ith slice to the $(i + 1)$th slice can be expressed as:

$$E'_{i+1} = \lambda_i E_i \tag{8.8}$$

Obviously, if $E_i \leq 0$, indicated there is no residual sliding force for the ith slice. Stability coefficient and slide thrust of the slope can be calculated as follows:

$$F_s = \frac{\sum_{i=1}^{n-1}\left(R_i \prod_{j=1}^{n-1} \lambda_j + R_n\right)}{\sum_{i=1}^{n-1}\left(T_i \prod_{j=1}^{n-1} \lambda_j + T_n\right)} \tag{8.9}$$

$$E_i = \lambda_{i-1} E_{i-1} + K_s T_i - R_i \tag{8.10}$$

$$R_i = W_i \cos \overline{\omega_i} \tan \varphi_i + c_i L_i \tag{8.11}$$

$$T_i = W_i \sin \overline{\omega_i} + D_i \tag{8.12}$$

$$D_i = \gamma_w A_w \sin \overline{\omega_i} \tag{8.13}$$

where F_s is stability coefficient of the slope; R_i is the sliding resistance for the ith slice; K_s is the designed security coefficient of the slope; T_i is the sliding force of the ith slice; W_i is the weight of the ith slice, in which submerged weight should be used when it is under saturated condition; c_i is the cohesion of the sliding surface of the ith slice, in which saturated cohesion should be used for the saturated slope; L_i is the length of the sliding surface for the ith slice.

Equations (8.9) and (8.10) reflect the comprehensive influence of the groundwater on the stability of the slope, such as the change of the mechanical parameters of sliding surface, buoyancy, seepage force, etc. From the two formulas, we can get two criterions on the slope stability. If $F_s \geq K_s$, the slope is stable, otherwise is unstable. If $E_n \leq 0$, the slope is stable, otherwise is unstable.

8.5.1.3 Stability Seepage Force of the Landslide

1. The basic concept

Landslide stability analysis is based on Mohr–Coulomb theory. The main two methods for landslide stability analysis are the total stress method (Su-analysis) with the consideration of the undrained shear strength (Eq. 8.14), and effective stress method ($\bar{c}$ and $\bar{\varphi}$ analysis) with the consideration of the drainage shear strength (Eq. 8.15).

$$K_f = \frac{\sum (N_i \tan \varphi_i + c_i L_i)}{\sum (W_i \sin \alpha_i)} \quad (8.14)$$

$$K_f = \frac{\sum (\overline{N}_i \tan \overline{\varphi}_i + \overline{c}_i L_i)}{\sum (W_i \sin \alpha_i)} \quad (8.15)$$

Look from the quantity, the difference between the two methods is that pore water pressure of the sliding zone is considered in the $\bar{c}$ and $\bar{\varphi}$ analysis method. The pressure is equal to h_w, the height the slide body beneath infiltrate surface, multiplied by γ_w, the weight of per unit volume water. For any vertical slice of the slope with the width of l_i, the effective normal stress $\overline{N}_i$ is equal to the total normal pressure, N_i minus pore water pressure, or as $\overline{N}_i = N_i - \gamma_w h_w l_i$.

However, from the intrinsic view, the seepage effect and declined water table are considered in the $\bar{c}$ and $\bar{\varphi}$ analysis method. The pore water pressure is deducted, thus the sliding resistance is produced by the effective stress completely in the stability analysis. This method is more suitable for the stability evaluation of accumulated layer landslide under long-term changes of water table. Taking the Three Gorges Project as an example, Most of the slopes influenced by the changes of reservoir water table are accumulated layer landslides. Continuous infiltration surface, i.e., a unity of underground water table, is normally formed by the control of reservoir water table. Long-term stability is the main goal of the landslide stability analysis. Thus, the effective stress method should be used for this.

According to Bernoulli equation, the total water head of the seepage flow of the saturated soil water is the sum of position head, pressure head, and flow velocity head. The sum of the position head and the pressure head is called the piezometric head. In general, the velocity head can be neglected in landslide because of the large soil seepage resistance and the velocity flow. Total stress can be considered to be the sum of the effective stress and buoyancy of the sliding body skeleton. However, the sharply decline of the water table and also its movements from the range of 175–145 m exist in Three Gorges reservoir area. The total head (equivalent to the piezometric head) of different parts of the main landslide section would decrease along the flow direction. The ratio between the head loss, Δh of the two points and the penetration length, L, is named as the hydraulic slope, I. Therefore, the seepage force which is the external force of water flow on soil slope should be considered, namely. Seepage force is a kind of volume force. It is proportional to the hydraulic grade, and has the same direction as the seepage.

2. Calculation formulas for the landslide stability coefficient and thrust under the seepage pressure

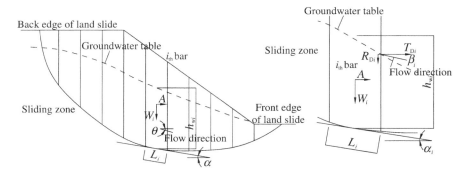

Fig. 8.3 A calculation model of accumulated layer landslides: Sweden slice method (*circular sliding* surface)

In general, the shapes of sliding zone of reservoir landslide are circular arc and broken lines with the corresponding calculation models:

(1) Sliding surface as a single plane or arc surface (Fig. 8.3) landslide stability calculation:

$$K_f = \frac{\sum((W_i(\cos\alpha_i - A\sin\alpha_i) - N_{wi} - R_{Di})\tan\varphi_i + c_i L_i)}{\sum(W_i(\sin\alpha_i + A\cos\alpha_i) + T_{Di})} \quad (8.16)$$

where $N_{wi} = \gamma_{wi} h_{wi} L_i$ is pore water pressure, which is approximately equal to the area beneath the infiltration plane $h_{wi} L_i$ multiplied by the weight of per unit volume water γ_w.

Component force of the parallel sliding slice resulted from the seepage force is expressed as:

$$T_{Di} = \gamma_w h_{wi} L_i \sin\beta_i \cos(\alpha_i - \beta_i)$$

Component force of the vertical sliding slice resulted from the seepage force is expressed as:

$$R_{Di} = \gamma_w h_{wi} L_i \sin\beta_i \cos(\alpha_i - \beta_i)$$

where W_i is the weight of the *i*th slice, kN/m; c_i is the cohesion of the *i*th slice, kPa; φ_i is the internal friction of the *i*th slice, °; L_i is the sliding length of the *i*th slice, m; α_i is the first slide angle of the *i*th slice, °; β_i is the groundwater flow direction of the *i*th slice, °; A is coefficient of the earthquake acceleration (gravity acceleration, g); K_j is the stability coefficient.

It's very difficult to determine the pore water pressure of the landslide. The effective stress can be assumed as:

$$\overline{N}_i = N_i - \gamma_{wi}h_{wi}l_i = (1 - r_U)W_i \cos \alpha_i$$

where r_U is the pore pressure ratio, defined as the ratio of the total pore water pressure and total uplifting pressure, can be expressed as:

$$r_U = \frac{\text{Underwater area of the sliding body} \times \text{unit water weight}}{\text{Total volume of the sliding body} \times \text{unit sliding body weight}}$$
$$\approx \frac{\text{Underwater area of the sliding body}}{\text{Total area of the sliding body} \times 2}$$

In general, the ratio of unit water weight and unit sliding body weight is reduced to 0.5. Thus, r_U is greatly simplified in calculation by the total underwater area of the slices.

Accordingly, Eq. (8.16) can be simplified as:

$$K_f = \frac{\sum \left(((W_i(1 - r_U)\cos \alpha_i - A \sin \alpha_i) - R_{Di}) \tan \varphi_i + c_i L_i\right)}{\sum (W_i(\sin \alpha_i + A \cos \alpha_i) + T_{Di})}$$

where W_i is the weight of the ith slice, kN/m. Different slope layers which may have various weights must be considered in the calculation. The slop body weight of above and under the seepage surface should also be distinguished, with the natural weight above the water table and saturated weight under the water table, and can be expressed as:

$$W_i = (\gamma_i h_{1i} + \gamma_{sat} h_{wi}) b_i$$

where γ_i and γ_{sat} are the natural and saturated weight of the slope respectively; h_{1i} and h_{wi} are the heights of the vertical slice above and under the seepage surface respectively. It should be noted that due to considering of the static pore water pressure, submerged weight γ' is not used in the formula. If submerged weight is considered, i.e., $\gamma' = \lambda_{sat} \gamma_w$, the pore water pressure N_{wi} should be merged in Eq. (8.16).

Accordingly, Eq. (8.16) can be transformed as:

$$K_f = \frac{\sum ((W_i(\cos \alpha_i - A \sin \alpha_i) - R_{Di}) \tan \varphi_i + c_i L_i)}{\sum (W_i(\sin \alpha_i + A \cos \alpha_i) + T_{Di})} \quad (8.17)$$

Calculation the landslide thrust

Shen Runzhi et al. proposed a calculation formula of the landslide thrust based on arc method. According to the characteristics of the slop and the prevention and treatment engineering, two calculation methods for thrust are proposed.

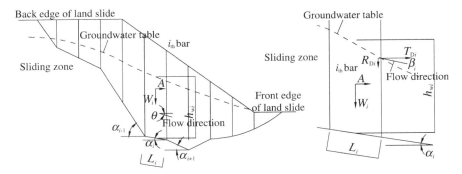

Fig. 8.4 Landslide model II: transfer coefficient method (*broken line* surface)

Shear resistant: the sliding body is relatively intact and intensity, and the shearing strength of the sliding slice is low. The landslide thrust is general in rectangular distribution.

$$H_s = (K_s - K_f) \times \sum (T_i \times \cos \alpha_i) \tag{8.18}$$

Bending resistance: sliding body integrity is poor. The landslide thrust is general in triangular distribution.

$$H_m = (K_s - K_f)/K_s \times \sum (T_i \times \cos \alpha_i) \tag{8.19}$$

where H_s and H_m are the landslide thrust, kN; K_s is the designed safety coefficient; T_i is the tangent force component of the vertical slice weight in sliding zone.

(2) The sliding surface is as broken lines shape (Fig. 8.4)

(1) Calculation of the landslide stability

$$K_f = \frac{\sum_{i=1}^{n-1} \left[((W_i((1-r_U)\cos \alpha_i - A \sin \alpha_i) - R_{Di}) \tan \varphi_i + c_i L_i) \prod_{j=i}^{n-1} \psi_j \right] + R_n}{\sum_{i=1}^{n-1} \left[(W_i(\sin \alpha_i + A \cos \alpha_i) + T_{Di}) \prod_{j=i}^{n-1} \psi_j \right] + T_n}$$

(8.20)

where $R_n = (W_n((1-r_U)\cos \alpha_n - A \sin \alpha_n) - R_{Dn}) \tan \varphi_n + c_n L_n$

$$T_n = W_n(\sin \alpha_n + A \cos \alpha_n) + T_{Dn}$$

$$\prod_{j=i}^{n-1} \psi_j = \psi_i \psi_{i+1} \psi_{i+2} \cdots \psi_{n-1}$$

8.5 Project Types and Instances of Bedrock Fissure Water

where ψ_j is the transfer coefficient passed from residual sliding force of the ith slice to the $(i+1)$th slice, and can be expressed as:

$$\psi_j = \cos(\alpha_i - \alpha_{i+1}) - \sin(\alpha_i - \alpha_{i+1}) \tan \varphi_{i+1}$$

The notes are the same as above.

(2) The landslide thrust

The calculation formula for landslide thrust based on the transfer coefficient method can be expressed as:

$$P_i = P_{i-1} \times \psi + K_s \times T_i - R_i \tag{8.21}$$

where sliding force is

$$T_i = W_i \sin \alpha_i + A \cos \alpha_i + \gamma_w h_{wi} L_i \tan \beta_i \cos(\alpha_i - \beta_i);$$

sliding resistance is

$$R_i = (W_i(\cos \alpha_i - A \sin \alpha_i) - N_{wi} - \gamma_w h_{wi} L_i \tan \beta_i \sin(\alpha_i - \beta_i)) \tan \varphi_i + c_i L_i;$$

transfer coefficient is

$$\psi = \cos(\alpha_{i-1} - \alpha_i) - \sin(\alpha_{i-1} - \alpha_i) \tan \varphi_i.$$

Considering the pore water pressure ratio, sliding resistance R_i can be calculated by the following formula:

$$R_i = (W_i((1 - r_U) \cos \alpha_i - A \sin \alpha_i) - \gamma_w h_{wi}) \tan \varphi_i + c_i L_i$$

(3) Working condition of landslide prevention design discussion

In general, dead weight is the basic load of the landslide, and rainstorm and earthquake are the exceptional loads. Taking the prevention and treatment engineering of Three Gorges Reservoir as an example, the load type is closely related to the changes of the reservoir water table.

After the buildup of the Three Gorges Reservoir, the water table before dam is at 145–175–145 m between in flood season (October to early April of the following year), and has a 30 m variation range. When the reservoir works in flood season (middle of June to the end of September), the reservoir storage table must be reduced to flood limit water table of 145 m to held the occurred flood. For once a flood in 5a, 20a, 100a and 1000a, water tables before dam must keep at 147.2 m, 157.5 m, 166.7 m and 175 m respectively. The reservoir water table would quickly decrease to 145 m after the flood peak to prevent flood occur again. Thus from the safety consideration, water table sharply declines from 175–145 m can a working condition for checking.

For the working condition, the dead weight would serve as a temporary base load which would be ended in 2009, and dead weight added with the rainstorm would also be the temporary exceptional load. These two kinds of load would not serve as the designed working conditions of the landslide prevention and control engineering affected by the reservoir for the Three Gorges Reservoir area.

Dead weight added with reservoir water table of 175 m should be the basic load of the Three Gorges Reservoir area. The second phase water table of 135 and 156 m for the transition are not as a design conditions into consideration. Corresponding with this, dead weight and rainstorm added with water table of 175 m, dead weight added with water table range of 175–145 m, and dead weight and rainstorm added with water table range of 175–145 m should serve as the working conditions of landslide design for engineering design and check. Because the reservoir is in weak seismic activity area, the general working condition of landslide control engineering does not consider the earthquake.

Due to the rapid development of many immigrant towns in the reservoir area in recent years, the construction land shortage is serious. Many houses, even the tall buildings, have to be built on landslide. Thus building load should be as a kind of working condition, or converted into the added dead weight, in these landslide areas. Specially, the load or overload (even several times) of the track should be considered at prevention and control engineering in riverside road areas.

8.5.1.4 Calculation of Slope Stability

The Three Gorges Project is famous in the world due to large storage capacity, long returning flow, and high water table fall. It could be ensured after putting completed reservoir into use that the normal water table in the front of dam is 175 m and limited water table for flood prevention is 145 m, indicating that the amplitude of variation is 30 m. The hydrogeological conditions in reservoir area are strongly affected because of water cycle fluctuation, resulting in deteriorated rock and soil mechanics features of slope body and large hydrostatic pressure and seepage force, which in turn cause revival of old landslide and buckling failure of current stable slope. As a result, these problems may bring a security risk on human life and property, and operation of the reservoir. Taking Guantangkou landslide of Wanzhou, Chongqing as an example, the effects of groundwater in bedrock area on slope stability are detailed introduced and further evaluated.

Case 1: Landslide of Guantangkou in Wanzhou, Chongqing

1. Basic characteristics of Guantangkou Landslide

   (1) Geological conditions

   Shown as Fig. 8.5, landslide area is located at right bank slope of Zhuxi River in Wanzhou, and it has a landscape of medium hills, river valley, and bank slope. It is also a part of the Taibai rock old collapse body slide. The elevation at right is higher than that at left and top elevation ranges from 212 to 222 m, 80 m higher than Zhuxi River. Landslide distributes along the rolling erosion gully. The terrain shows a step type, and the surface slope at top is steeper than that at bottom. The top landslide, which is demarcated by Guoben branch road is a slope with a slope angle of 25°–35°. The bottom landslide is a rank-2 slope with a small slope angle of 5°–10°. And it extends to Zhuxi River, which becomes incongruous. The table-board elevation of rank-1 step from south to north ranges from 160 to 177 m, and width ranges from 140 to 260 m. The table-board elevation of rank-2 slope ranges from 140 to 160 m, and width ranges from 80 to 150 m. A slope or scarp connects the slope of rank-1 and rank-2 with a slope angle of 10°–25° and scarp height of 3–10 m. Band slope of front exit is along with the river with a height of 8–16 m, and a slope angle of 20°–50°.

   Landslide group of Guantangkou shows a type of a short boot and its elevation of south is higher than that in north. Back edge elevation ranges from 212 to 222 m, front exit elevation ranges from 133 to 141 m and the difference is 80 m. Furthermore, the length between south and north is about 500 m, width between east and west is about 800 m, and thickness ranges from 17.0 to 49.1 m. The total

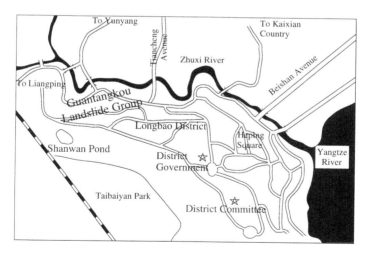

Fig. 8.5 Location of the Guantangkou landslide in Wanzhou

area is about 0.4 km² and total volume is about 1280×10^4 m³. The whole sliding body is mainly composed of quaternary block clamp silty clay, silty clay, and gravel rock. The rock composition is mainly sandstone and mudstone. According to the trial pit water penetration test, silty clay proves to be a good waterproof performance. Groundwater of slippery body is mainly recharged by atmospheric precipitation and surface water, and flow path is short, discharging at low point of the terrain. Thus, the groundwater occurrence condition is poor and inhomogeneity in the accumulated layer landslide.

(2) Selection of calculation scheme

The landslide of Guantangkou is located at mid-front of Taibai rock history sliding body, and it is still stable as a whole with some extent signs of deformation. The bottom edge deformation mainly occurred at shallow parts of the Guantangkou slope according to the survey. The II–II′ and V–V′ profiles are selected to be calculated combing with the analysis of landslide stability. The methods of Fillennius, Bishop, Janbu, and transfer coefficient are selected to evaluate the stability, because the landslide zone shows circular arc and break line shapes. Landslide thrust is obtained using transfer coefficient method and checked using other methods.

(3) Selection of parameters of soil physical and mechanical properties in sliding zone

Some parameters of soil physical and mechanical properties at selected profiles in sliding zone are shown in Table 8.2. These parameters are obtained through the exploration of the hole and shaft in soil sample in indoor test and in situ shear test.

Natural weight of side body is 20.3 kN/m³, and saturated weight is 20.6 kN/m³. Values of shear strength parameters are determined after comprehensive judgements and inverse analysis listed in Table 8.3.

Table 8.2 Physical and mechanical parameters of the Guantangkou landslide

Natural state				Saturated state			
Peak		Salvage value		Peak		Salvage value	
c (kPa)	φ (°)	c (kPa)	φ (°)	c (kPa)	φ (°)	c (kPa)	φ (°)
38.14	17.17	29.93	13.49	27.46	13.82	19.86	9.59

Table 8.3 Stability analysis and calculation parameter selection

	Natural state		Saturated state	
	c (kPa)	φ (°)	c (kPa)	φ (°)
The sliding zone	30	15	21	13.5
Trailing shallow edge	27	12.9	20	12.5

8.5 Project Types and Instances of Bedrock Fissure Water

(4) Calculation model and load combination of landslide.

1. Calculation model

Form line of sliding slope and sliding slice are both simplified into broken lines when calculating by Calslope procedure, where slide unit width is 1 m, which could be applied into two-dimensional calculation.

2. Load combination

   a. Self-weight: there is no concentrated load but self-weight in sliding body.
   b. Acting force of groundwater

   The geological exploration indicated that seepage field was formed in landslide under the action of rainfall infiltration. So, buoyancy produced by pore water pressure on sliding zone should be taken into account, as well as the influence of rainstorm on landslide stability.
   c. Water table of the Three Gorges Reservoir

Both storage table and reservoir regulation have various influences on the landslide stability after the completion of the Three Gorges Project. Therefore, the following two aspects should be taken into account when calculating, that is, ① the effect of normal storage table (i.e., 175 m) on landslide stability; and ② calculation of landslide stability when water table drops (i.e., from 175 m down to 145 m).

3. Working condition combination

   a. Working condition 1: dead weight and heavy rain $K_s = 1.25$ (overall), $K_s = 1.20$ (trailing shallow edge).
   b. Working condition 2: dead weight and water table at 175 m, $K_s = 1.15$.
   c. Working condition 3: dead weight, water table from 175 to 145 m and rainstorm $K_s = 1.15$.

Working conditions of each computation section are shown in Figs. 8.6 and 8.7.

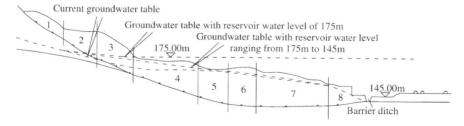

Fig. 8.6 Sketch map of landslide stability calculation at II–II' profile of Guantangkou slope, Wanzhou

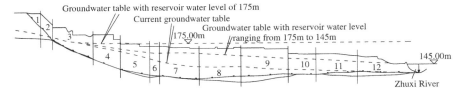

Fig. 8.7 Sketch map of landslide stability calculation at V–V' profile of Guantangkou, Wanzhou

2. Data for calculation

Quantitative analysis results on landslide stability of Guantangkou landslide are listed in Tables 8.4 and 8.5.

3. Results of stability analysis and calculation

Calculation results of stability analysis of Guantangkou landslide under the working conditions of 1–3 are shown in Tables 8.6 and 8.7.

4. Analysis on calculation of landslide thrust.

It is essential to divide slide body into slices when calculating thrust (Figs. 8.6 and 8.7). Residual sliding force is calculated under three conditions based on basic loads, self-weight, and aquiferous groups. The residual slide force is zero in the cases of basic loads. In order to make a trend analysis and comparison, though the residual sliding force is determined by calculated values in theory, the value is considered as zero in real designing. In this book, the slices method is adopted.

(1) Calculation results of landslide thrust at V–V' profile

The landslide thrust (kN) is shown in Table 8.8 when water table reaches 175 m and security coefficient equals to 1.15. Landslide thrust under the following two conditions are, respectively, calculated with seismic acceleration of 0, 0.05g and 0.1g: saturated slide zone and no groundwater table; mere hydrostatic pressure with groundwater table.

Table 8.9 indicates the value of landslide thrust with security coefficient of 1.15 when water table drops from 175 to 145 m. The working condition of considering groundwater table and seepage pressure is introduced after comparing with Table 8.8.

Calculation results listed in Table 8.9 shows that as for No. 8 slice, landslide thrust equals to zero when seismic acceleration equals to zero under saturated sliding zone condition. Landslide thrust equals to 1334 kN under only hydrostatic pressure condition. Landslide thrust equals to 3138 kN under seapage pressure condition. Similarly, landslide thrust equals to 8936 kN when seismic acceleration equals to 0.1g under saturated sliding zone condition. Landslide thrust equals to 24,237 kN under only hydrostatic pressure condition. Landslide thrust equals to 26,041 kN under seepage pressure condition. These results show an increasing tendency with the order increase of the above-mentioned conditons.

8.5 Project Types and Instances of Bedrock Fissure Water 343

Table 8.4 Calculation of landslide stability at II–II′ profile of Guantangkou landslide in Wanzhou

Profile number	Soil slice	Boundary	X Coordinate (m)	Slope crest (m)	Slip band (m)	Water table (m)			Natural state			Saturated state			Imposed load (kN)
						Cond. 1	Cond. 2	Cond. 3	c (kPa)	φ (°)		c (kPa)	φ (°)		
II–II′		X1	5	213	213	213	213	213							
	1	X2	37.44	200.05	190.02	190.02	188.91	180.98	27	12.9		20	12.5		0
	2	X3	64.79	196.68	174.29	174.36	177.99	177	27	12.9		20	12.5		0
	3	X4	93	176.9	161.41	166.91	175.4	173.07	30	15		21	13.5		0
	4	X5	144.93	167.99	143.2	165.17	167.99	166.14	30	15		21	13.5		0
	5	X6	168.55	166.85	137.26	164.08	166.85	164.42	30	15		21	13.5		0
	6	X7	191.05	166.2	135.22	161.88	166.2	162.31	30	15		21	13.5		0
	7	X8	247.6	154.06	136.37	153	154.06	153.11	30	15		21	13.5		0
	8	X9	279.02	145.9	140.4	140.8	145.9	140.4	30	15		21	13.5		0

Table 8.5 Calculation of landslide stability at V–V′ profile of Guantangkou Landslide in Wanzhou

Profile number	Soil slice	Boundary	X Coordinate (m)	Slope crest (m)	Slip band (m)	Water table (m)			Natural state			Saturated state			Imposed load (kN)
						Cond. 1	Cond. 2	Cond. 3	c (kPa)	φ (°)		c (kPa)	φ (°)		
V–V′															
	1	X1	17.49	218.79	218.79	218.79	215	218.79							
	2	X2	37.16	216.73	198.08	198.08	198.08	198.08	27	12.9		20	12.5		0
	3	X3	55.25	198	183.44	183.44	183.44	183.44	27	12.9		20	12.5		0
	4	X4	111.59	188.13	155.96	173.25	176.95	174.31	27	12.9		20	12.5		0
	5	X5	151	181.3	142.56	165.9	175	169.1	30	15		21	13.5		0
	6	X6	192.28	171.86	133.97	155.7	171.86	164.54	30	15		21	13.5		0
	7	X7	206.28	171.78	132.65	152.25	171.78	163.29	30	15		21	13.5		0
	8	X8	263.67	168.51	128.5	144.38	168.51	161.85	30	15		21	13.5		0
	9	X9	323.21	168.93	131.09	143.22	168.93	158.68	30	15		21	13.5		0
	10	X10	390.61	166	133.49	140.02	166	154.88	30	15		21	13.5		0
	11	X11	430.6	159.8	133.3	138.77	159.8	152.51	30	15		21	13.5		0
	12	X12	489.27	156.27	135.59	136.59	156.27	149.02	30	15		21	13.5		0
		X13	580	133.5	133.5	133.5	133.5	133.5	30	15		21	13.5		0

8.5 Project Types and Instances of Bedrock Fissure Water

Table 8.6 Calculation results of landslide stability coefficient when water tables dropped from 175 to 145 m at II–II′ profile

Woking conditions	$a = 0g$				$a = 0.05g$			
	Fillennius	Bishop	Janbu	Transfer coefficient	Fillennius	Bishop	Janbu	Transfer coefficient
Saturated slide band	1.444	1.4961	1.4844	1.55	1.1506	1.1825	1.1824	1.22
Hydrostatic pressure	1.0102	1.0581	1.046	1.09	0.8016	0.8305	0.8267	0.85
Hydrostatic and osmotic pressure	0.9674	1.0142	1.0035	1.05	0.7642	0.7916	0.7886	0.82

Table 8.7 Calculation results of landslide stability coefficient when water tables drops from 175 to 145 m at V–V′ profile

Woking conditions	$a = 0g$				$a = 0.05g$			
	Fillennius	Bishop	Janbu	Transfer coefficient	Fillennius	Bishop	Janbu	Transfer coefficient
Saturated slide band	2.5495	2.6311	2.5035	2.83	1.7379	1.7842	1.751	1.86
Hydrostatic pressure	1.8267	1.905	1.8328	2.03	1.2418	1.2849	1.2689	1.33
Hydrostatic and seepage pressure	1.7567	1.8327	1.766	1.96	1.1968	1.2388	1.2247	1.29

Table 8.8 Landslide thrust with normal water table of 175 m at V–V′ profile (unit:kN)

Saturated slide band			Hydrostatic pressure		
$a = 0g$	$a = 0.05g$	$a = 0.1g$	$a = 0g$	$a = 0.05g$	$a = 0.1g$
2090	2276	2462	2180	2367	2554
5072	5582	6092	5179	5689	6199
11438	13504	15570	13069	15146	17223
14381	18104	21826	18523	22274	26025
12940	18504	24068	20414	26031	31649
10715	16767	22819	19217	25330	31443
1878	10712	19547	15678	24613	33548
0	0	8936	4942	16420	27898
0	0	0	0	9265	23684
0	0	0	0	6827	22842
0	0	0	0	2030	19565
0	0	0	0	0	18179

Table 8.9 Calculation results of landslide thrust when water tables drops from 175 to 145 m at V–V' profile (unit:kN)

Saturated slide band			Hydrostatic pressure			Seepage pressure		
$a = 0g$	$a = 0.05g$	$a = 0.1g$	$a = 0g$	$a = 0.05g$	$a = 0.1g$	$a = 0g$	$a = 0.05g$	$a = 0.1g$
2090	2276	2462	2090	2276	2462	2090	2276	2462
5072	5582	6092	5092	5602	6111	5112	5622	6132
11438	13504	15570	12704	14879	16954	13408	15484	17559
14381	18104	21826	17857	21603	25350	19006	22753	26499
12940	18504	24068	19117	24725	30334	20687	26295	31904
10715	16767	22819	17701	23804	29906	19314	25417	31519
1878	10712	19547	13148	22064	30981	14888	23804	32721
0	0	8936	1334	12786	24237	3138	14589	26041
0	0	0	0	3843	18324	0	5874	20255
0	0	0	0	594	16563	0	2628	18597
0	0	0	0	0	12356	0	0	14438
0	0	0	0	0	10096	0	0	12278

8.5 Project Types and Instances of Bedrock Fissure Water

However, calculation results in Tables 8.8 and 8.9 show that under the same security coefficient (K_s = 1.15), landslide thrust under water table of 175 m, in slice 8 for an example: $a = 0$, $P = 0$ (saturation), $P = 4942$ kN (hydrostatic pressure); $a = 0.1g$, $P = 8936$ kN (saturation), $P = 27898$ kN (hydrostatic pressure), the landslide thrust are higher than that of the water tables dropped from 175 to 145 m. It indicates that landslide stability does not always decrease when water table of the Three Gorges Reservoir drops from 175 to 145 m. The thin landslide with smooth sliding zone is helpful for landslide stability due to the decrease of sliding body underwater and small hydraulic grade, indicating that the decreasing rate of landslide buoyancy is higher than the increasing rate of seepage pressure.

(2) Calculation results of landslide thrust at II–II′ profile

Table 8.10 shows the landslide thrust when water table of the Three Gorges Reservoir equals to 175 m. Table 8.11 shows the landslide thrust when water table of the Three Gorges Reservoir drops from 175 to 145 m. The conclusion is the same as that at V–V′ profile.

5. Evaluation of landslide stability

1. The stability under current situation

Above stability analysis and calculation indicates that the landslide of Guantangkou is stable except the western parts, i.e., II–II′ profile under current self-weight and storm rain situation. In the process of calculation, the effect of hydrostatic pressure on slide is considered, however, accurately, the trailing edge cracks on the Shalong road is closed of emulsified asphalt and drainage ditches are built. A poor infiltration capacity of slide is found in the process of landslide investigation. stability coefficient at II–II′ profile is low due to hydrostatic pressure. The western part on Guantangkou landslide is unstable under self-weight and rainstorm conditions. It is possible that landslide will become unstable if drainage condition is poor when raining heavily.

Table 8.10 Landslide thrust with normal water table of 175 m at II–II′ profile (unit:kN)

Saturated slide band			Hydrostatic pressure		
$a = 0g$	$a = 0.05g$	$a = 0.1g$	$a = 0g$	$a = 0.05g$	$a = 0.1g$
882	1066	1251	882	1066	1251
3828	4531	5234	3910	4614	5317
6024	7360	8696	6723	8064	9405
8142	10741	13340	11287	13908	16529
8072	11406	14740	12686	16053	19420
5235	9251	13268	11229	15289	19349
0	1718	7281	5149	10777	16406
0	0	4265	2171	8060	13948

Table 8.11 Landslide thrust when water tables drops from 175 to 145 m at II–II' profile (unit:kN)

Saturated slide band			Hydrostatic pressure			Seepage pressure		
$a = 0g$	$a = 0.05g$	$a = 0.1g$	$a = 0g$	$a = 0.05g$	$a = 0.1g$	$a = 0g$	$a = 0.05g$	$a = 0.1g$
882	1066	1251	882	1066	1251	882	1066	1251
3828	4531	5234	4029	4734	5438	4121	4825	5530
6024	7360	8696	6723	8064	9405	6985	8326	9667
8142	10741	13340	11021	13641	16260	11873	14492	17111
8072	11406	14740	12305	15669	19034	13327	16692	20057
5235	9251	13268	10700	14755	18811	11808	15863	19919
0	1718	7281	4319	9941	15564	5448	11070	16693
0	0	4265	1136	7016	12897	1849	7730	13611

2. Landslide stability under reservoir table and scheduling conditions of the Three Gorges Reservoir

The landslide stability of Guantangkou landslide proves to be the worst according to the above calculation when water table reaches 175 m. Front part of the Guantang landslide is the anti-sliding segment with smooth topography and water table is lower than 175 m resulting in a poor infiltration capacity. When water table drops from 175 to 145 m, negative influence of the increasing groundwater seepage pressure on landslide is weaker than positive influence of decreasing buoyancy on landslide. Therefore, landslide stability of Guantangkou increased comparing with that when water table is 175 m. The landslide condition is the worst under water table of 175 m and heavily rainfall conditions. Water table is 175 m, stability coefficient at II–II′ profile equals to 1.03, which indicated a unstable status. However, stability coefficient at V–V′ profile ranges from 1.5 to 1.6, showing a good global stability.

8.5.2 Groundwater of the Tunnel Project

Fissure water of bedrock is one of the most common groundwater in China. In engineering projects of tunnel, dam and other large underground constructions, the enrichment of bedrock fissure water directly pose a threat to the construction and operation management, rising a number of special engineering problems, such as landslide, roof caving, water gushing or inrush, and even the underground debris flow and tunnel deformation, etc. According to the statistics of the 415 tunnels of the Nan-Kun railway, landslides occurred in 15 %, and water burst occurred in 93.5 % of these tunnels. More than 60 collapses occurred in 11 faults of the GuanJiao tunnel during the construction of the Qinghai–Tibet railway. The short-term maximum fissure water discharge of the 9th fault was 38,000 t/d and 29 massive landslides occurred during the construction of the Dayaoshan tunnel. The reasons for collapses may be that, bedrock fissure water immersed and softened the structural plane, soft layer and the fracture zone, reducing its strength and taking away the fillings of the weak surfaces. The rock mass would disintegrate quickly, inducing or worsening the collapses. The occurrence of bedrock fissure water mainly depends on the bedrock type (Table 8.12). However, due to uneven burial distribution of bedrock fissure water, and complexity of its occurrence and transport, there was little hydrogeological research for bedrock area, which should be taken into consideration seriously. As engineering example, the bedrock fissure water of new Qidaoliang tunnel area was introduced.

Table 8.12 Structure types of fractured rock

Structure types	Fissure surface	Integrity coefficient	Permeability	Quantitative model
Massive structure	Few fractures, microcracks	>0.9	Hardly impermeable	Continuous medium
Blocky structure	Generally developed joints	0.6–0.8	Poor permeability	Noncontinuous medium
Stratified structure	Relatively developed fractures	0.3–0.6	Obvious anisotropy permeability	Noncontinuous medium
Cataclastic structure	Developed fractures	0.1–0.3	High permeability	Noncontinuous medium
Unconsolidated structure	Developed fractures	<0.1	Strong permeability	Seem-continuous Medium

8.5.2.1 Geologic Features of Bedrock Fissure Water in New Qidaoliang Tunnel Area

Qidaoliang is the watershed of Qilihe district of Lanzhou and Lintao county of the Dingxi region, and belongs to the temperate zone with semiarid climate. North slope of the area is dankness, but the southern slope is dry and rainless. Annual precipitation is 500 mm and the evaporation is 1500 mm. Tunnel site belongs to low mountainous area. Topography steep, exposed bedrock, erosional and structural fissures are developed at north part of the watershed. It is advantageous for precipitation to infiltrate bedrock and form the fissure water. The topography is gradual at south of the watershed, and the overlying eolian loess that contains no water is 3–15 m. Due to the presence of mudstone at the bottom of the loss, there may be some fissure and pore water in weathered zone of the mudstone and overlying eolian loss. If fractures are developed at the bottom of the loess, the bedrock fissure water would formed by the infiltrated loess water. Bedrock fissure water in the area is mainly distributed in sandstone and conglomerate of the Hekou group. Due to the various rock structure character and size, and terrains of the area, the groundwater is uneven distribution with various water abundances.

Bedrock fissure water of the tunnel site is mainly recharged by precipitation, which seeps into ground through the small thickness of quaternary overburden and bedrock weathering zone. Then the water flows along the bedrock fissures forming the fracture phreatic water, and may be accumulated at the local parts of tectonic fracture zone. The bedrock fissure water finally discharges at depression springs of the slope. The small scale of confined water may also be formed at some local regions with certain constructs, terrains, impermeable and permeable layers and attitude of rock formations. Dynamic of groundwater table was affected by climate, hydrogeological conditions. The depth of water table increased in dry seasons, resulting in decreased water discharge. The situation was opposite in wet seasons with decreased depth of water table and increased water discharge.

8.5 Project Types and Instances of Bedrock Fissure Water

8.5.2.2 The Partition of Bedrock Fissure Water, According to the Seepage Characteristics

Surrounding rock of New Qidaoliang tunnel was characterized as the dual pore-fracture medium, and fissure water was the basic groundwater type. Thus the distribution of groundwater is uneven, with various water abundances of the tunnel surrounding rock.

1. Southern entrance—F4 fault

    (1) Engineering geological conditions

    The lithology of this section was mainly characterized as weathering conglomerate, conglomerate, silicide crystalline limestone, phyllite, and F4 fault. Weathering conglomerate and conglomerate were flat and integrity, with not very developed fault and joint. Limestone and phyllite section have the steeply dipping occurrence and folds along the tendency. Limestone is hard and relatively complete with developed fissures. Phyllite has the thin plate structure which is softness and poor integrity. F4 fault, the thrust fault with steep dip angle, is the regional major fault with the trend of 315° and dip angle of 74°. The width of the fault and the affected zone, filled with fault gouge, breccia, rock fragments and phyllite, etc., is about 120 m.

2. Seepage analysis

Entrance section of the tunnel is in vadose zone, recharged by infiltration of the precipitation and surface water. Thus, it is wet in the entrance section of the tunnel because of the fissure water infiltration in rainy season, and is desiccation in the dry season with not a large number of water burst in generally. Due to no incision of large geological structural plane, Limestone is only with the joint and fissure water which are not too much. Phyllite which is dry and has no water acts as an aquiclude. Only crack ooze water and moist occur in the partial phyllite section. F4 fault and the influence zone with the characteristics of rich water, full water and water resistance, mainly filled with the fault mud, form the aquitard and divide the hanging wall and footwall into two different hydrogeological units. Hanging wall is the medium-thickness layered rock of the limestone with embed phyllite or interbed limestone and phyllite, the resisted bedrock fissure water is enriched at the hanging wall. Lithology of the footwall is sandstone embed mudstone or interbed limestone and phyllite. Due to the control of lithology and tectonic interface, the enrichment zone of interval structure crevice water was developed.

According to the observation statistics of water discharge after the tunnel excavation, the water burst and seepage phenomena were only in the section of XK21 + 045–XK20 + 825. Vault, left and right side of the tunnel wall occurred different scales of the water burst and seepage, but most water burst points distributed at the right tunnel wall. The lithology of this segment is crystalline limestone characterized as the hard brittle rock with developed fissures, forming the

Table 8.13 Entrance—statistical water discharge of F4 fault in tunnel

Mileage	Lithology	Rock structural types	Structural planes	Rich-water area	Measured water discharge (t/d)		Dynamic	Predicted water discharge (t/d)	Deviation (t/d)
					Maximum (location)	Average			
XK21 +440– +270	Weathering conglomerate, conglomerate and fault fracture zone (F5)	Layered and unconsolidated structure	Developed fractures	Vadose zone with basically no water	No water	No water	Unstable and obviously affected by precipitation	No water	
XK21 +270– XK20 +800	Siliceous limestone with interbed phyllite	Block-layer structure	Developed fractures	Abundant fissure water	48.22 (XK20 + 911, Right Wall)	42.50/300 m	Stable and unconspicuous affected by precipitation	50/300 m	−7.5/300 m
XK20 +800– +690	F4 fault	Unconsolidated structure	Developed fractures	Squeezing fissures with basically no water	1.227 (XK20 + 691, Right Wall)	0.41/110 m	Stable and unconspicuous affected by precipitation		

8.5 Project Types and Instances of Bedrock Fissure Water

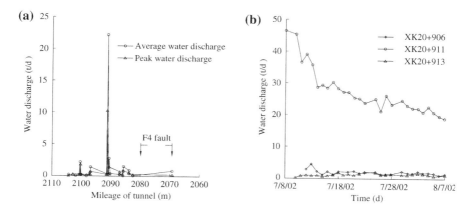

Fig. 8.8 Tunnel entrance water discharge curves F4 fault. **a** water discharge variation on mileage. **b** water discharge variation time

bedrock fissure water. Comparison of the designed and forecast and actual observation water discharge contrast was shown in Table 8.13 and Fig. 8.8.

2. F4 fault—F3 fault

1. Engineering geological conditions

This section is located in the Shuichi river valley, the southern and northern parts of which are the F3 and F4 faults zone. The basic bedrock lithology is sandstone interbed mudstone which is medium thickness-thick layer with rock breakage, developed fractures and local small fold with an asymmetric syncline. F3 fault is a normal fault with steep dip angle of 60°, strike of NW to SE and tendency of SW. The stratums of both hanging wall and footwall of the fault were Hekou group of lower cretaceous. The fault was filled with broken sandstone and mudstone with developed fissures and poor integrity.

2. Seepage analysis

Fracture development is closely controlled by lithology and fold structure. The characteristics of fissures development are obviously different in sandstone and mudstone, and fissures in sandstone are interpenetration with each other. Due to the strong rigidity, the sandstone in general with tension fractures which have high connectivity, become a relative aquifer. The mudstone is a relative aquitard due to its soft plastic character with fine and close of joints and poor water conductivity. The fracture water of sand and mudstone is relatively in layers, so that the interlayer fissure water is under pressure. Because of the fissure water in layers, the amount of fissure water is decreased with depth increasing. Sandstone aquifers are separated by mudstone layer, resulting in no hydraulic connections and transfluence of two aquifers. F3 fault was a small fault with the hanging wall of sandstone interbeded mudstone layer. The groundwater was recharged rainfall and surface water infiltration. The main recharge source is Shuidi river that infiltrates under gravity

Table 8.14 Statistical water discharge of F4–F3 fault section

Mileage	Lithology	Rock Structural types	Structural planes	Rich water area	Measured water discharge (t/d) Maximum (location)	Measured water discharge (t/d) Average	Dynamic	Predicted water discharge (t/d)	Deviation (t/d)
XK20 +690– +370	Sandstone with interbed mudstone	Layered structure	Developed fractures	Abundant fissure water	24.19 (XK20 +536, Right wall)	122.47/300 m	Unstable and obviously affected by precipitation	938/500 m	−440.33/300 m
XK20 +370– +320		Unconsolidated structure	Developed fractures	Abundant fissure water	4.46 (XK20 +363, Vault)	19.33/50 m	Stable and unconspicuous affected by precipitation		

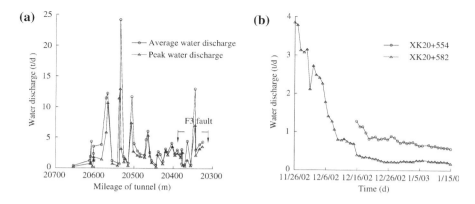

Fig. 8.9 Water discharge curves of F4 fault—F3 fault section. **a** Water discharge variation on mileage. **b** Water discharge variation time

through the surface fissures and weathered zone. Thus hanging wall of F3 fault had the rich water. Water containing and strong transport properties of the F3 fault, not only resulted in the drainage of fissure water in sand and mudstone layers, but also resulted in the strongly discharge of fissure water in the weathered zone with sharply declined water table.

According to the excavation of tunnel water discharge after observation statistics, the design water discharge forecast and actual observation contrast are shown in Table 8.14 and Fig. 8.9.

8.5.2.3 Conclusions

The fissure water seepage of bedrock brings great influence on construction of large underground engineering. Understanding its seepage features has realistic meaning. However, the study and governance of bedrock fissure water are very difficult due to the complex groundwater seepage in fractured rock. According the process of tunnel excavation of new Qidaoliang tunnel, the analysis results of water seepage observation of the tunnel indicated the following conclusions:

1. Due to random distribution and anisotropy connectivity of the vast fractures in rock mass, the occurrence and status of bedrock fissure water are various, resulting in the heterogeneity of seepage.
2. Groundwater flow in fractured rock mass is closely related to the fracture occurrence, and anisotropy of the fissures caused the anisotropy of fissure water seepage.
3. Most water burst points distributed at right side wall of the downlink of the New Qidaoliang tunnel, suggesting that the connectedness of the fissures in rock mass is strong and river recharge has a certain influence on the tunnel water discharge.

4. During the excavation of Qidaoliang tunnel, when water discharge was mainly from storage resource, initial water discharge was very large characterized by sudden flood water. However, the water discharge decreased quickly with the passage of time, rarely has the stable status (see dynamic curves of water discharge of XK20 + 911 in Fig. 8.8 and XK20 + 582 in Fig. 8.9). When water discharge was mainly from the recharge resource of surrounding rock, the water discharge tended to remain a relatively stable flow (see dynamic curves of water discharge of XK20 + 906 and XK20 + 913 in Fig. 8.8, and XK20 + 554 in Fig. 8.9).

8.6 Exercises

1. How to categorize the bedrock groundwater according to the hydraulics and medium types? Is it perfect?
2. What are the burial characteristics of bedrock fissure water? What are the laws involved in groundwater occurrence and migration?
3. The fracture-pore dual-medium model is divided into two types; what are they and the assumptions in them?
4. Review literatures and understand the new progress of groundwater seepage models of fractured rock mass.
5. How many numerical simulation softwares can be used to depict the bedrock fissure water? What is the process?
6. List the engineering examples associated with bedrock fissure water, and briefly expounds the influences of bedrock groundwater.

Chapter 9
Numerical Simulation of Engineering Groundwater

9.1 Basic Principle

Finite Difference Method (FDM), Finite Element Method (FEM), Boundary Element Method, and Finite Volume Method are the common methods of numerical calculation for engineering groundwater. These calculation methods have been deduced in detail in related textbooks and monographs. In this section, only the basic principle of the common methods will be introduced.

9.1.1 Finite Difference Method

The basic idea of FDM is to replace the continuous seepage area with the set of a finite number of discrete points within seepage area. Derivative is replaced by difference quotient approximately at these discrete points. Differential equations and their definite conditions are transformed into difference equations with unknown quantities of approximation of unknown functions at discrete points. Then, the equations will be solved and the approximation of the solution for differential equations at discrete points will be obtained eventually.

The basic principle of FDM is to represent the derivative of the hydraulic head function at a point approximately by the head values of that point, adjacent points, and the spaces between. The spacing of these points may be equal or not equal, which are equivalent, respectively, to the finite difference grids with a uniform or nonuniform grid. These points can be located on one side or both sides of the point, which forms different finite difference formulas of the derivatives. The finite difference approximation of derivatives of head function can be established in several ways; the most common method among which is being elicited through Taylor expansion.

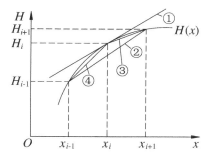

Fig. 9.1 Schematic diagram of finite difference approximation to first order derivatives ①—$K = \frac{\partial H}{\partial x}\big|_i$ ②—$K = \frac{H_{i+1}-H_{i-1}}{2\Delta x}$ ③—$K = \frac{H_{i+1}-H_i}{\Delta x}$ ④—$K = \frac{H_i-H_{i-1}}{\Delta x}$ (*Note K is slope*)

In the groundwater seepage equation, there are first order and second order derivatives. In FDM, differential is replaced by difference quotient in most time. The calculation method of the first difference is shown in Fig. 9.1.

One point i is taken from x-axis, whose coordinate is $x_i = i\Delta x$. Point $(i-1)$ and $(i+1)$, whose coordinates are $x_{i-1} = (i-1)\Delta x$ and $x_{i+1} = (i+1)\Delta x$, are taken from the left and right side Δx, away from point i. The head function $H(x)$ is expanded by Taylor series in the center of point i.

$$H_{i+1} = H_i + \Delta x \cdot \frac{dH}{dx}\bigg|_i + \frac{(\Delta x)^2}{2!} \cdot \frac{d^2H}{dx^2}\bigg|_i + \frac{(\Delta x)^3}{3!} \cdot \frac{d^3H}{dx^3}\bigg|_i + \frac{(\Delta x)^4}{4!} \cdot \frac{d^4H}{dx^4}\bigg|_i + \cdots \tag{9.1}$$

$$H_{i-1} = H_i - \Delta x \cdot \frac{dH}{dx}\bigg|_i + \frac{(\Delta x)^2}{2!} \cdot \frac{d^2H}{dx^2}\bigg|_i - \frac{(\Delta x)^3}{3!} \cdot \frac{d^3H}{dx^3}\bigg|_i + \frac{(\Delta x)^4}{4!} \cdot \frac{d^4H}{dx^4}\bigg|_i - \cdots$$

where H_i represents the head value of i point; $\frac{dH}{dx}\big|_i$ represents value of head derivative at point i.

So,

$$\frac{H_{i+1}-H_i}{\Delta x} = \frac{dH}{dx}\bigg|_i + \frac{\Delta x}{2!} \cdot \frac{d^2H}{dx^2}\bigg|_i + \frac{(\Delta x)^2}{3!} \cdot \frac{d^3H}{dx^3}\bigg|_i + \frac{(\Delta x)^3}{4!} \cdot \frac{d^4H}{dx^4}\bigg|_i + \cdots \tag{9.2}$$

$$\frac{dH}{dx}\bigg|_i = \frac{H_{i+1}-H_i}{\Delta x} + O(\Delta x) \tag{9.3}$$

where $O(\Delta x)$ represents remainder, which is infinitesimal, the same order as Δx when $\Delta x \to 0$. If the remainder is ignored, the finite difference of the first order derivative can be expressed approximately as follows:

9.1 Basic Principle

$$\left.\frac{dH}{dx}\right|_i = \frac{H_{i+1} - H_i}{\Delta x} \qquad (9.4)$$

This is the forward-difference approximation to first order derivatives, which has the first truncation error.

Similarly,

$$\left.\frac{dH}{dx}\right|_i = \frac{H_i - H_{i-1}}{\Delta x} + O(\Delta x) \qquad (9.5)$$

where $O(\Delta x)$ represents remainder, which is infinitesimal, the same order as Δx when $\Delta x \to 0$. If the remainder is ignored, the finite difference of the first order derivative can be expressed approximately as follows:

$$\left.\frac{dH}{dx}\right|_i = \frac{H_i - H_{i-1}}{\Delta x} \qquad (9.6)$$

This is the backward-difference approximation to first order derivatives, which has the first truncation error.

$$\left.\frac{dH}{dx}\right|_i = \frac{H_{i+1} - H_{i-1}}{2\Delta x} + O[(\Delta x)^2] \qquad (9.7)$$

where $O[(\Delta x)^2]$ represents remainder, which is infinitesimal, the same order as Δx when $\Delta x \to 0$. If the remainder is ignored, the finite difference of the first order derivative can be expressed approximately as follows:

$$\left.\frac{dH}{dx}\right|_i = \frac{H_{i+1} - H_{i-1}}{2\Delta x} \qquad (9.8)$$

This is the central-difference approximation to first order derivatives, which has the second truncation error.

From the three finite difference equations of the first order derivative, we can see that unilateral differences (forward-difference and backward-difference) both have first truncation error. Central-difference has second truncation error and the less the Δx and Δt are, the less the truncation errors are.

Similarly, for second order derivative, there is

$$\left.\frac{d^2H}{dx^2}\right|_i = \frac{H_{i-1} - 2H_i + H_{i+1}}{(\Delta x)^2} + O[(\Delta x)^2] \qquad (9.9)$$

where $O[(\Delta x)^2]$ represents remainder, which is infinitesimal, the same order as Δx when $\Delta x \to 0$. If the remainder is ignored, the finite difference of the first order derivative can be expressed approximately as follows:

$$\left.\frac{\mathrm{d}^2 H}{\mathrm{d}x^2}\right|_i = \frac{H_{i-1} - 2H_i + H_{i+1}}{(\Delta x)^2} \qquad (9.10)$$

This is the central-difference approximation to second order derivatives, which has the second truncation error.

Those differential equations for derivatives are based on the independent variable x. If head is seen as the function of space variable y or time variable t, the corresponding differential equations can be obtained.

Specific solving process about finite difference of one-dimensional, two-dimensional, and three-dimensional with pressure/no pressure steady/unsteady seepage refer to the related professional books.

9.1.2 Finite Element Method

Finite element method is a numerical method for solving the definite problem of partial differential equations. Similar with the differential method, finite element method is used to solve groundwater flow problems by converting the definite problem to algebraic equations through regional subdivision and interpolation method. Based on the different ways of building algebraic equations, finite element method is divided into Galerkin method and Ritz method.

Taking Biarritz finite element method for example, the method is based on the variational principle and partial interpolation. The so-called variational principle is to convert the solution of the definite problem of partial differential equations describing groundwater flow to extremum problems of one functional. The essence of this method is to divide the seepage area into line, plane, and body elements. Then, according to actual situation, using interpolation method of some form, the head approximate expression is established with the head values of the nodes at the elements. In the end, the interpolation of the whole element aggregation is formed and the solution of functional extremum problems is converted to the solution of algebraic equations.

So, the most critical step of Ritz method is to find one functional extremum problem corresponding to the definite solution of groundwater flow. When subdivision and interpolation of Ritz method and Galerkin method are the same, the algebraic equations formed are the same.

Taking the two-dimensional triangular grid for example (Fig. 9.2), we will analyze the basic principles of unit interpolation.

The first step is spatial-temporal discrete, do mesh generation, and divide the calculated seepage area into a series of triangles. Pay attention to the shape and homogeneity of the elements in the process of division, and then number the element nodes.

One element e is taken from D casually. The three nodes of this element are numbered i, j, k, whose coordinates are (x_i, y_i), (x_j, y_j), (x_k, y_k) with counterclockwise

9.1 Basic Principle

Fig. 9.2 Element meshing

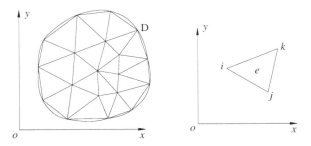

order. The values of head function at the three nodes are H_i, H_j, H_k. There are several methods to stipulate the approximation of head value within the element. One simple and commonly used method is to replace the head curved surface with plane, namely using the linear interpolation of junction head bit H_i, H_j, H_k as approximate (trial solution) solution of head distribution at the triangular element. So, the following equation can be set:

$$h_i^e(x, y, t) = \beta_1^e + \beta_2^e x + \beta_3^e y$$

where β_1^e, β_2^e, and β_3^e are undetermined functions.

The value of $h_i^e(x, y, t)$ at node i, j, k are $H_i(t), H_j(t), H_k(t)$, namely

$$\begin{cases} H_i(t) = \beta_1^e + \beta_2^e x_i + \beta_3^e y_i \\ H_j(t) = \beta_1^e + \beta_2^e x_j + \beta_3^e y_j \\ H_k(t) = \beta_1^e + \beta_2^e x_k + \beta_3^e y_k \end{cases} \tag{9.11}$$

Solving linear equations based on Cramer's rule, there are

$$\begin{cases} \beta_1^e = \dfrac{A_1}{A} \\ \beta_2^e = \dfrac{A_2}{A} \\ \beta_3^e = \dfrac{A_3}{A} \end{cases} \tag{9.12}$$

$$A = \begin{vmatrix} 1 & x_i & y_i \\ 1 & x_j & y_j \\ 1 & x_k & y_k \end{vmatrix} \quad A_1 = \begin{vmatrix} H_i & x_i & y_i \\ H_j & x_j & y_j \\ H_k & x_k & y_k \end{vmatrix} \quad A_2 = \begin{vmatrix} 1 & H_i & y_i \\ 1 & H_j & y_j \\ 1 & H_k & y_k \end{vmatrix} \quad A_3 = \begin{vmatrix} 1 & x_i & H_i \\ 1 & x_j & H_j \\ 1 & x_k & H_k \end{vmatrix}$$

Solving the equations,

$$\begin{cases} a_i = x_j y_k - x_k y_j & a_j = x_k y_i - x_i y_k & a_k = x_i y_j - x_j y_i \\ b_i = y_j - y_k & b_j = y_i - y_k & b_k = y_k - y_i \\ c_i = x_k - x_j & c_j = x_i - x_k & c_k = x_j - x_i \end{cases} \quad (9.13)$$

Represent the area of triangle with Δ^e, A can be expressed as

$$A = 2\Delta^e$$

$$\begin{cases} \beta_1^e = \dfrac{1}{2\Delta^e}[a_i H_i + a_j H_j + a_k H_k] \\ \beta_2^e = \dfrac{1}{2\Delta^e}[b_i H_i + b_j H_j + b_k H_k] \\ \beta_3^e = \dfrac{1}{2\Delta^e}[c_i H_i + c_j H_j + c_k H_k] \end{cases} \quad (9.14)$$

Eventually,

$$h_i^e(x,y,t) = \frac{1}{2\Delta^e}[(a_i + b_i x + c_i y)H_i + (a_j + b_j x + c_j y)H_j + (a_k + b_k x + c_k y)H_k]$$

Taking the two-dimensional triangular grid for example (Fig. 9.2), we will analyze the basic principles of unit interpolation.

Setting

$$N_i^e = \frac{1}{2\Delta^e}(a_i + b_i x + c_i y) \quad N_j^e = \frac{1}{2\Delta^e}(a_j + b_j x + c_j y) \quad N_k^e = \frac{1}{2\Delta^e}(a_k + b_k x + c_k y)$$

$$h_i^e(x,y,t) = N_i^e(x,y)H_i + N_j^e(x,y)H_j + N_k^e(x,y)H_k$$

where $h_i^e(x,y,t) = N_i^e(x,y)H_i + N_j^e(x,y)H_j + N_k^e(x,y)H_k$ are basis functions at element e.

For the promoter region (usually, we call the polygon region consisted of triangle elements around the node as promoter region at the node), basic function of every node is structured at the seepage region. Triangulation and linear interpolation method are used to construct the basis function. Then, finite element equations can be structured with Galerkin method and Ritz method as

$$[G]\{H\} + [\mu^*]\left\{\frac{\mathrm{d}H}{\mathrm{d}t}\right\} = \{E\} + \{B\} \quad (9.15)$$

where $[G]$ is water-mediated matrix; $\{H\}$ is column matrix of unknown head; $[\mu^*]$ is water-storage matrix; $\left\{\frac{\mathrm{d}H}{\mathrm{d}t}\right\}$ is column matrix of first order derivative of node head to time; $\{E\}$ is the resource of discharge/recharge column matrix; $\{B\}$ is boundary column matrix.

9.1 Basic Principle

Process finite difference to the column matrix of first order derivative of node head to time

$$\left\{\frac{dH}{dt}\right\} = \frac{1}{\Delta t}\left(\{H^{t+\Delta t}\} - \{H^t\}\right) \tag{9.16}$$

$$[G]\{H^{t+\Delta t}\} + \frac{1}{\Delta t}[\mu^*]\left(\{H^{t+\Delta t}\} - \{H^t\}\right) = \{E\} + \{B\} \tag{9.17}$$

$$\left([G] + \frac{1}{\Delta t}[\mu^*]\right)\{H^{t+\Delta t}\} = \{\mu^*H\} + \{E\} + \{B\} \tag{9.18}$$

$$\{\mu^*H\} = \frac{1}{\Delta t}[\mu^*]\{H^t\} \tag{9.19}$$

This format is fully implicit. Under the given initial and boundary conditions, we can calculate each coefficient matrix or vector of the equations. So, the head $\{H^{\Delta t_1}\}$ of every node at the end of the first Δt_1 can be solved. Then, it will be seen as the initial head value $\{H^t\}$ at the second Δt_2. A new coefficient matrix (vector) will be solved.

9.1.3 Boundary Element Method

Boundary element method is also called boundary integral equation method. The basic idea is to convert the definite problem of partial differential equations to integral equations at boundary with Green theorem. Then, the integral equations are made discretization by subdivision and interpolation method of elements (unit). By solving the integral equations, the head value H at the second boundary node and the hydraulic gradient at the first boundary nodeare obtained. In order to find out the head value of any point within seepage area, integral calculation needs to be done. Since the boundary integral equations become the discrete objects, this method can deal with the problem by reducing the dimension with one-dimension lower.

9.2 Numerical Simulation of Foundation Pit Dewatering

Principles and procedures of numerical simulation of engineering groundwater has been deduced in related textbooks and monographs. In this section, taking the foundation pit dewatering of Yishan Road Station, Shanghai Metro Line 9 for example, we will explain the basic processes of numerical simulation of foundation pit dewatering by FDM.

9.2.1 Analysis of Prototype

9.2.1.1 Project Overview

Yishan Road Station, located on Yishan Road, west from Zhongshan Road, east to Kaixuan Road, is a terminal and reentrant station of the first phase of project of Shanghai Metro Line 9. The station, 297.40 m in length and 21.2 m in width of the standard segment, is a four underground island platform station. The subsidiary structure of the station includes five entrances and three air shafts (Fig. 9.3).

The thickness of the underground wall of the station's main structure is 1.2 m, the depth of the underground wall of the standard section is 48 m, and the depth of the underground wall of the end well is 51 m. The depth of the deepest part of the standard section in the excavation of foundation pit is 27.855 m and the depth of the deepest part of the end well is 29.718 m. Five steel supports and four reinforced concrete supports are set up.

Shanghai seven decorating confluence is located in the central of the south side of the station. It has a 17-story concrete frame structure, which is 14 m away from the envelope edge of the station. The north side of the station locates the good quality decorative furnishings and Treasure Island City commercial building, which are 13 m away from the outside edge of the foundation pit. The east side of Yishan Road Station of Metro Line 9 locates the elevation in operation of the first phase of the Pearl Line and Yishan Road Station completed of the second phase of Metro Pearl Line. The nearest distance between the station and the elevated pile of the first phase of the Pearl Line is only 7 m. The nearest distance between underground station and the second phase of Metro Pearl Line is 23 m. The west of the station locates the Zhongshan Road Viaduct, the nearest distance between the center of which and the foundation pit is approximately 25 m.

9.2.1.2 Hydrogeological Conditions

1. Buried depth of groundwater

According to "*Engineering Geological Survey Report*" of the research area and the regional data collected, the groundwater of Shanghai region mainly contains the phreatic water in shallow clay layer, micro-confined groundwater in shallow silty layer, and the confined groundwater in deep silty and sand layer (Fig. 9.4).

The buried depth of the phreatic water in shallow layer is 0.3–1.5 m under the ground and the annual average groundwater depth is 0.5–0.7 m under the ground.

Micro-confined groundwater in shallow layer (layer ④$_2$) and the confined groundwater in deep layer (layer ⑦) are generally lower than the phreatic level. The buried depth of micro-confined water in layer ④$_2$ and ⑤$_{2-2}$ is generally 3–6 m. The buried depth of confined water in layer ⑦ is generally 4–12 m. Phreatic and confined water level changes with seasons, weather, tides, and other factors.

2. Hydraulic parameters of formation

9.2 Numerical Simulation of Foundation Pit Dewatering

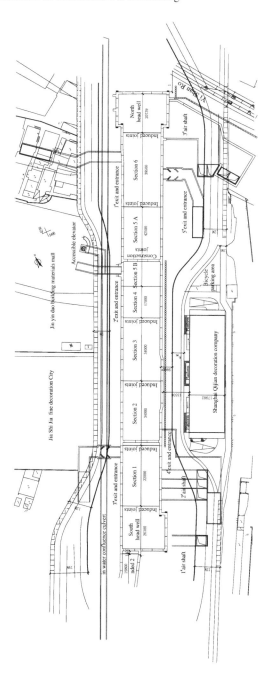

Fig. 9.3 Plan view of the total site

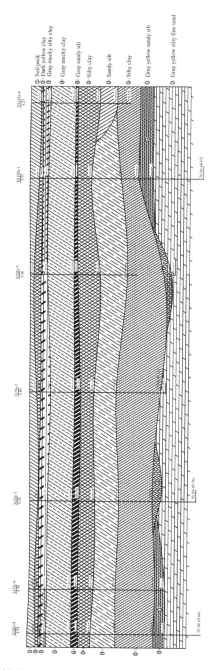

Fig. 9.4 Cross section of 2-2 stratigraphic prototype in generalized computing

9.2 Numerical Simulation of Foundation Pit Dewatering

Table 9.1 Coefficients in laboratory permeability test

Sequence	Soil	Permeability coefficient (cm/s)	
		K_I	K_{II}
②₁	Clay	5.72×10^{-7}	2.17×10^{-6}
③	Mucky silty clay	7.39×10^{-7}	1.21×10^{-6}
④₁	Mucky clay	8.23×10^{-8}	1.18×10^{-7}
④₂	Sandy silt	5.30×10^{-5}	2.99×10^{-4}
⑤₁₋₂	Clay	4.65×10^{-7}	1.67×10^{-6}
⑤₂₋₂	Sandy silt	1.00×10^{-5}	2.03×10^{-4}
⑤₃₋₁	Silty clay	3.75×10^{-7}	2.37×10^{-6}
⑤₃₋₂	Silty clay	1.51×10^{-6}	6.66×10^{-6}

The coefficients of layer ②, ③, ④₁, ④₂, ⑤₁₋₁, ⑤₁₋₂, and ⑤₃ in laboratory permeability test provided by *Engineering Geological Survey Report* are shown in Table 9.1.

According to preliminary survey and injection test data of the near engineering site, the permeability coefficient and static water level of related soil layers are shown in Table 9.2.

From the two tables mentioned above, the permeability coefficient obtained from water injection test is larger than the coefficient from laboratory test. This phenomenon occurs because the horizontal bedding and thin silty sand intercalation of the soil layer strength and the ability of permeability, while permeability coefficients obtained in laboratory permeability test are less with the limitation of soil quality and testing boundary conditions.

Hydrogeological parameter inversion of the study area and the seepage model of foundation pit dewatering belong to the third type of seepage. In the third type of

Table 9.2 Permeability coefficient and static water level of related soil layers

Test depth (m)	Sequence	Permeability coefficient (cm/s)	Buried depth of static water level (m)	Elevation of static water level (m)
3.50–5.00	③	1.06×10^{-5}	2.03	2.49
4.00–5.50		2.82×10^{-5}	1.65	3.07
8.00–9.50	④₁	9.15×10^{-6}	2.08	2.64
10.00–11.50		9.20×10^{-6}	2.39	2.13
18.00–19.50	④₂	2.53×10^{-4}	2.86	1.77
21.00–22.50	⑤₁₋₂	9.33×10^{-6}	2.85	1.87
23.00–24.50		9.21×10^{-6}	3.35	1.17
28.00–29.50		1.01×10^{-5}	3.59	1.13
37.00–38.50	⑤₂₋₂	2.74×10^{-5}	4.20	0.52
45.00–46.50	⑦₁	7.02×10^{-5}	8.64	−3.72
51.00–52.20		1.11×10^{-4}	5.86	−1.34

groundwater seepage model, envelope structure (or watertight curtain) deepens into the middle and lower part of the dewatering aquifer. Most of the confined aquifers inside or outside the pit are separated by the envelope structure (or watertight curtain) with the bottom of the aquifer unseparated. The characteristics of the groundwater flow are as follows: because of the block of the envelope structure (or watertight curtain), the groundwater inside and outside of the upper pit is not continuous, but the bottom aquifer is continuously connected. The flow boundary is very complex and the groundwater presents three-dimensional flow regime. In addition, because of the significant drop of confined aquifer level, the groundwater from the upper unconfined aquifer recharges the confined aquifer through the aquitard. In this case, the seepage resulted from foundation pit dewatering is called third type of seepage.

9.2.2 Three-Dimensional Numerical Modeling of FDM

9.2.2.1 Calculation Range

1. Plane range

The research of the hydrogeological parameter inversion is an important step of envelope design, foundation pit dewatering design and numerical calculation. In order to analyze pumping test results rationally and inverse hydrogeological parameters of formation scientifically, range of the calculation model should be set reasonably to eliminate the boundary effect of groundwater seepage computation effectively. In hydrogeological parameter inversion, the calculation range is set at least 1000 × 1000 m outside the core of the pumping-well group, and the calculation range of the three-dimensional analysis of foundation pit dewatering is 2108.6 × 2027.35 m in the plane.

2. Depth range

The calculation range in the vertical is selected 150 m below the ground.

The length of the calculation region is 2108.6 m and the width is 2027.35 m. The study area was divided into 270,976 units (116 rows, 146 columns, and 16 floors). Three-dimensional finite difference numerical model is shown in Fig. 9.5.

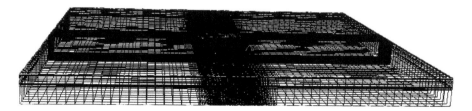

Fig. 9.5 Planar graph of three-dimensional numerical model inversed by hydrogeological parameters

9.2 Numerical Simulation of Foundation Pit Dewatering

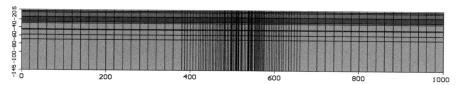

Fig. 9.6 Profile of three-dimensional numerical model inversed by hydrogeological parameters

9.2.2.2 Strata

The strata to be calculated and inversed were generalized according to *Engineering Geological Survey Report*. The strata and division generalized are shown in Fig. 9.6 and Table 9.3.

9.2.2.3 Water-Proof Curtain

This project uses diaphragm wall as water-proof curtain. The three-dimensional numerical model of the end well and standards section are shown in Fig. 9.7.

9.2.2.4 Pumping Well and Observation Well

The layout of pumping test and pumping and observation wells in foundation pit dewatering are shown in Figs. 9.8, 9.9, 9.10, and 9.11. The layouts of pumping and observation wells in three-dimensional numerical model are shown in Figs. 9.12 and 9.13.

9.2.2.5 The Initial and Boundary Conditions

The initial conditions are given in accordance with the results of field observations; the surrounding boundary conditions are taken as constant water head.

9.2.2.6 Calculation Cases

According to the hydrogeological parameters designed and set initially by the pumping test, three-dimensional finite difference analysis was conducted. By the fitting of logging level and monitoring data of land subsidence of controlling points, the hydraulic parameters and physico-mechanical parameters of the soil model were inversed. Then, the confined water level and land subsidence of the points 10, 20, 40, and 80 m away from the pit when the confined water levels decreased to the design depth were analyzed.

Table 9.3 The initial parameter values selected in hydrogeological parameter inversion modeling of pumping test

Layer	Soil	Thickness	Elevation	Water content (%)	Specific weight (kN/m³)	Permeability coefficient (cm/s)		Compression modulus (MPa)
						K_V	K_H	
②₁	Clay	1.25	0.72	33.4	18.4	5.72×10^{-7}	2.17×10^{-6}	4.21
③	Mucky silty clay	3.47	−2.75	41.1	17.5	7.39×10^{-7}	1.21×10^{-6}	3.53
④₁	Mucky clay	9.88	−12.63	49.6	16.7	8.23×10^{-8}	1.18×10^{-7}	2.18
④₂	Sandy silt	1.86	−14.59	30.7	18.4	5.30×10^{-5}	2.99×10^{-4}	10.47
⑤₁₋₂	Silty clay	9.12	−23.07	34.8	18.0	4.65×10^{-7}	1.67×10^{-6}	4.43
⑤₂₋₂	Sandy silt	7.87	−35.02	33.5	17.7	1.00×10^{-5}	2.03×10^{-4}	15.94
⑤₃₋₁	Silty clay	9.75	−30.71	34.2	18.0	3.75×10^{-7}	2.37×10^{-6}	4.81
⑤₃₋₂	Silty clay	14.81	−45.37	31.8	18.2	1.51×10^{-6}	6.66×10^{-6}	5.69
⑥₄	Silty clay	2.4	−47.86	22.1	19.6			8.50
⑦₁	Sandy silt	3.74	−47.86	24.5	19.0	7.02×10^{-5}	1.11×10^{-4}	14.25
⑦₂	Fine sand	Undrilled	Undrilled	24.4	19.2	3.02×10^{-3}	3.11×10^{-3}	14.40

9.2 Numerical Simulation of Foundation Pit Dewatering

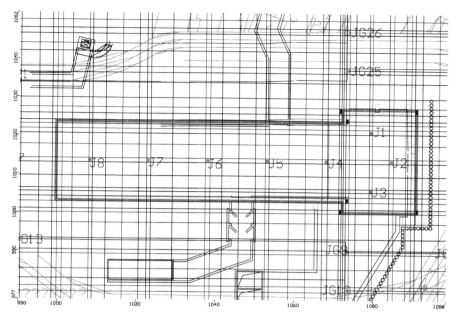

Fig. 9.7 No. Z3 foundation pit dewatering wells and the water-proof curtain of the standard section

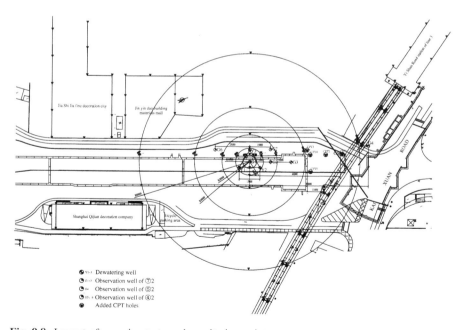

Fig. 9.8 Layout of pumping tests and monitoring points

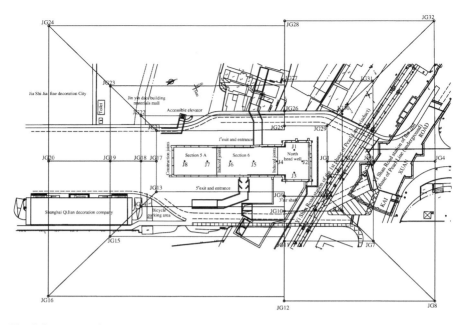

Fig. 9.9 Layout of groundwater dewatering wells and water level controlling points of No. Z3 foundation pit

Diaphragm walls are used in building envelope of No. Z3 foundation pit. The excavation depth of foundation pit is 27 m. The depth of the end well and diaphragm walls of the standard section are all 62 m. The depths of the eight dewatering wells are 60 m. The buried depths of filter tubes are 53–60 m.

9.2.3 Three-Dimensional Numerical Simulation of FDM

9.2.3.1 Unsteady Flow Mathematical Model

Three-dimensional unsteady flow mathematical model of confined aquifer used in the calculation is as follows:

$$\begin{cases} \frac{\partial}{\partial x}\left(K_{xx}\frac{\partial h}{\partial x}\right) + \frac{\partial}{\partial y}\left(K_{yy}\frac{\partial h}{\partial y}\right) + \frac{\partial}{\partial z}\left(K_{zz}\frac{\partial h}{\partial z}\right) - W = \mu_s \frac{\partial h}{\partial t} & (x,y,z) \in \Omega \\ h(x,y,z,t) = H_0 & (x,y,z) \in \Gamma_1 \\ h(x,y,z,t)|_{t=t_0} = h_0(x,y,z) & (x,y,z) \in \Omega \end{cases} \quad (9.20)$$

where K_{xx}, K_{yy}, K_{zz} are the Hydraulic conductivity along the x, y, z axis direction, cm/s; h is the head value of point (x, y, z) at time t, m; W is the resource of discharge/recharge, 1/d; μ_s is the specific storage in point (x, y, z), 1/m; t is the time, h; Ω is

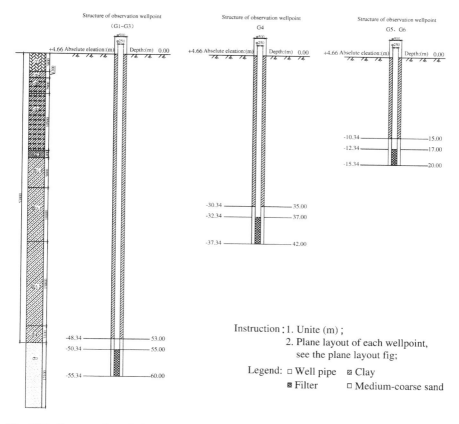

Fig. 9.10 Cross section of observation wells in hydrogeological parameter inversion model

three-dimensional time domain; Γ_1 is the first boundary condition; H_0 is the constant water head, m.

9.2.3.2 Three-Dimension Numerical Parameter Inversion

The pumping process of pumping well Y1 belongs to unsteady flow actually. The water level of the observation wells changed over time. The water levels of observation well G1, G2, and G3 were used as the observation level. Estimation-correction method was used to inverse hydrogeological parameters.

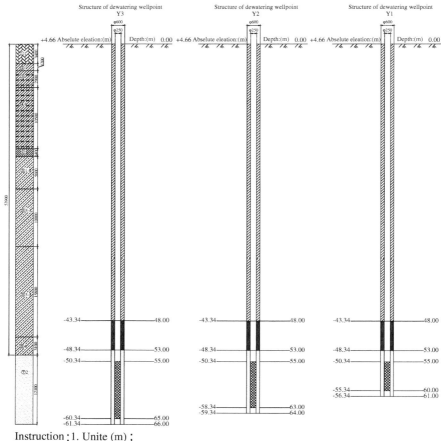

Fig. 9.11 Cross section of pumping wells in hydrogeological parameter inversion model

Using the model above, the hydrogeological parameter inversion focused on single-well Y1 pumping tests. Estimation-correction method was used in calculation. The hydrogeological parameters were fixed through analysis with the water level observations of observation well G1, G2, and G3 in order to minimize the calculation error. The calculation was divided into 23 stress periods and every stress period was divided into 10 time steps.

Through the trial calculation of the aquifer parameters and the adjustment of boundary conditions, fixed flux boundary was selected as the optimal boundary of ⑦$_2$. The optimal set of parameters was as follows:

9.2 Numerical Simulation of Foundation Pit Dewatering

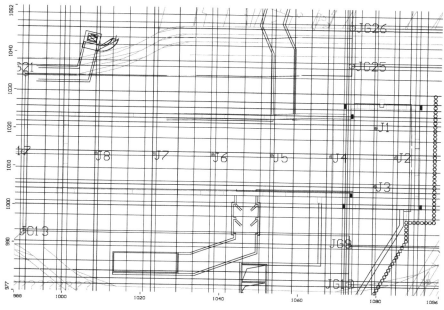

Fig. 9.12 Finite difference grid of dewatering wells and the near area in No. Z3 foundation pit

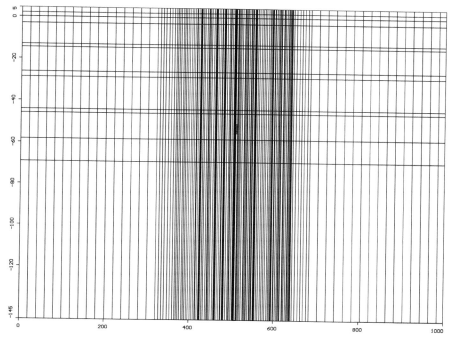

Fig. 9.13 The location of Y1 pumping filter tube used in hydrogeological parameter inversion model

Table 9.4 Inversion parameters of single-well pumping test

Layer	Soil	Initial permeability coefficient of model (cm/s)			Specific storage (1/m)
		K_{xx}	K_{yy}	K_{zz}	
⑦$_2$	Fine sand	8.3×10^{-3}	8.3×10^{-3}	3×10^{-4}	1.75×10^{-6}

$$K_{xx} = 8 \times 10^{-3} \text{cm/s}, K_{yy} = 8 \times 10^{-3} \text{cm/s}, K_{zz} = 3 \times 10^{-3} \text{cm/s},$$
$$\mu_s = 1.75 \times 10^{-6} (1/\text{m}).$$

The inversion parameters determined eventually are shown in Table 9.4.

9.2.3.3 Three-Dimensional Numerical Simulation of Dewatering

After several trials, the water level near ⑦ layer of foundation pit was fallen down to 27 m in buried depth. Analysis was conducted according to the adjusted water quantity after the optimization and adjustment of the water quantity of dewatering well. The results are shown in Figs. 9.14, 9.15, and 9.16.

Water-proof curtains of 62 m in deepness were used in the standard section and end wells of foundation pit, and the length of the outer edge of the filter tube of

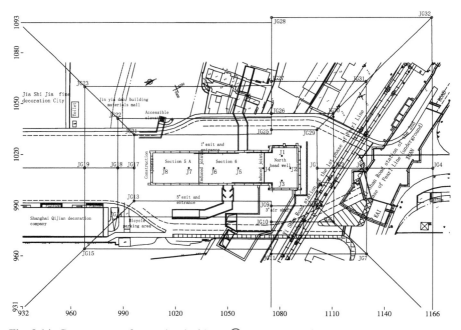

Fig. 9.14 Contour map of water level of layer ⑦$_2$ near No. Z3 foundation pit (t = 5 d)

9.2 Numerical Simulation of Foundation Pit Dewatering

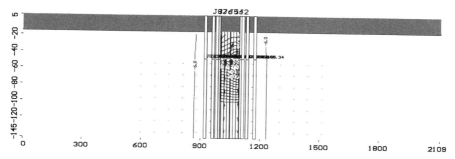

Fig. 9.15 Contour map of water level of layer ⑦$_2$ of No. Z3 foundation pit (Partially vertical, $t = 5$ d)

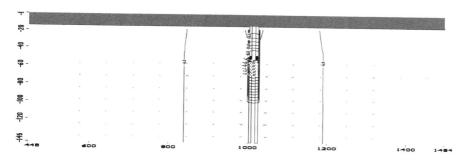

Fig. 9.16 Contour map of water level near end well of layer ⑦$_2$ of No. Z3 foundation pit (Lateral, Part of end well, $t = 5$ d)

pumping wells inside the curtain was 2 m. The water yield of pumping wells could be reduced effectively. After pumping for 5 d, the water level inside the pit dropped to the design antihypertensive elevation and stabilized. The maximum water level of the places 10, 20, 40, and 80 m away from the outside pit dropped to -6.377 m, equivalent to 1.037 m in drawdown.

9.2.4 Settlement Calculation

Since layer ⑦$_2$ is silt, its contribution to land subsidence is considered as instantaneous elastic deformation mainly, and the additional consolidation settlement is mainly in the layer ⑤$_{1-2}$–⑦$_1$, where produce level fluctuation. By the numerical analysis of the third type of seepage around foundation pit support (curtain of diaphragm wall), the fall of water level of controlling points (such as Metro Line 3) outside the pit are given. According to the national standard *Code on Geotechnical Investigations for Metro and Light Rail Transit* (GB50307—1999) line 8.5.7, the additional land subsidence caused by dewatering can be calculated.

According to the code, the additional loads caused by groundwater decrease can be calculated as follows:

$$\Delta P = \gamma_w (h_1 - h_2) \quad (9.21)$$

where ΔP is the additional load caused by groundwater decrease, kPa; h_1 is the water head height before dewatering, m; h_2 is the water head height after dewatering, m; γ_w is specific weight of water, kN/m^3.

In the calculation, specifically, the buoyancy of groundwater is reduced in the use of the actual head multiplied by moisture content.

Additional land subsidence caused by dewatering can be calculated by layer-wise summation method as

$$S = \sum_{i=1}^{n} S_i = \sum_{i=1}^{n} \frac{\Delta P_i}{E_i} H_i \quad (9.22)$$

where S is the total additional settlement of the surface caused by dewatering, m; S_i is the additional settlement of the layer i, m; ΔP_i is the additional loads caused by dewatering of the layer i, kPa; E_i is the compression modulus of the layer i, kPa; H_i is the thickness of the layer i, m.

E_i in the equation above is elastic modulus to sand. To clay and silt, it can be calculated by the following equation:

$$E_s = \frac{1 + e_0}{a_v} \quad (9.23)$$

where e_0 is the initial void ratio of soil layer; a_v is the volume compressibility of soil layer (in MPa^{-1}), which should be selected from the stress section from effective self-weight stress to the sum of effective self-weight and additional stress.

In the place 10 m away from the curtain, the land subsidence caused by consolidation is calculated according to the maximum water level drawdown of the soil layers above. The accomplishments are shown in Table 9.5.

Table 9.5 The calculation of land subsidence by layer-wise summation method (10 m away from curtain)

Layer	Thickness	Moisture content (%)	Compression modulus (MPa)	Water level drawdown (m)	Layer-wise consolidation settlement (mm)
⑤$_{1-2}$	6.42	0.348	4.43	1.69	−3.67
⑤$_{3-1}$	9.74	0.342	4.81	1.55	−10.73
⑤$_{3-2}$	13.54	0.318	5.69	1.06	−8.02
⑦$_1$	1.78	0.245	14.25	1.04	−0.32
Total settlement (m)					−22.75

9.2 Numerical Simulation of Foundation Pit Dewatering

Table 9.6 Calculation of consolidation degree with dewatering of 113 d

Layer	Consolidation degree
⑤$_{1-2}$	0.189
⑤$_{3-1}$	0.189
⑤$_{3-2}$	0.189
⑦$_1$	0.198

According to dewatering of 113 d, the consolidation degrees of soil layers 10 m away from the curtain are shown in Table 9.6.

According to dewatering of 113 d, the amount of land subsidence of the points 10 m away from the curtain is 4.29 mm.

9.2.5 Effects and Analysis

In order to verify the effects of dewatering and measure the water level drawdown inside and outside the pit when dewatering dropped to the design drawdown, equal drawdown tests were conducted in sites. Dropping to the design drawdown by single-wells and multi-wells, the dropping development of water level inside and outside the pit were inspected.

The monitoring data when the well Y3-1, Y3-5, Y3-6 of foundation pit Z3 dropping to design drawdown are shown in Table 9.7. When the water level inside foundation Z3 reached to the design drawdown, the water level drops of piezometers are 0.60, 1.00, and 1.75 m, which is closer to pre-analysis of decompression.

Table 9.7 Accomplishments of equal drawdown tests

Time	Well flow (m³/h)				Water level drawdown (m)				
	Y3-6	Y3-3	Y3-1	Sum	Inside		Outside		
					Y3-4	Y3-2	Y4-5	Y3	G3-1
9:00					10.7	10.65	10	10.5	10.8
9:30	7	7			23.25	23.1	10.1	10.7	11.35
10:00	8	7	2	17	26.1	24.8	10.2	10.9	11.9
10:30	8	9	4	21	26.35	25.05	10.25	11	11.75
11:00	8	5	4	17	26.6	25.1	10.25	11.1	12.1
11:30	5				25.8	25.25	10.3	11.25	12.2
12:00	4				25.85	25.35	10.35	11.3	12.35
13:00	8	13	8	29	25.85	25.5	10.4	11.5	12.4
14:00	9	14	8	31	25.9	25.8	10.5	11.5	12.45
15:00	8	14	8	30	26	25.9	10.6	11.5	12.55
16:00	9	14	8	31	26.05	25.6	10.75	11.55	12.6
17:00	9	14	7	30	26.05	25.8	10.8	11.6	12.6
Drawdown (m)					15.3	15.25	0.6	1	1.75

9.3 Case Study

1. Please derive the first order and second order derivatives of head function by difference quotient, when taken point $(i + 1)$ and $(i + 2)$ adjacent to point i, shown as Fig. 9.17 in textbook.
2. If the coordinates value of i, j, k are (3,5), (7,3), and (8,7) shown as Fig. 9.18 in textbook, and the value of head function at these three nodes are 10, 7, and 12; please calculate the approximation of the head value within this element.

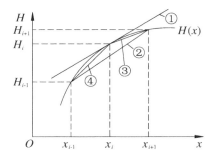

Fig. 9.17 Schematic diagram of finite difference approximation to first order derivatives

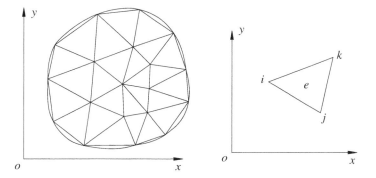

Fig. 9.18 Element meshing

9.4 Exercises

1. What are the mechanisms of FEM, FDM, and BEM, respectively? And how about the difference?

Chapter 10
Groundwater Pollution and Corrosivity Assessment

10.1 Groundwater Quantity Analysis

Water quantity analysis can be categorized into simple analysis, full analysis, special analysis, and professional analysis based on different purposes. The current special analysis in engineering geological investigation mainly refers to groundwater corrosivity assessment to building materials.

10.1.1 Groundwater Quantity Analysis Representation Methods

The results of groundwater quantity analysis can be represented by indexes and chemical formulas. The specific methods are as follows:

1. Ion content indexes
   Salts that dissolve in water exist in cations and anions, such as Na^+, Ca^{2+}, Cl^-, SO_4^{2-}. Ion content in groundwater is measured in mmol/L, mg/L, meq/L, while in the sea it would be mol/L and g/L. Ultra trace elements are measured in µg/L.
2. Molecule content indexes
   Gases and colloids, such as CO_2, SiO_2, which dissolve in groundwater, can be measured in mmol/L, mg/L.
3. Comprehensive indexes
   Comprehensive indexes include pH value, acidity and alkalinity, hardness, and salinity, which all represent the chemical properties of groundwater.

   (1) pH value
   Groundwater pH value reflects groundwater acid-base properties, which depend on hydrolytic factors of acids, alkalis, and salts. pH value also has some relativity to electrode potential, which could influence the migration intensity of chemical

Table 10.1 Groundwater types according to pH value

Types	Strong acidic water	Acidic water	Weak acidic water	Neutral water	Weak alkaline water	Alkaline water	Strong alkaline water
pH values	<4.0	4.0–5.0	5.0–6.0	6.0–7.5	7.5–9.0	9.0–10.0	>10.0

elements. Thus, pH value is an important index in groundwater analysis and chemical equilibrium calculation. Generally, the pH value of groundwater is about 4–10. Table 10.1 shows different groundwater types according to their pH value.

(2) Acidity and alkalinity

Acidity is mainly caused by uncombined CO_2, inorganic acid, strong-acid weak-base salt, and organic acid. Total acidity is tested in strong alkali titration using phenolphthalein indicator, and hence it is also called phenolphthalein acidity. Another kind of acidity is methyl orange acidity, which uses phenolphthalein as indicator.

Likewise, the alkalinity also includes total alkalinity (methyl orange alkalinity) and phenolphthalein alkalinity.

Acidity and alkalinity are measured in mmol/L and meq/L.

(3) Hardness

The hardness of water depends on the content of Ca^{2+}, Mg^{2+} and other metal ions, excluding alkali metal ions. Hardness can be divided as follows:

(a) Total hardness: Total content of $Ca(HCO_3)_2$, $Mg(HCO_3)_2$, chloride, sulfate, and nitrate in groundwater.
(b) Temporary hardness (carbonate hardness): Carbonate precipitation content after groundwater boiling.
(c) Permanent hardness (non-carbonate hardness): Calcium salt and magnesium salt content that stay in groundwater after boiling.
(d) Negative hardness (sodium-potassium hardness): Content of sodium and potassium carbonate, bicarbonate, hydroxide in groundwater.

$$\text{Total hardness} = \text{Temporary hardness} + \text{Permanent hardness}$$
$$= \text{Carbonate hardness} + \text{Non-carbonate hardness}$$
$$\text{Negative hardness} = \text{Total alkalinity} - \text{Total hardness}$$

Hardness is measured in mmol/L, mg/L, meq/L, and H^0.

Table 10.2 shows the classification of groundwater according to hardness. Table 10.3 shows the scaling factors between hardness units.

(4) Salinity

Salinity is the total content of ions, molecules, and compounds in groundwater, which means that salinity includes all dissolved and colloidal substances, but not free gas.

10.1 Groundwater Quantity Analysis

Table 10.2 Groundwater types according to hardness

Types	Extremely soft water	Soft water	Light hard water	Hard water	Extremely hard water
Germanic degree (H^0)	<4.2	4.2–8.4	8.4–16.8	16.8–25.2	>25.2
meq/L	<1.5	1.5–3.0	3.0–6.0	6.0–9.0	>9.0
mg/L	<42	42–48	48–168	168–252	>252

Notes 1 "mg/L" measured by CaO
2 1 mmol/L = 0.3566 H^0

Table 10.3 Scaling factors between hardness units

	mg/L	H^0	meq/L
mg/L	1	2.8/E	1/E
H^0	E/2.8	1	
meq/L	E	2.8	1

Note E refers to the equivalent weight of calculated matter

Salinity can be measured by many ways. First, it can be represented by the quantity of evaporation residue of groundwater. The summation of all cations and anions can represent the quantity of evaporation residue theoretically. Besides, salinity can be measured by ion exchange method.

Salinity is measured in g/L and mg/L.

Table 10.4 shows the classification of groundwater based on salinity.

(5) KypmoBa formula

Groundwater chemical type can be represented by major anion and cation percentage of meq, as follows:

Trace elements (g/L) gas (g/L) salinity (g/L) $\frac{\text{Anion}(\text{meq} > 10\%)}{\text{Cation}(\text{meq} > 10\%)} \cdot \text{temp}\,(^\circ\text{C})$

where the percentage of meq is the meq of certain ions dividing the total meq of all ions.

The ions with (meq/L %) > 25 % can be named as groundwater type.

Take the groundwater of Tanggu, Tianjin for example:

$$F_{0.05} M_{1.05} \frac{HCO^3_{53.4} Cl_{39.6}}{Na_{95.16}} \cdot T_{15}$$

The groundwater type is HCO_3–Cl–Na.

Table 10.4 Groundwater types according to salinity

Types	Freshwater	Low salinity water (light salty water)	Medium salinity water (salty water)	High salinity water (saline)	Brine
Salinity (g/L)	<1	1–3	3–10	10–50	>50

10.1.2 Groundwater Quantity Analysis Contents

Analytic contents of groundwater quantity can be different based on engineering requirement and accuracy. In engineering geological investigation, only two items need to be analyzed, basic ion balance analysis and special analysis on erosion CO_2.

10.1.2.1 Ion Balance Analysis

If the contents of Na^+ and K^+ have been measured, the total anion of meq per liter ($\sum A$) is theoretically equal to the total anion of meq per liter ($\sum C$). However, because of analysis error, they are not equal in most cases. Analysis error can be calculated by the following formula, and its limited range is ±2 %.

$$x\% = \frac{\sum C - \sum A}{\sum C + \sum A} \times 100\% \tag{10.1}$$

If the contents of Na^+ and K^+ are not measured, $\sum A$ is usually larger than $\sum C$, and the limited range of x is shown in Table 10.5.

10.1.2.2 Erosion CO_2 Analysis

For natural groundwater, the measured and theoretical values should be close to each other. The latter can be calculated by the following formula:

$$[HCO_3^-]^3 + \left(2[Ca^{2+}] - [HCO_3^-]_0\right)[HCO_3^-]^2 + 1/Kf[HCO_3^-]$$
$$- 1/Kf\left(2[CO_2]_0 + [HCO_3^-]_0\right)$$
$$= 0$$

where $[Ca^{2+}]_0$, $[HCO_3^-]_0$, and $[CO_2]_0$ are the concentration of Ca^{2+}, HCO_3^- and free CO_2, mmol/L; $[HCO_3^-]$ is the equilibrium concentration of HCO_3^- after adding $CaCO_3$; K is the equilibrium constant and f is the activity coefficient, which can be looked up in Tables 10.6 and 10.7.

Table 10.5 Limited range for groundwater analysis error

Total content of anion and cation (mg/L)	Limited range (%)
>300	3
<300	5

Table 10.6 Equilibrium constant under different temperatures

Temp (°C)	0	5	10	15	20	25	30
K	0.0160	0.0152	0.0171	0.0189	0.0222	0.0260	0.0328

Table 10.7 Activity coefficient under different ionic strengths (μ)

μ	f	μ	f	μ	f	μ	f
0.001	0.809	0.012	0.522	0.032	0.381	0.055	0.307
0.002	0.745	0.014	0.499	0.034	0.372	0.060	0.297
0.003	0.703	0.016	0.480	0.036	0.364	0.065	0.286
0.004	0.668	0.018	0.463	0.038	0.357	0.070	0.277
0.005	0.641	0.020	0.449	0.040	0.350	0.075	0.269
0.006	0.616	0.022	0.434	0.042	0.343	0.080	0.261
0.007	0.597	0.024	0.421	0.044	0.337	0.085	0.254
0.008	0.579	0.026	0.410	0.046	0.331	0.090	0.247
0.009	0.562	0.028	0.0400	0.048	0.325	0.095	0.241
0.10	0.547	0.030	0.390	0.050	0.320	0.100	0.235

Note $\mu = 1/2 \sum C_i Z_i^2$, where C_i is the ion concentration and Z_i is the ionic charge

10.1.3 Water Sample Requirements

10.1.3.1 Basic Principles for Water Sample Gathering

The water sample must represent the natural condition of groundwater, which means it should be freshwater gathered from drillings, observation holes, wells, and test pit. Spring samples should be taken from spring vent.

10.1.3.2 General Requirements for Water Sample Gathering

1. The containers can be plastic bottle, or drum or glass bottle with ground glass stopper, which should be cleaned with distilled water. Flush the containers, including their stoppers, more than thrice with the water which will be gathered. Then pour water into the containers slowly. A space with a height of 10–12 mm should be left in the top of containers. Seal these containers with paraffin or sealing wax in time and label them with sampling notes. Send these samples with inspection application sheets to lab.
2. If the water sample is unstable, the stabilizer should be added immediately after sampling, and impurity must be kept off. Specific methods are shown in Table 10.8.
3. Water samples should be prevented from frost or direct sunlight, and stored as required. Tests should be taken before expiration.
4. The amount of water samples is determined by test contents, as follows:

Table 10.8 Sampling methods for unstable groundwater

Unstable components for special analysis	Sampling amount (L)	Processing methods	Notes
Corrosive CO_2	0.25–0.30	Add 2–3 g marble powder	Take simple and complete analysis samples at the same time
Total sulfide	0.30–0.5	Add 10 mL 1:3 Tin acetate solution or 2–3 mL 25 % Zinc acetate solution, and 1 mL 4 % sodium hydroxide solution	Weigh the sample (with bottle)
Cu, Pb, Zn	1.0	Add 5 mL 1:1 hydrochloric acid solution	Hydrochloric acid should not contain the ions tested; sand mixing is absolutely forbidden
Fe	0.5	Freshwater: add 15–25 mL acetic acid-acetate liquid	Turbid water should be filtrated rapidly
		Alum water or acidic water: add 5 mL 1:1 sulfuric acid solution and 0.5–1.0 g ammonium sulfate	
Dissolved oxygen	0.3	Add 1–3 mL alkaline potassium iodide solution and 3 mL manganese chloride solution, then shake it violently and seal it	Weigh the sample bottle before sampling; water should be full in the bottle; record total volume of additives and temperature
		When the water sample contains a lot of organic matter and reducing substances, add 0.5 mL bromine (or potassium permanganate solution), and place it for 24 h after shaking. Then add 0.5 mL salicylic acid solution, and then perform the above procedures	
Cyanide	0.5	Add 2 g solid sodium hydroxide in 1 L water	Keep cold and send for analysis as soon as possible
Phenol	0.5	Add 2 g solid sodium hydroxide in 1 L water	Keep cold and send to analysis as soon as possible
N	1.0	Add 0.7 g concentrated sulfuric acid	Keep cold and send to analysis as soon as possible
Ra	2–3	Add 4–6 g concentrated sulfuric acid	

(continued)

Table 10.8 (continued)

Unstable components for special analysis	Sampling amount (L)	Processing methods	Notes
U	0.5–1.0 (Blue ray method)	Add hydrochloric acid solution	Colorimetric method needs 2–3 L water sample
Rn	0.1	Sampling with vacuumed glass dilator; dilator can be replaced by glass bottle with ground glass stopper	No air should in sample bottle; record the sampling time in detail; avoid stirring water samples

Simple analysis: 500–1000 mL;
Complete analysis: 2000–3000 mL.
Special analysis: see Table 10.8.
The amount of water sample should be above 20–30 % of the amount needed.

10.2 Groundwater Pollution

Nowadays, more and more large engineering constructions inevitably influence the vulnerable groundwater environment system. On the one hand, groundwater quantity, quality, and deposit condition directly connect to the construction safety and quality, and service safety. On the other hand, engineering construction could in turn affect groundwater hydrodynamic conditions and pollution. This means the relationship of groundwater and engineering construction is interacting and counteracting, and they both should be handled carefully. Thus, it is necessary to study groundwater pollution problems.

10.2.1 Concepts of Groundwater Pollution

According to the *PRC Water Pollution Control Law (1984)*, water pollution can be defined as: "A water quality deterioration phenomenon that results from chemical, physical, biological or radioactive change caused by intervention of certain substance or energy, which may affect the effective use of water, cause health hazard, and destroy ecological balance."

Groundwater pollution is a kind of water pollution. Groundwater, surface water, and precipitation can be converted into each other. Thus, affected by human activities, pollutants, microorganism, and heat energy can transmit into groundwater, which may lead to groundwater quality deterioration that show up as ineffective use of water, damage to human health, and destruction of ecological balance and environment. This phenomenon is called "groundwater pollution."

10.2.2 Pollutants, Pollution Sources, and Pollution Paths or Ways

Pollutants are substances or energy that causes groundwater pollution.
Pollution sources are the place or point that the pollutants come from.
Pollutants from pollution source get into pumping or studying area through pollution paths in many ways.

10.2.2.1 Pollutants

Groundwater pollutants can be divided into three types, as follows:

1. Inorganic pollutants:
   Common pollutants for major components are NO_3^-, Cl^- and total dissolved solid. Trace nonmetal components mainly include arsenic, phosphate, and fluoride, while trace metal components include Cr, Hg, Cd, Zn, Fe, Mn, and Cu etc.
2. Organic pollutants:
   So far, phenolic compound, cyanide, and pesticide have been detected in groundwater.
3. Pathogen pollutants:
   Common pathogen pollutants in polluted groundwater are nonpathogenic *E. coli*, pathogenic *S. typhimurium*, *M. tuberculosis*, and hepatitis bacteria.

   Major pollutants in industrial sewage are shown in Table 10.9.

10.2.2.2 Pollution Sources

Groundwater pollution sources can be divided into four types, as follows:

1. Domestic pollution: including municipal sewage and domestic solid waste.
2. Industrial pollution: including industrial sewage, refuse, waste and corrupt matter, waste gas, and radioactive material.
3. Agricultural pollution: including pesticide, chemical fertilizer, insecticide, backwater in sewage irrigation, and animal waste.
4. Environmental pollution: including natural salty aquifers, seawater, and soluble substances of dewatering formation in mining area.

10.2.2.3 Pollution Paths and Ways

The pollution paths of groundwater pollutants are complex. However, they can be divided into four types: intermittent infiltration type, continuous infiltration type, leaky flow type, and underground runoff type, specifically shown in Table 10.10.

10.2 Groundwater Pollution

Table 10.9 Sewage pollutants of major industry departments

Department	Industry	Major pollutants
Metallurgical industry	Ferrous metallurgy (mineral processing, sintering coking, steelmaking, steel rolling)	Suspended matter, acidity, phenol, cyanide, oil, chemical oxygen-demanded substance, biochemical oxygen-demanded substance, chroma, sulfide, polycyclic aromatic hydrocarbon
	Nonferrous metallurgical (mineral processing, sintering, smelting, electrolysis, refining)	Suspended matter, Cu, Zn, Pb, Hg, Ag, As, Cr, fluoride, chemical oxygen-demanded substance, acidity
Chemical industry	Basic chemical industry (acid, alkali, inorganic, and organic material)	Hg, As, Cr, phenol, cyanide, sulfide, benzene, aldehyde, alcohol, oil, suspended matter, fluoride, acid, alkali, chemical oxygen-demanded substance
	Fertilizer industry (synthetic ammonia, nitrogen fertilizer, phosphate fertilizer)	Suspended matter, chemical oxygen-demanded substance, As, acid, alkali, fluoride, ammonia, total phosphorus
	Chemical fiber industry	Chemical oxygen-demanded substance, soluble solid, total organophosphorus, biochemical oxygen-demanded substance, acid, alkali, suspended matter, Zn, Cu, SO_2
	Synthetic rubber industry	Aniline, alkene, total organic phosphorus, chemical oxygen-demanded substance, biochemical oxygen-demanded substance, oil, Cu, Zn, Cr, acid, alkali, polycyclic aromatic hydrocarbon
	Plastic industry	Chemical oxygen-demanded substance, Hg, organochlorine, As, acid, alkali, Pb, polycyclic aromatic hydrocarbon
	Pesticide, pharmaceutical and painting industry	Organochlorine, organophosphorus, chlorobenzene, chloral, NaClO, acidity, chemical oxygen-demanded substance, biochemical oxygen-demanded substance, suspended matter, oil, polycyclic aromatic hydrocarbon

(continued)

Table 10.9 (continued)

Department	Industry	Major pollutants
Light industry	Paper industry	Suspended matter, alkali, biochemical oxygen-demanded substance, chloral, phenol, sulfide, Hg, lignin
	Textile and dyeing industry	Acid, alkali, sulfide, suspended matter, biochemical oxygen-demanded substance, chemical oxygen-demanded substance, total organic carbon
	Food industry	Chemical oxygen-demanded substance, biochemical oxygen-demanded substance, suspended matter, acid, alkali, Escherichia coli, total bacteria
	Leather industry	Acid, alkali, Cr, sulfide, biochemical oxygen-demanded substance, chemical oxygen-demanded substance, total organic carbon, suspended matter, nitrate
Machinery industry	Electronic industry	Acid, Cr, Cd, Zn, Cu, Hg, suspended matter
	Agricultural machinery, universal machine building, machining industry	Acid, alkali, cyanide, Cr, Cd, Zn, Cu, Ni, oil, suspended matter
Petroleum industry	Refining, distillation, cracking	Biochemical oxygen-demanded substance, chemical oxygen-demanded substance, oil, phenol, cyanide, benzene, polycyclic aromatic hydrocarbon, aldehyde, alcohol, suspended matter
Building material industry	Cement industry, asbestos industry, glass industry	Suspended matter, acid, alkali, phenol, cyanide
Mining industry	Coal mining, nonferrous metal mining, ferrous metal mining	Acid, alkali, suspended matter, heavy metal, radioactive material

The ways in which groundwater is polluted can be summarized as direct pollution and indirect pollution.

1. Direct pollution:
   Groundwater pollutant comes from the pollution source directly without property change. This is the major way for groundwater pollution, and pollution source and path can be found easily.
2. Indirect pollution:
   Pollutant in a low level or even does not exist at pollution source, and is generated during polluting. This kind of pollution is a complex slow varying process, for example, human activities cause increase in groundwater hardness.

10.2 Groundwater Pollution

Table 10.10 Types of groundwater pollution paths (from Handbook of Engineering Geology, 1992)

Types	Pollution paths	Pollution source	Polluted aquifer
Intermittent infiltration type	Precipitation leaching to solid wastes	Industrial or living solid waste landfills	Unconfined aquifer
	Precipitation leaching to dewatering formation in mining area	Soluble minerals of dewatering formation	Unconfined aquifer
	Irrigation and precipitation leaching to farmland	Pesticide, chemical fertilizer and soluble salt residues on farmland surface	Unconfined aquifer
Continuous infiltration type	Leaking from sewage gathering and chemical warehouse	Sewages and chemical liquid	Unconfined aquifer
	Leaking from polluted surface water	Polluted surface water	Unconfined aquifer
	Leaking from sewage lines	Sewages	Unconfined aquifer
Leaky flow type	Leaky caused by groundwater exploitation	Polluted phreatic water and natural salty water	Unconfined and confined aquifers
	Leaky flow through hydrogeological window	Polluted phreatic water and natural salty water	Unconfined and confined aquifers
	Between wells	Polluted phreatic water and natural salty water	Unconfined and confined aquifers
Underground runoff type	Through karst pipelines	Sewages and polluted surface water	Unconfined aquifer
	Through wastewater treatment wells	Sewages	Unconfined and confined aquifers
	Seawater intrusion	Seawater	Unconfined and confined aquifers

10.2.3 Investigation and Monitoring of Groundwater Pollution

To find out whether groundwater has been polluted and leads to damage to buildings and underground facilities, investigation and monitoring of groundwater pollution will be taken, which can provide bases for environment assessment and treatment.

10.2.3.1 Environment Investigation of Groundwater Pollution

1. Water quality analysis projects of environment investigation including pH value, Eh, electrical conductivity, oxygen consumption, Mn, Ni, phenol, total nitrogen, ammonia nitrogen, phosphate, cyanide, biochemical oxygen-demanded substance, chemical oxygen-demanded substance, total organic carbon, soluble solvent (oil), bacteria analysis, heavy metal analysis, and so on.
2. The requirements of water sample collection have been discussed in Sect. 10.1.3. Measuring points setting principles are as follows: pollution source, pollution path, and way should be considered; 2–3 water samples should be taken from one measuring point; unpolluted water samples should be taken for comparison.

10.2.3.2 Environment Monitoring of Groundwater Pollution

1. Water quality analysis projects of environment monitoring shown in Table 10.11.
2. The requirements of water sample collection have been discussed in Sect. 10.1.3. Measuring points setting principles are as follows: Section lines of point pollution, line pollution, and block pollution should be parallel and vertical to groundwater flow; 1–3 samples should be taken from each section line; unpolluted water samples also should be taken as comparison; time factor should be considered.

10.3 Groundwater Corrosion Evaluation

Some ingredients in groundwater can corrode concrete and metal. Thus, groundwater corrosion evaluation should be taken in the condition that buildings contact groundwater for a long time.

10.3.1 Groundwater Corrosive Effects to Concrete

It is confirmed by a number of experiments that there are three forms of corrosion: decomposition corrosion (DC), crystallization corrosion (CC), and decomposition-crystallization corrosion (DCC). Corrosion forms depend on the chemical components of groundwater and the type of cement. Distinguishing standards are shown in Table 10.12.

10.3 Groundwater Corrosion Evaluation

Table 10.11 Water quality analysis projects and processing methods (from Handbook of Engineering Geology, 1992)

Test items	Water sample types			Sample container: plastic (P) or glass (G)	Preservation methods	Maximum preservation time
	Groundwater	Lake water	River water			
Necessary projects						
Temperature	+	+	+		In situ determination	
pH	+	+	+		In situ determination	
Electrical conductivity	+	+	+		In situ determination	
Dissolved oxygen	+	+	+		In situ determination	
Chloride	+	+	+	P or G	None	7 days
Total alkalinity	+	+	+	G	4 °C	24 h
Suspended matter	−	−	+	P or G	4 °C	7 days
Ammonia nitrogen (AsN)	+	+	+	P or G	pH_2	1–7 days
Nitrogen, potassium nitrate and nitrite (AsN)	+	+	+	P or G	pH_2	1–7 days
BOD_5 (20 °C)	−	−	+	G	4 °C	4–24 h
Fluoride	+	−		P	4 °C	7 days
Ortho phosphate (dissolution reaction, as P)	−	+	+	P (>100 µg/L)	4 °C	24 h
				G (<100 µg/L)		
Coliform group	+	+	+	G	4 °C	6–24 h
Selective projects						
TOC	+	+	+	G	Adding H_2SO_4	7 days
COD	+	+	+	G	Adding H_2SO_4, pH_2	7 days
Anionic detergent	+	+	+	G	Adding $HgCl_2$	24 h
Nonionic detergent	−	−	+	G	4 °C	24 h
Transparency	−	+	+		In situ determination	
Sulfate	+	+	+	P or G	4 °C	7 days
Ca	+	+	+	P	None	7 days
Mg	+	+	+	P	None	7 days

(continued)

Table 10.11 (continued)

Test items	Water sample types			Sample container: plastic (P) or glass (G)	Preservation methods	Maximum preservation time
	Groundwater	Lake water	River water			
Volatile suspended solid	−	−	+	P or G	4 °C	7 days
N (Kjeldahl method)	−	+	+	P or G	Adding H_2SO_4, pH_2	24 h
Total chromium	−	−	+	P or G	Adding HNO_3, pH_2	6 months
Hexavalent chromium	−	−	+	P or G	Adding HNO_3, pH_2	6 months
Ni	−	−	+	P or G	Adding HNO_3, pH_2	6 months
Zn	−	−	+	P or G	Adding HNO_3, pH_2	6 months
Cu	−	−	+	P or G	Adding HNO_3, pH_2	6 months
As	+	−	+	P or G	Adding HNO_3, pH_2	6 months
B	+	−	+	P or G_S^*	Adding HNO_3, pH_2	6 months
Cyanide	−	−	+	P or G	4 °C	24 h
Si (reaction)	−	+	−	Only P	Adding NaOH, pH_{12}, in situ filtration at 4 °C	7 days
Total iron	+	+	+	P or G	Adding HNO_3, pH_2	6 months
Mn	+	+	+	P or G	Adding HNO_3, pH_2	6 months
K	+	+	+	P	Adding HNO_3, pH_2	6 months
Na	+	+	+	P	Adding HNO_3, pH_2	6 months
Total phosphorus	−	+	+	P or G	4 °C	24 h
Streptococcus faecalis	+	+	+	G	4 °C	6–24 h
Chlorophyll a	−	+	−	P or G		
Phytoplankton	−	+	−	P	Lugd's iodine, acidified by acetic acid	
Primary yield	−	+	−		Preserved in clear or dark bottles	

(continued)

10.3 Groundwater Corrosion Evaluation

Table 10.11 (continued)

Test items	Water sample types			Sample container: plastic (P) or glass (G)	Preservation methods	Maximum preservation time
	Groundwater	Lake water	River water			
Dissolved CO_2	+	+		P	4 °C	24 h
Permanganate	+		+	G	4 °C	24 h
Se	+	+	+	P or G	Adding HNO_3, pH_2	6 months
H_2S	+	+	+	G	Adding zinc acetate and NaOH 2 mL/L respectively	24 h
Ba	+			P or G	Adding HNO_3, pH_2	6 months
Phenol	+		+	Only G	Adding $CuSO_4$ 1 g/L, then adding H_3PO_4 till pH = 4	24 h
Li	+	+	+	P or G	Adding HNO_3, pH_2	6 months
Polycyclic aromatic hydrocarbons	+	+	+			
Testing projects with global meaning						
Heavy metal						
Gr				P	Adding HNO_3, pH_2	6 months
Hg				G	Adding HNO_3, $pH_{1\pm0.2}$	6 months in glass bottle or 2 weeks in plastic bottle
Pb				P	Adding HNO_3, pH_2	6 months
Organic chloride						
DDT				Glass bottle with PTFE lining and lid		
DDE				Glass bottle with PTFE lining and lid		
DDD				Glass bottle with PTFE lining and lid		

(continued)

Table 10.11 (continued)

Test items	Water sample types			Sample container: plastic (P) or glass (G)	Preservation methods	Maximum preservation time
	Groundwater	Lake water	River water			
Dieldrin				Glass bottle with PTFE lining and lid		
BHC				Glass bottle with PTFE lining and lid		
Polychlorinated Biphenyls				Glass bottle with PTFE lining and lid		

Note G_S represents C-glass

10.3.1.1 Decomposition Corrosion

Decomposition corrosion is the breakdown effect caused by acidic water leaching calcium and $CaCO_3$ from concrete. It has two basic types: ordinary acidic corrosion and carbonate corrosion.

Ordinary acidic corrosion is the reaction between H^+ of groundwater and $Ca(OH)_2$ of concrete. The chemical equation is

$$Ca(OH)_2 + 2H^+ \rightleftharpoons Ca^{2+} + 2H_2O$$

The degree of reaction depends on the pH value of groundwater, which means the lower the pH value is, the larger reaction degree is.

Carbonate corrosion is the reaction between corrosive CO_2 in groundwater and $CaCO_3$ of concrete. First of all, the reaction between lime and CO_2 on the surface of concrete produces a $CaCO_3$ layer. Then further reaction produces calcium bicarbonate, which can easily dissolve in water. These reactions lead to concrete breaking. The chemical equation is

$$CaCO_3 + CO_2 + H_2O \rightleftharpoons Ca^{2+} + 2HCO_3^-$$

This is a reversible reaction. The dissolving of calcium carbonate requires a certain amount of free CO_2 to maintain balance, which is called balance carbon dioxide. The reducing of free CO_2 will lead to precipitating of $CaCO_3$, while the increase of free CO_2 will make the reaction to move forward to reach a new balance. The part of CO_2 that reacts with $CaCO_3$ is called corrosive CO_2.

10.3 Groundwater Corrosion Evaluation

Table 10.12 Distinguishing standards for groundwater-concrete corrosion form

Corrosion form	Corrosion index		Gravelly soil				Sandy soil				Clayey soil			
		Cement type	A		B		A		B		A		B	
			Regular	Sulfate erosion resisted	Regular	Sulfate erosion resisted	Regular	Sulfate erosion resisted	Regular	Sulfate erosion resisted	Regular	Sulfate erosion resisted	Regular	Sulfate erosion resisted
Decomposition corrosion	DC index pH_s		Groundwater is corrosive when $pH < pH_s$, $pH_s = \frac{[HCO_3^-]}{0.15[HCO_3^-] - 0.25} - K_1$								None			
			$K_1 = 0.5$		$K_1 = 0.3$		$K_1 = 1.3$		$K_1 = 1.0$					
	pH		<6.2		<6.4		<5.2		<5.5					
	Free CO_2 (mg/L)		Corrosive when free $[CO_2] > a[Ca^{2+}] + b + K_2$											
			$K_2 = 20$		$K_2 = 15$		$K_2 = 80$		$K_2 = 60$					
Crystallization corrosion	$[SO_4^{2-}]$ (mg/L)	$[Cl^-]$ (mg/L) <1000	>250	>3000	>250	>4000	>300	>3500	>300	>3500	>400	>4000	>400	>5000
		$[Cl^-]$ (mg/L) 1000–6000	>100+0.15 $[Cl^-]$		>100+0.15 $[Cl^-]$		>150+0.15 $[Cl^-]$		>150+0.15 $[Cl^-]$		>250+0.15 $[Cl^-]$		>250+0.15 $[Cl^-]$	
		$[Cl^-]$ (mg/L) >6000	>1050		>1050		>1100		>1100		>1200		>1200	
Decomposition-crystallization corrosion	Weak base sulfate ions [Me]		Corrosive when [Me] >1000 mg/L or [Me] > $K_3 - SO_4^{2-}$								None			
			$K_3 = 7000$		$K_3 = 6000$		$K_3 = 9000$		$K_3 = 8000$					

Notes: 1. A represents Portland cement; B represents Portland cement mixed with volcano ash, sand, and slag
2. Indexes *a* and *b* can be found in Table 10.13

The distinguishing standards of decomposition corrosion are shown in Table 10.12. There are three indexes.

1. Decomposition corrosion index pH_s.
   It is the main index of decomposition corrosion that can be calculated as follows:

$$pH_s = \frac{[HCO_3^-]}{0.15[HCO_3^-] - 0.25} - K_1$$

   where $[HCO_3^-]$ is the concentration of HCO_3^-, measured in meq/L; K_1 is found in Table 10.12. When $pH \geq pH_s$ the groundwater does not have decomposition corrosivity, otherwise it has this kind of corrosion.

2. Acidic corrosion index pH.
   When pH value of groundwater is lower than that shown in Table 10.2, it has acidic corrosivity.

3. Carbonate corrosion index.
   The carbonate corrosion index is the concentration of free CO_2. If the concentration of free CO_2 is larger than $[CO_2]_3$, the groundwater has carbonate corrosivity. $[CO_2]_3$ is calculated by the following formula:

$$([CO_2]_3) = a[Ca^{2+}] + b + K_2$$

   where $[Ca^{2+}]$ is the concentration of Ca^{2+}, mg/L; a and b can be found in Table 10.13; K_2 can be found in Table 10.12.

The groundwater has decomposition corrosion if any of the above three indexes meets the condition.

10.3.1.2 Crystallization Corrosion

Crystallization corrosion mainly refers to sulfuric acid corrosion, which is the reaction between sulfate in groundwater and concrete. This kind of reaction can produce gypsum and aluminum sulfate crystals in the pores of the concrete. The volume of gypsum crystal is 1–2 times larger than cement and 2–5 times for aluminum sulfate crystal, which will lead to the decrease of concrete strength, even to breaking. Gypsum is the intermediate product of aluminum sulfate. The reaction formula is

$$4CaO \cdot Al_2O_3 \cdot 12H_2O + 3CaSO_4 \cdot nH_2O$$
$$\rightarrow 3CaO \cdot Al_2O_3 \cdot 3CaSO_4 \cdot 3H_2O + Ca(OH)_2$$

10.3 Groundwater Corrosion Evaluation

Table 10.13 Values of coefficients a and b (from Handbook of Design and Construction of Underground Engineering, 1999)

[HCO$_3^-$] (meq/L)	Total amount of Cl$^-$ and SO$_4^{2-}$ (mg/L)										
	0–200		201–400		401–600		601–800		801–1000		>1000	
	a	b	a	b	a	b	a	b	a	b	a	b
1.4	0.01	16	0.01	17	0.07	17	0.00	17	0.00	17	0.00	17
1.8	0.04	17	0.04	18	0.03	17	0.02	18	0.02	18	0.02	18
2.1	0.07	19	0.08	19	0.05	18	0.04	18	0.04	18	0.04	18
2.5	0.04	21	0.09	20	0.07	19	0.06	18	0.06	18	0.05	18
2.9	0.13	23	0.11	21	0.09	19	0.08	18	0.07	18	0.07	18
3.2	0.16	25	0.14	22	0.11	20	0.10	19	0.09	18	0.08	18
3.6	0.2	27	0.17	23	0.14	21	0.12	19	0.11	18	0.10	18
4	0.24	29	0.2	24	0.16	22	0.15	20	0.13	19	0.12	19
4.3	0.28	32	0.24	26	0.19	23	0.17	21	0.16	20	0.14	20
4.7	0.32	34	0.28	27	0.22	24	0.20	22	0.19	21	0.17	21
5	0.36	36	0.32	29	0.25	26	0.23	23	0.22	22	0.19	22
5.4	0.40	38	0.36	30	0.29	27	0.26	24	0.24	23	0.22	23
5.7	0.44	41	0.40	32	0.32	28	0.29	25	0.27	24	0.25	24
6.1	0.48	43	0.43	34	0.36	30	0.33	26	0.30	25	0.28	25
6.4	0.54	46	0.47	37	0.4	32	0.36	28	0.38	27	0.31	27
6.8	0.61	48	0.51	39	0.44	33	0.40	30	0.37	29	0.34	29
7.1	0.67	51	0.55	41	0.48	35	0.44	31	0.41	30	0.38	30
7.5	0.74	53	0.60	43	0.53	37	0.48	33	0.45	31	0.41	31
7.8	0.81	55	0.65	45	0.58	38	0.53	34	0.49	33	0.44	33
8.2	0.88	58	0.70	47	0.63	40	0.58	35	0.53	34	0.48	34
8.6	0.96	60	0.76	49	0.68	42	0.63	37	0.57	36	0.52	36
9	1.04	63	0.81	51	0.73	44	0.67	39	0.61	38	0.56	38

It should be noticed that crystallization corrosion is not isolated. It is always accompanied with decomposition corrosion, which can enhance crystallization corrosion. In addition, sulfuric acid corrosion is relevant to the concentration of Cl$^-$ and the location of buildings. For example, changes in water level will enhance this kind of corrosion. In recent years, in order to prevent the damaging effects of high content SO_4^{2-} to the cement, sulfate-resisting cement is used in construction.

The concentration of SO_4^{2-} is the crystallization corrosion index. If the concentration of SO_4^{2-} in groundwater is larger than the corresponding value in Table 10.12, the groundwater has crystallization corrosivity. Crystallization corrosion of Portland cement is also related to the concentration of Cl$^-$, while the sulfate-resisting cement is not.

10.3.1.3 Decomposition-Crystallization Corrosion

Decomposition-crystallization corrosion refers to the corrosion of weak base sulfate ions, which means the chemical reaction between cement and high concentration $Mg^{2+}, Fe^{2+}, Fe^{3+}, Ca^{2+}, Zn^{2+}, NH_4^+$, etc. This reaction can decrease the strength of concrete until damage. For example, the reaction between $MgCl_2$ in groundwater and $Ca(OH)_2$ crystal in concrete, which can produce $Mg(OH)_2$ and soluble $CaCl_2$, will damage the concrete.

The decomposition-crystallization corrosion index is the concentration of weak base sulfate ions $[M_e]$, which is mainly used in industrial wastewater corrosion evaluation. The groundwater has decomposition-crystallization corrosivity if $[M_e] > 1000\,\text{mg/L}$ and meets the following condition:

$$[M_e] > K_3 - [SO_4^{2-}]$$

where $[M_e]$ is the concentration of one or some of the $Mg^{2+}, Fe^{2+}, Fe^{3+}, Ca^{2+}, Zn^{2+}, NH_4^+$ ions, mg/L; $[SO_4^{2-}]$ is the concentration of SO_4^{2-}, mg/L; K_3 can be found in Table 10.12.

10.3.2 Groundwater Corrosive Effects to Steel

If the pH value of groundwater is low and there are dissolved oxygen, free sulfuric acid, H_2S, CO_2, and other heavy metal sulfates in it, it can corrode rebar, iron pipe and other kinds of steels violently. Therefore, groundwater corrosive effects should be considered when designing steel structures that are immersed in groundwater. Especially, sulfide mines often form acidic mine water, which is harm to mining equipment.

Groundwater corrosion of iron is mainly related to hydrogen ion concentration. When the pH value of groundwater is less than 6.8, it has corrosivity, and if the pH value is less than 5, the corrosive effect will be strong.

The dissolved oxygen and iron material can have oxidation, which leads to rust. The company of CO_2 will aggravate this reaction.

The corrosion of free H_2SO_4 in groundwater is also caused by hydrogen ions. In order to prevent corrosion of iron material by sulfuric acid, the concentration is preferably not more than 25 mg/L.

The dissolved CO_2 or H_2S makes the groundwater an electric conductor, which can lead to electrochemical reaction, aggravating the reaction. The formulas are as follows:

$$CO_2 + H_2O \rightleftharpoons H_2CO_3 \rightleftharpoons H^+ + HCO_3^-$$
$$H_2S \rightleftharpoons H^+ + HS^-$$

10.3 Groundwater Corrosion Evaluation

At this time, iron releases charges which are accepted by hydrogen, that is:

$$Fe \rightarrow Fe^{2+} + 2e$$
$$2H^+ + 2e \rightarrow H_2 \uparrow$$

Heavy metal sulfates such as $CuSO_4$ can also aggravate corrosion. This is because iron and copper can form microcell reaction, which is iron releases charges that are accepted by hydrogen:

$$Fe \rightarrow Fe^{2+} + 2e$$
$$Cu^{2+} + 2e \rightarrow Cu$$

At present, there is no uniform distinguishing standard for groundwater-metal corrosion, but the above conclusions must be considered in underground engineering designing.

10.4 Case Study

Analysis report of water quality is shown in Table 10.14.
Some descriptions of the water quality project are presented below as well.

Engineering: well water in the southwest of Wulitun	Drilling No. EA_2	
Date of water taken: 9 May 1988		
Engineering No.	Exploration well No.	Date of analysis: 16 May–2 June 1988
Laboratory No. 85015	Depth of water taken	Date of presentation 13 June 1988
Temperature	28 °C	Water source confined water of Quaternary
Water temperature	18 °C	Room temperature °C

Table 10.14 Water quality analysis report

Odor		Odorless			Taste			Tasteless	
Color and chroma		Colorless			Transparency			Transparent	
Suspended matter		None							
Project		Content in per liter of water			Hardness	Project	Germanic degrees	Project	mg/L
		mg	mN	mN%		Total hardness	12.18		
cation	$Na^+ + K^+$	21	0.84	16.18		Temporary hardness	8.94	Residue on ignition	
	Ca^{++}	60.14	3	57.8		Permanent hardness	3.24	loss on ignition	
	Mg^{++}	16.45	1.35	26.02		Negative hardness	0	Al^{+++}	
	Fe^{+++}	0	0			Project	Bacteria index/value	M_n^{++}	
	Fe^{++}	0	0			Escherichia coli		Cu^{++}	Not detected
	NH_4^+	0	0			Sum of bacteria	60/L	P_b^{++}	Not detected
	Sum	97.59	5.19	100				Z_n^{++}	Not detected
					Project	mg/L		$C_r...$	0.002
Anion	Cl'	33.62	0.95	18.31				A_s	Not detected
	SO_4^-	25.72	0.54	10.40	Free CO_2	13.20		F^-	0.15
	HCC	194.50	3.19	61.46	Erosive CO_2			I^-	
	CO_3^-				Consumed oxygen	0.66		CN^-	Not detected
	NO_3^-	31.80	0.51	9.83	Dissolved oxygen			PO_4^-	
	NO_2^-	0.01	–		H_2S	0		Phenol	Not detected
	Sum	285.65	5.19	100	Soluble S_1O_2	20.6		Hg	Not detected
					Dried leavings	311			
Sum								Total mineralization	306.74
								pH	7.2
								Total alkalinity	3.19 mN/L
								Acidity	mN/L

Conclusion and decision

Note when testing water with iron, add 1:1 nitric acid to make water sample pH < 2 and obtain total iron content 0.04 mg/L

Person in charge:
Analyst:
Requirements:

1. Please write Kurllov's (KypmoBa) formula of the water sample and determine the chemical type of groundwater.
2. Based on the analysis report of water quality above, please assess the corrosion of groundwater to concrete, steel and iron.
   Note mN = meq (milliequivalent)

10.5 Exercises

1. What indexes are commonly used in the groundwater quality analysis?
2. How many types of concrete corrosion are there by groundwater? What are they?

Bibliography

An GF, Gao DZ (2001) 3D-FEM Application to the prediction of creep settlement of soft clay considering elastic-visco-plastic constitution. J Tongji Univ 29(2):195–199 (in Chinese)
Andrus RD, Stokoe KH, Roesset JM (1991) Liquefaction of gravelly soil at pence ranch during the 1983 Borah Peak, Idaho Earthquake. In: First international conference on soil dynamics and earthquake engineering V, Karlsruhe, Germany, September
Andrus RD, Stokoe KH, Chung RM (1997) Draft guidelines for evaluating liquefaction resistance using shear wave velocity measurements and simplified procedures, NISTIR 6277. National Institute of Standards and Technology, Gaithersburg, MD 121 pp
Bahar R, Cambou B (1995) Forecast of creep settlements of heavy structures using pressure meter tests. Comput Geotech 17:507–521
Beijing Disaster Reduction Association (1998) Urban sustainable development and disaster prevention. Meteorological Press, Beijing (in Chinese)
Cai YQ, Qian L (2001) Seismic liquefaction and stability analysis of Qiantang River embankment. J Hydraul Eng 1:57–61 (in Chinese)
Cai X, Guo XW, Shen PL et al (2001) Three dimensional consolidation analysis of soft soil foundation and its application. J Hohai Univ 29(5):27–32 (in Chinese)
Cao H, Zhong CH, Zhang YJ (2004) The study methods and its progress about soft clay creep. J Zhuzhou Inst Technol 18(2):102–106 (in Chinese)
Chen ZH (1993) Consolidation theory of unsaturated soil based on the theory of mixture (I). Appl Math Mech 14(8):687–698 (in Chinese)
Chen Z (2001) The impact of urban transit on ecological environment. China Railw Sci 22(3):126–131 (in Chinese)
Chen CX, Tang ZH (1990) Numerical method on groundwater flow problems. China University of Geosciences Press, Wuhan (in Chinese)
Chen ZH, Xie DY, Wang YS (1993) Experimental studies of laws of fluid motion, suction and pore pressures in unsaturated soil. Chin J Geot Eng 15(3):9–20 (in Chinese)
Chen YH, Hong BN, Gong DY et al (2002) Modified Cambridge visco-elastic-plasticity model considering rheology of soil. J Hohai Univ 30(5):44–47 (in Chinese)
Dai RL, Chen H, Yu YY (2001) Subsoil character of pile foundation and settlement analysis of high-rise buildings in Shanghai. Chin J Geot Eng 23(5):627–630 (in Chinese)
Dai FL, Shang HX, Lin GS et al (2002) Study of high temperature creep deformation for crystallizing materials. Chin J Theor Appl Mech 34(2):186–191 (in Chinese)
Dong MG, Fan HB, Hu ZP (2003) Numerical simulation technology of soft clay rheological test. Soil Eng Found 17(3):42–45 (in Chinese)
Escario V, Saez J (1986) The shear strength of partly saturated soils. Geotechnique 36(3):453–456
Fredlund DG (1976) Density and compressibility characteristics of air-water mixtures. Can Geotech J 13(4):386–396
Fredlund DG, Dakshanamurthy V (1982) Prediction of moisture flow and related swelling or shrinking in unsaturated soils. Geotech Eng 13:15–49

Fredlund DG, Hasan JU (1979) One-dimensional consolidation theory: unsaturated soils. Can Geotech J 16:521–531
Fredlund DG, Morgenstern NR (1977) Stress state variables for unsaturated soils. J Geotech Eng Div ASCE 103(5):447–466
Fredlund DG, Rachardjo H (1993) Soil mechanics for unsaturated soils. Wiley, New York
Gan KM, Fredlund DG, Rahardjio H. Determination of the shear strength parameters of an unsaturated soil using the direct shear test. Can Geotech J 25(3):500–510
Gao WH (1998) Time dependent analysis and stiffness calculation of ground model in rheological soft-clay. Rock Soil Mech 19(4):25–30 (in Chinese)
Gao HY, Feng QM (1997) Response analysis for buried pipelines through settlement zone. Earthq Eng Eng Vib 17(1):68–75 (in Chinese)
Gong SL (1998) Effects of urban construction on the land subsidence in Shanghai. Chin J Geol Hazard Control 9(2):108–111 (in Chinese)
He KQ, Wang SQ, Wang RL et al (2007) Action law of groundwater and evaluation of Huanglashi slope stability. Hydrogeol Eng Geol 34(6):90–94 (in Chinese)
Hu RL, Yue ZQ, Wang LC (2004) Review on current status and challenging issues of land subsidence in China. Eng Geol 76:65–77
Huang Y, Ye WM, Tang YQ et al (2001) Dynamic analysis of effective stress for seismic subsidence of pile foundations. Eng Mech 18(4):123–129 (in Chinese)
Handbook of design and construction of underground engineering (1999) Influence and prevention of groundwater on underground engineering (Xiao MY, Zeng JL eds). China Architecture & Building Press, Beijing
Handbook of engineering geology (1992) Groundwater, 3rd edn. China Architecture & Building Press, Beijing
Handbook of excavation engineering (1997) Design and construction of engineering dewatering (Liu JH, Hou XY eds). China Architecture & Building Press, Beijing
Handbook of hydrogeological investigation of water supply (1977) Calculation on hydrogeology, Beijing: China Geology Press
Ishihara K (1985) Stability of natural deposits during earthquakes. In: Proceedings of 11th internatioanl conference on soil mechanics and foundation engineering, San Fransisco, CA, A. A. Balkema, Rotterdam, vol 1, pp 321–376
Jiang PN (1989) On the engineering properties of unsaturated soils. Chin J Geot Eng 11(6):39–59
Jiang J, Chen LZ (2001) One-dimensional settlement due to long-term cyclic loading. Chin J Geot Eng 23(3):366–369 (in Chinese)
Lancellotta R (1997) A general nonlinear mathematical model for soil consolidation problems. Int J Eng Sci 35:1045–1063
Lenards GA (1962) Foundation engineering. McGraw-Hill, New York
Li JT (1989) Groundwater flow numerical simulation. Geology Publishing House, Beijing (in Chinese)
Li XY (1997) Three-dimensional nonlinear visco-elasticity plasticity analysis of saturated soft soil in Shanghai. Eng Mech (supp):443–352 (in Chinese)
Li JS, Chen CX (Trans), Bear J (1983) Dynamics of fluids in porous media. China Architecture and Building Press, Beijing
Li WX, Rong RD, Feng SQ (1987) Superficial sediments in the Yangtze River Estuary and the distribution characteristics of organics. Shanghai Geol 23(3):40–54 (in Chinese)
Li QF, Wang HM, Lu Y et al (2002) Study of land subsidence model and exiting issues. Shanghai Geol 84(4):11–15 (in Chinese)
Liao ZS (1998) Groundwater classification and basic concepts of fracture water. Geol J China Univ 4(4):473–477 (in Chinese)
Lin P, Xu ZH, Xu P et al (2003) Research on coefficient of consolidation of soft clay under compression. Rock Soil Mech 24(1):106–108 (in Chinese)
Liu HZ, Sun J (2001) Study on numerical simulation of factors of ground settlement during shield driving in soft ground. Mod Tunn Technol 36(6):24–28 (in Chinese)

Liu ZD, Lu SQ, Bao CG et al (1986) Shear strength of soil. Chin J Geot Eng 8(1):26–31 (in Chinese)

Liu Y, Zhang XL, Wan GF et al (1998) The situation of land subsidence and counter measures in Shanghai for recent years. Chin J Geol Hazard Control 9(2):13–17 (in Chinese)

Liu CH, Chen CX, Feng XT et al (2005) Effect of groundwater on stability of slopes at reservoir bank. Rock Soil Mech 26(3):419–422 (in Chinese)

Liu Y, Wan HP, Jiang YC et al (2005) Ground water action law and evaluation on dynamic stable of Huanglashi slope in three gorges region. Chin J Rock Mech Eng 24(19):3571–3576 (in Chinese)

Lv SW, Tang YQ, Ye WM (1998) Analysis of the engineering properties of soil in shallow stratum storing marsh gas. Shanghai Geol 67(3):50–55 (in Chinese)

Ma SZ, Jia HB, Meng GT (2002) The FEM simulation of dynamic penetration process of CPTU. Rock Soil Mech 23(4):478–481 (in Chinese)

Me YF, Xie DY, Wang SF (1995) Test study of the consolidation deformation law for saturated sand after vibration. Earthq Resist Eng 4:32–34 (in Chinese)

Men FL (1999) Preliminary investigations on rheological properties of clay and ground settlement in Shanghai city (I). J Nat Disaster 8(3):117–126 (in Chinese)

Men FL (1999) Preliminary investigations on rheological properties of clay and ground settlement in Shanghai city (II). J Nat Disaster 8(4):123–132 (in Chinese)

Meng QS, Wang N, Chen Z (2004) Pore water pressure mode of oozy soft clay under impact loading. Rock Soil Mech 25(7):1017–1022 (in Chinese)

Miao JF, Wu LG (1991) Viscoelastic theory of leakage flow with the consideration of three-dimensional compaction of soil skeletons due to pumping. J Tongji Univ 23(3):309–314 (in Chinese)

Ministry of Construction of the People's Republic of China (1999) Technical standard for water supply well (GB50296-99). China Planning Press, Beijing (in Chinese)

Ministry of Construction of the People's Republic of China (1999) Technical code for building and municipal dewatering engineering (JGJ/T111-98). China Architecture and Building Press, Beijing (in Chinese)

Ministry of Construction of the People's Republic of China (2001) Code for hydrogeological investigation of water supply (GB50027-2001). China Planning Press, Beijing (in Chinese)

Ministry of Construction of the People's Republic of China (2002) Code for design of building foundation (GB50007-2002). China Architecture and Building Press, Beijing (in Chinese)

Ministry of Construction of the People's Republic of China (2012) Technical specification for retaining and protection of building foundation excavation (JGJ120-2012). China Architecture and Building Press, Beijing (in Chinese)

Ministry of Metallurgy of the People's Republic of China (1999) Technical specification for foundation pits excavation (YB9258-97). China Architecture and Building Press, Beijing (in Chinese)

Neuman SP (1972) Theory of flow in unconfined aquifers considering delayed response of water table. Water Resour Res 8(4):1031–1045

Neuman SP (1975) Analysis of pumping test data from anisotropic unconfined aquifers considering delayed gravity response. Water Resour Res 11(2):329–342

Prickett TA (1965) Type-curve solution to aquifer test under water-table condition. Ground Water 3(3):5–14

Qian JZ, Yang LH, Li RZ et al (2003) Advances in laboratory study of groundwater flow in fractures system. J Hefei Univ Technol (Nat Sci) 26(4):510–513 (in Chinese)

Ran QQ, GuXY (1998) A coupled model for land subsidence computation with consideration of rheological property. Chin J Geol Hazard Control 9(2): 99–103 (in Chinese)

Research group of major natural disasters of the State Scientific and Technological Commission (1994) Major natural disasters and reduction countermeasures in China. Science Press, Beijing (in Chinese)

Seed HB (1968) Landslides during earthquakes. J Soil Mech Found Div ASCE 94(SM5)

Seed HB (1968) Seismic response of horizontal soil layers. ASCE 94 (SM.4)
Seed HB, Idriss IM (1971) Simplified procedure for evaluating soil liquefaction procedure. J Soil Mech Found Div ASCE 97(SM9):1249–1273
Shearer TR (1998) A numerical model to calculate land subsidence applied at Huangu in China. Eng Geol 49:85–93
Shi WH (1999) Analysis for the effects construction engineering and water resources' development on Shanghai land subsidence. Shanghai Geol 72(4):50–54 (in Chinese)
Shanghai Tongji Geotechnical Construction Industrial Co. (2006) Pumping test report of Baosteel's NO.3 hot rolling whirling pool foundation pit dewatering. Shanghai (in Chinese)
Shanghai Tongji Geotechnical Construction Industrial Co. (2006) Dewatering design scheme of Baosteel's NO.3 hot rolling whirling pool foundation pit dewatering. Shanghai (in Chinese)
Shanghai Guanglian Foundation Engineering Company (2006) Dewatering operation scheme of Baosteel's NO.3 hot rolling whirling pool foundation pit dewatering. Shanghai (in Chinese)
Shanghai Tongji Geotechnical Construction Industrial Co. (2006) Pumping test scheme and preliminary dewatering design scheme of steel rolling area whirling pool foundation pit for Pusteel's moving engineering. Shanghai (in Chinese)
Shanghai Tongji Geotechnical Construction Industrial Co. (2006) Pumping test report and dewatering design scheme of steel rolling area whirling pool foundation pit for Pusteel's moving engineering. Shanghai (in Chinese)
Shanghai Tongji Geotechnical Construction Industrial Co. (2007) Group well pumping test report and dewatering operation scheme of steel rolling area whirling pool foundation pit for Pusteel's moving engineering. Shanghai (in Chinese)
Si LX (1957) Design and construction of wellpoint system. Shanghai Science Press, Shanghai
Simoni L, Salomoni V, Schrefler BA (1999) Elastoplastic subsidence models with and without capillary effects. Comput Method Appl M 171:491–502
Soil Mechanics Department of Hohai University (East China Institute of Water Conservancy) (1984) Geotechnical principles and calculation: vol one. China Water & Power Press, Beijing (in Chinese)
Tang YQ, Xu C (1997) Problems of environmental geology of Shanghai urban development in the 21st century. Undergr Space 17(2):95–98 (in Chinese)
Tang YQ, Ye WM, Huang Y (2000) Prediction of ground settlement caused by subway tunnel construction in Shanghai. In: Proceeding of the international conference on engineering and technological sciences
Tang YQ, Ye WM, Zhang QH (1993) Some problems in tunneling by shield-driven method for Shanghai metro (Part I). Undergr Space 13(2):93–99 (in Chinese)
Tang YQ, Ye WM, Zhang QH (1993) Some problems in shield-driven tunneling of Shanghai metro (Part II). Undergr Space 13(3):171–177 (in Chinese)
Tang YQ, Ye WM, Zhang QH (1995) Study of ground settlement caused by subway tunnel construction in Shanghai. Undergr Space 15(4):250–258 (in Chinese)
Tang YQ, Ye WM, Zhang QH (1996) Marsh gas in soft stratum at the estuary of the Yangtse River and safety measures of construction of the tunnel. J Tongji Univ 24(4):465–470 (in Chinese)
Tang YF, Feng ZL, Lin H et al (2000) FEM analysis method of consolidation-nonlinear rheology on soft engineering. Build Struct 30(11):47–50 (in Chinese)
Tang YQ, Huang Y, Ye WM (2002) Discussion on several problems of construction of deep foundation ditch project. Constr Technol 31(1):5–6, 11 (in Chinese)
Tang B, Chen XP, Zhang W (2003) The general development of consolidation theory under considering its rheological characteristics. Soil Eng found 17(3):87–90 (in Chinese)
Tang YQ, Huang Y, Ye WM et al (2003) Critical dynamic stress ratio and dynamic strain analysis of soils around the tunnel under subway train loading. Chin J Rock Mech Eng 22(9):1566–1570 (in Chinese)
Tang YQ, Wang YL, Huang Y et al (2004) Dynamic strength and dynamic stress-strain relation of silt soil under traffic loading. J Tongji Univ 32(6):701–704 (in Chinese)

Tang YQ, Song YH, Zhou NQ et al (2005) Experimental research on troubles of EPB shield construction in sandy soil. Chin J Rock Mech Eng 24(1):52–56 (in Chinese)

Tang YQ, Zhang X, Wang JX et al (2005) Earth pressure balance shield tunneling-induced disturbance in silty soil. J Tongji Univ 33(8):1031–1035 (in Chinese)

Tang YQ, Zhang X, Wang JX et al (2005) Subsidence caused by metro tunnel excavation with shield method in Shanghai. In: Proceeding of the seventh international symposium on land subsidence, vol I, pp 257–269

Tang YQ, Zhang X, Zhou NQ et al (2005c) Microscopic study of saturated soft clay's Behavior under cyclic loading. J Tongji Univ 33(5):626–630 (in Chinese)

Tang YQ, Yan XX, Wang JX et al (2007) Model test study of influence of high-rise building on ground subsidence. J Tongji Univ 35(3):13–18 (in Chinese)

Tang YQ, Zhang X, Zhao SK et al (2007) Model of pore water pressure development in saturated soft clay around a subway tunnel under vibration load. China Civil Eng J 40(4):67–70 (in Chinese)

Tang YQ, Yang P, Shen F et al (2007c) Microscopic study of dark green silty's behavior after melted. J Tongji Univ 35(1):6–9 (in Chinese)

Tsuchida H (1970) Prediction and countermeasure against the liquefaction in sand deposits. Abstract of the Seminar in the Port and Harbor Research Institute

Walker AJ, Steward HE (1989) Cyclic undrained behavior of nonplastic and low plasticity silts. Technical report NCEER-89-0035, National Center for Earthquake Engineering Research, SUNY at Buffalo, 26 July 1989

Wang Y, Su BY, Xu ZY (1996) Comment on the models of seepage flow in fractured rock masses. Adv Water Sci 7(3):276–282 (in Chinese)

Wang SL, Shi GH, Wang DC (2000) Study and application of bed rock crevice water three-dimensional numerical model. Water Resour Hydropower Northeast China 18(4):36–38 (in Chinese)

Wang H, Luo GY, Li YH et al (2000) Structural analysis of water collection of faults. Hydrogeol Eng Geol 27(3):12–15 (in Chinese)

Wang HM, Tang YQ, Yan XX (2006) An unequal time-interval GM(1,1) model for predicting ground settlement in Shanghai soft soil engineering. J Eng Geol 14(3):398–400 (in Chinese)

Wang JX, Hu LS, Wu LG et al (2009) Hydraulic barrier function of the underground continuous concrete wall in the pit of subway station and its optimization. Environ Geol 57(2):447–453

Wang JX, Wu LG, Zhu YF et al (2009) Mechanism of dewatering-induced ground subsidence in deep subway station pit and calculation method. Chin J Rock Mech Eng 28(5):1010–1019 (in Chinese)

Wang JX, Guo TP, Wu LG et al (2010) Mechanism and application of interaction between underground wall and well in dewatering for deep excavation. Chin J Undergr Space Eng 6(3):564–570 (in Chinese)

Wang JX, Wu LG, Hu LS et al (2010) Pumping test and its application to deep foundation pit under complex more flow hydrogeological conditions. Chin J Rock Mech Eng 29(z1):3082–3087 (in Chinese)

Wheeler SJ (1988) A conceptual model for soils containing large gas bubbles. Geotechnique 38:389–397

Wheeler SJ (1988) The undrained shear strength of soils containing large gas bubbles. Geotechnique 38:399–413

Wu LG et al (2003) Design and execution of dewatering and theory of seepage in deep excavation. China Communications Press, Beijing (in Chinese)

Wu LG, Li G, Fang ZC et al (2009) Dewatering case history for excavation. China Communications Press, Beijing

Wuhan Surveying-Geotechnical Research Institute Co., Ltd. of MCC (2005) Geotechnical engineering investigation report of Baosteel's NO.3 hot rolling whirling pool foundation pit. Shanghai (in Chinese)

Wuhan Surveying-Geotechnical Research Institute Co., Ltd. of MCC (2005) Hydrogeology investigation report of Baosteel's NO.3 hot rolling whirling pool foundation pit dewatering. Shanghai (in Chinese)

Wuhan Surveying-Geotechnical Research Institute Co., Ltd. of MCC (2005) Hydrogeology investigation report of steel rolling area sedimentation tank sites for Pusteel's moving engineering. Shanghai (in Chinese)

Xue YQ (1986) Groundwater dynamics. Geological Publishing House, Beijing (in Chinese)

Yan XX, Gong SL, Zeng ZQ et al (2002) Relationship between building density and land subsidence in Shanghai urban zone. Hydrogeol Eng Geol 6:21–25 (in Chinese)

Yang HJ, Han WF, Chen WW et al (2003) Seepage features of fissure water in original rock and practical case study. Chin J Rock Mech Eng 22(z2):2582–2587 (in Chinese)

Ye WM, Tang YQ, Zhang XL et al (2000) Study of the subarea of geological disaster to underground engineering in Shanghai. J Tongji Univ 28(6):726–730 (in Chinese)

Ye WM, Tang YQ, Zhang XL et al (2001) Superficial geological environment and disasters in Shanghai area. Tongji University Press, Shanghai (in Chinese)

Yin YP (2003) Seepage pressure effect on landslide stability at the Three Gorges Reservoir area. Chin J Geol Hazard Control 14(3):1–8 (in Chinese)

Zhang AG, Wei ZX (2002) Past, present and future research on land subsidence in Shanghai City. Hydrogeol Eng Geol 29(5):72–75 (in Chinese)

Zhou NQ, Tang YQ, Wang JX et al (2006) Response characteristics of pore pressure in saturated soft clay to the metro vibration loading. Chin J Geot Eng 28(12):2149–2152 (in Chinese)